C0-CFL-594

Advances in

Photochemistry

Volume 9

Advances in Photochemistry

Volume 9

Editors

J. N. PITTS, JR., *Department of Chemistry,*
University of California, Riverside, California

GEORGE S. HAMMOND, *Vice Chancellor of Natural Sciences,*
University of California, Santa Cruz, California

KLAUS GOLLNICK, *Institut fur Organische Chemie,*
Universitat Munchen, Munchen, West Germany

An Interscience® Publication

John Wiley & Sons, New York • London • Sydney • Toronto

REF
QD
601
A1 A28
v. 9

Copyright © 1974, by John Wiley & Sons, Inc.

All rights reserved. Published simultaneously in Canada.

No part of this book may be reproduced by any means, nor transmitted, nor translated into a machine language without the written permission of the publisher.

Library of Congress Catalog Card Number 63-13592
ISBN 0-471-69092-9

Printed in the United States of America

10 9 8 7 6 5 4 3 2 1

506494

Foreword

Volume 1 of this series, Advances in Photochemistry, appeared in 1963. The stated purpose of this series was to explore the frontiers of photochemistry through the medium of chapters written by pioneers who are experts. Frontiers in photochemistry are at times full of confusion and for this reason they are fascinating!

In the first eight volumes the editors have not acted as censors. We have followed the principle set forth in the original introduction to this series, that is, "the authors are free to make any statement they wish which cannot be proven wrong." We have solicited articles from experts who have strong personal points of view concerning fundamental aspects of photochemistry and spectroscopy. We have asked these authors to present their points of view through critical discussion and evaluation of existing data. In no sense have the articles been simply literature surveys although in some cases they have fulfilled both purposes.

We trust that during the past few years, when photochemistry has truly come of age, that this series has proven to be stimulating and has served our original purpose of bringing together points of views of photochemists and spectroscopists who only a short five to ten years ago did not even use the same vocabulary.

Two of us welcome the addition of Dr. Klaus Gollnick as an editor and hope that this will give even better coverage to the "European photochemical scene." We intend to continue the policies as established in the previous volumes.

The present editors wish to express their deep gratitude to Professor W. Albert Noyes, Jr., who served with two of us as co-editors of the first eight volumes of this series. Professor Noyes' wise guidance and his willingness to read and work with manuscripts was invaluable to the success of the series. The field of photochemistry has profitted in many ways and for many years from the inspiration and hard work contributed by Professor Noyes. It has been a privilege and source of personal satisfaction to have worked with him in this enterprise.

James N. Pitts, Jr.
George S. Hammond
Klaus Gollnick

Acknowledgment

The editors are deeply indebted to Mrs. Dolores V. Tanno for her painstaking effort in final typing this volume of the Series.

Contents

*Permanent address: Department of Chemistry, University of British Columbia, Vancouver, Canada

†Permanent address: Geophysical Fluid Dynamics Lab, Princeton University, P. O. Box 308, Princeton, NJ 08540

Application of Electron Spin Resonance Spectroscopy to Photochemistry

J. K. S. WAN
Department of Chemistry
Queen's University
Kingston, Ontario, Canada

CONTENTS

I. INTRODUCTION

Most, if not all, photochemical systems involve the formation and reactions of electronically excited molecules, free radicals, or atomic transients. The study of a photochemical reaction involves the elucidation of the primary process and of all secondary reactions which follow. The determination of the mechanism of a photochemical reaction requires ideally the identification of each elementary reaction occurring, together with a quantitative assessment of the role which each step plays in the overall reaction. In recent years an increasing number of photochemists have focused their attention on the relationships between molecular and electronic structures of excited molecules and free radicals and their chemical reactivities in photochemical systems.

Historically the most popular method for studying photochemical reactions is the "product analysis" method which involves a process of deduction based on observations of the overall behavior. However, such a method cannot provide unequivocal evidence for the existence of free radical intermediates, let alone information on their structures. The development of flash photolysis since the 1940s has proved to be a general and perhaps the most powerful of all techniques in photochemistry (1). Coupled with UV-visible spectroscopy, the flash method permits direct observation of transient intermediates and their kinetic behavior. Difficulty is sometimes encountered in specific systems with the optical detection technique. For example, the identification of CH_3 radicals in the flash photolysis of gaseous $(CH_3)_2Hg$ has been complicated by the fact that the transitions in the near UV and visible region are forbidden due to the geometry of the radical (2). When the flash method is applied to condensed phase systems, high resolution absorption spectra of the transients are difficult to obtain. Great care must also be taken to differentiate possible spurious absorption due to impurities and solvent radicals. In recent years the flash photolysis method has been coupled to various specifically designed detection systems including time-resolved mass spectrometry (3), photoconductivity (4), rapid scan infrared spectroscopy (5), and rapid scan electron spin resonance spectroscopy (6).

Since 1945 the development of electron spin resonance (ESR) spectroscopy, more satisfactorily known as electron paramagnetic resonance, has proved exceedingly useful for the study of paramagnetic molecules. It concerns magnetic dipole transitions between Zeeman energy levels of any system that possesses electronic magnetic moments. Systems that usually respond to ESR study include atoms with an odd number of electrons, ions having partly filled inner electron shells, and other molecular species that possess electronic angular momentum. It should be noted that some free radicals having unpaired electrons may not invariably show ESR phenomena and some diatomic molecules containing no unpaired electrons can be readily studied by ESR. The best known example of the first case is the $^2\pi_{1/2}$ state of NO which has no magnetic moment although it is a doublet state with an unpaired electron. An example of the second case is molecular oxygen in the $^1\Delta_g$ state which has no unpaired electron but possesses orbital angular momentum. However, similar exceptions are rare among polyatomic systems, and for practical purposes we consider that the technique of ESR can be potentially applied to studies of all polyatomic radicals.

In recent years the state of the art in ESR instrumentation has matured enough that the technique is being used more and more by photochemists to elucidate the structure and reactivity of free radicals and organic triplet molecules in photochemical systems. Indeed, a renewed and increasing interest in the field of free radical chemistry during the last decade can be attributed mainly to the rapid development of electron spin resonance spectroscopy. The technique is very versatile as it can be applied to gaseous, liquid, or solid systems over a large range of temperatures. In the past most photochemical systems studied by ESR were in the condensed phase and usually involved reactions in the steady state. Nevertheless, ESR studies have contributed significantly to photochemistry. It is clearly not feasible to present here a detailed review of all aspects of the enormous amount of work which has been done in this field. Rather, we shall aim at general illustrations of the application of ESR to a variety of photochemical systems and discussions of some of the current development in the study of fast free radical reactions.

II. SOME REQUISITE THEORETICAL BACKGROUND IN ESR SPECTROSCOPY

The first experimental observation of electron spin resonance was made on an iron compound by Zavoisky in Russia in 1944. A year later Purcell and Bloch and their collaborators observed independently nuclear magnetic resonance in condensed matter. While the celebrated Purcell and Bloch discoveries have sparked an explosive wave of research on nuclear magnetic resonance throughout the world, little activity on electron spin resonance followed immediately after the first success. In fact, even Zavoisky did not continue working in the field of ESR, and the development of the technique was almost exclusively left to the physicists in the decade after the war. Indeed, the potential diversity of applications of ESR to chemistry could hardly have been foreseen in the earlier days.

The basic theory of the phenomenon of electron spin resonance was developing even before the first experimental observation. Formal and rigorous introduction of all aspects of the theory have been documented in many excellent monographs (7) on ESR published in the last decade. Obviously the kind of ESR information the photochemist seeks can be quite different from that required by the physicist. In general the photochemist is interested in the detection and identification of radicals and photoexcited molecules and in analyzing their molecular structure and quantitative kinetic behavior. The intention of this section is merely to briefly summarize several important points which should be remembered in interpreting ESR data of interest to the photochemist.

A. Magnetic Properties of Radicals, Molecules, and Excited Molecules

The photochemist's first concern is whether or not a particular transient in a photochemical reaction can be detected by ESR under certain experimental conditions. This requires the knowledge of the magnetic properties of the species in question. Atoms having odd numbers of electrons are paramagnetic and we will not consider them here.

Let us first consider the species known as free radical. Since Lavoisier coined the term "radical" at the end of the 18th century, it has become one of the most popular terms in chemistry. Thus a free radical has generally been defined as "a molecule, or part of a molecule in which the normal chemical binding has been modified so that an unpaired electron is left associated with the system."(8). It has also been widely recognized that free radicals usually have a high chemical reactivity, but "stable radicals" which can be isolated in well-defined liquid or crystalline form are well known.

For a radical to be detected by ESR, it must possess

a magnetic moment originating from its unpaired electron. First-order contributions to the magnetic dipole moment μ may arise from the electronic orbital and spin angular momenta as well as from the rotation of the whole radical. In general an orbital magnetic moment will occur only in a state with twofold or more orbital degeneracy. Since the Jahn-Teller theorem shows that degenerate states in nonlinear molecules are normally unstable, most polyatomic organic radicals have zero orbital magnetic moment, and their magnetic moments arise mainly from spin and can be expressed as

$$\mu = -g_e \beta_e \underline{S} \qquad (2\text{-}1)$$

where $\underline{S}$ is the dimensionless spin vector of magnitude 1/2 $\hbar$ which gives the intrinsic spin angular momentum of the unpaired electron; β_e is the Bohr magneton, and g_e the free electron g-factor which has the value 2.00232.

It should be noted that several aromatic hydrocarbon ions with an unpaired electron and threefold or higher axis of symmetry, such as $C_6H_6^+$ and $C_6H_6^-$ are known to have orbitally degenerate ground states. Linear radicals with orbital moments, such as NO, OH, SH, NCO, and NCS, are also well known. In the case of NO, it has a $^2\pi_{1/2}$ ground state, and the magnetic moment in this state is given by

$$\begin{aligned} -\mu &= \beta_e \Lambda + g_e \beta_e \underline{S} \\ &= (1 - 1/2 g_e)\beta_e \\ &= -0.00116\beta_e \qquad (2\text{-}2) \end{aligned}$$

and NO is often referred to as having zero moment. Another state of NO, $^2\pi_{3/2}$, lies 121.1 cm^{-1} above the ground state and has a moment $\mu \simeq -2\beta_e$. Thus at room temperature both states are populated and NO can be detected by ESR. As the temperature is lowered, the relative population in the paramagnetic $^2\pi_{3/2}$ state is greatly reduced, and the magnetic moment of NO tends almost to zero at very low temperatures.

We have so far implied that radicals are species with an odd number of electrons. We now turn to molecules having an even number of electrons. Organic molecules in their ground state with an even number of electrons almost invariably have no orbital magnetic moment. Spin magnetic moments can, however, arise in some ground state molecules with an even number of electrons. This usually occurs in molecules which have a triplet ground state. The most common example is the ground triplet state of molecular oxygen. In recent years an increasing number of interesting molecules including methylene and its derivatives, nitrene derivatives, the species CNN, NCN, and NCCCN have been studied by ESR and their triplet ground state confirmed (9). In addition, aromatic hydrocarbons with three- or six-fold symmetry, such as triphenylene and decacyclene, may form dianions, with a triplet ground state (10). For the present we shall regard a triplet state molecule as having two "odd electrons."

Molecules with triplet ground states may have low-lying metastable electronic excited states with different multiplicity, symmetry, and magnetic properties. For example, the triplet methylene is bent and paramagnetic while the excited singlet methylene is linear and diamagnetic. Their chemical reactivities are also known to be different (11). Another example is molecular oxygen in which the two low-lying singlet excited states $^1\Delta_g$ and $^1\Sigma g^+$ have significantly different chemical (12,13) and magnetic properties from those of the ground triplet state $^3\Sigma g^-$. Thus the $^3\Sigma g^-$ state has a spin magnetic moment; the $^1\Delta g$ excited state possesses no unpaired electrons, and its magnetic moment arises mainly from electronic orbital angular momentum; and the $^1\Sigma g^+$ excited state is diamagnetic.

Molecules with a singlet ground state may have an electronically excited triplet state accessible by photoexcitation via a radiationless intersystem crossing. Indeed, photochemical study of excited triplet states coupled with ESR spectroscopy is perhaps one of the most exciting chapters in the recent advances in photochemistry.

B. Electron Spin Interactions: A Source of Chemical Information

In addition to the capability of detecting very low concentration of free radicals as transients in a chemical reaction, ESR spectroscopy provides experimentally the following very important sources of chemical information on their structure. (a) The resonant position referred to as the g value is directly determined by the separation of the energy levels of the radical under investigation. Variation of the g value is interpreted in terms of the perturbation by the spin-orbit interactions, and may be used as a probe to study substitution effects and changes in molecular structure (14,15). (b) The width and shape of the ESR lines are determined by the relaxation of the spin energy state. Elucidation of the various interactions which contribute to the relaxation provides very valuable information on the behavior of the unpaired electron (16). (c) The hyperfine structure of ESR spectra is produced by the interaction of the electron spin magnetic moment with the magnetic moments of the nuclei contained by the molecular orbital of the unpaired electron. Well-resolved hyperfine patterns often help to "fingerprint" the radicals under investigation and also reveal information regarding the spin densities at the various sites in the molecule and their correlation with the chemical nature of bonding. (d) For systems with more than one unpaired electron, such as triplet states, the interaction between two unpaired electrons leads to fine structure in the ESR spectra and provides a wealth of information concerning the electronic structures and the dynamics of the triplet state molecules. We will not consider here interaction between the electron spin and the molecular rotational magnetic moment. Spin-rotation coupling is important mainly in the gas phase ESR study of simple radicals at very low pressures, and the interpretation of this coupling in ESR spectra is beginning to be understood (17,18).

1. Electrons in Magnetic Fields

In a magnetic field $\underline{H}$ the energy of interaction $\underline{W}$ with the electron spin moment is given by

$$\underline{W} = -\mu \cdot \underline{H} \qquad (2\text{-}3)$$

and for a field $\underline{H}$ along the z direction the degenerate eigenstates with $\underline{m}_s = \pm 1/2$ split into a lower level $\underline{m}_s = -1/2$ and a higher level $\underline{m}_s = 1/2$ separated by an energy $g_e \beta_e \underline{H}$. The relative population of these two states in thermal equilibrium and in the absence of other polarization effects is governed by the Maxwell-Boltzmann distribution. Transitions between these two energy states $\underline{m}_s = 1/2$, $\underline{m}_s = -1/2$ can be equally induced by applying an alternating field having its magnetic component perpendicular to the constant magnetic field and its frequency ν being equal to $g_e \beta_e \underline{H}/h$. The transition probability for a free spin has been treated both by the Heisenberg method and by the perturbation theory for the coupling of the spin with the rotating field (7a). Since the probabilities of these transitions are equal and the equilibrium population of the lower state is slightly higher than that of the upper state, there will be a net absorption of the energy of the rotating field.

Experimentally the possibility of observing the transitions depends on the population difference between the two spin states, and the lifetime of the upper state which is determined by the spontaneous emission probability and by the various relaxation processes available. ESR emissions have been observed (19) in several systems in which the upper spin state has a higher population. The lifetime of the upper state as determined by the probability of spontaneous emission between the two spin levels can be calculated from the Einstein coefficient $\underline{A}_{1/2 \rightarrow -1/2}$:

$$\underline{A}_{1/2 \rightarrow -1/2} = \frac{32\pi^3\nu^3}{3C^3\ h}\ g_e^2\beta_e^2 \qquad (2\text{-}4)$$

In a magnetic field of 10,000 gauss, the resonance frequency ν of the free electron ($g = 2.0023$) is 28,026 Mc/s and the spontaneous emission lifetime is $\sim 10^{13}$ sec. Obviously this lifetime is too long and the transitions will be saturated exceedingly easily. In other words, the populations of the two levels become essentially equal and no net transition can be observed. Fortunately there are a number of nonradiative relaxation mechanisms open to the upper spin level including interactions with other electrons, with nuclei having nuclear magnetic moments, and with the lattice. The latter process is often known as <u>spin-lattice relaxation</u>. The term "lattice" generally refers to the degrees of freedom of the system other than those directly related with spin. Spin lattice relaxation is possible because the spin is coupled to the thermal motions of the system. The spin-lattice relaxation time for most organic radicals in condensed phase is generally much smaller than one second. In addition, interactions may lead to fluctuation of the relative energies of the spin levels and contribute to a relaxation process usually called the spin-spin relaxation.

2. Electron Spin-Orbit Interaction: Variation in g-Factors

First-order effects due to spin-orbit coupling vanish for all polyatomic molecules with nondegenerate ground states. Second-order effects appear in two different ways. The inclusion of spin-orbit coupling in the molecular Hamiltonian results in the given molecular state being not an eigenstate of the total spin operator $\underline{\hat{S}}^2$, thus causing a mixing between singlet and triplet states and consequently a relaxation of the selection rule for the "spin forbidden" transition in electronic spectra (20).

In ESR spectroscopy, the second-order effect of the spin-orbit coupling causes the variation of the $\underline{g}$-factor from the free electron value and the elimination of the degeneracy of the triplet state of transition-metal ions.

Let us consider a paramagnetic species with no magnetic nuclei as being described by an effective spin Hamiltonian

$$\mathcal{H} = \beta_e \underline{\hat{S}} \cdot \underline{g} \cdot \underline{H} + \underline{\hat{S}} \cdot \underline{D} \cdot \underline{\hat{S}} \qquad (2\text{-}5)$$

Here $\underline{g}$ and $\underline{D}$ are tensors of second rank. The $\underline{g}$ tensor is given by

$$g_{ij} = g_e(\delta_{ij} - 2\lambda\ \Delta g_{ij}) \qquad (2\text{-}6)$$

where g_e is the free electron g value; δ_{ij} is the Kronecker's delta; λ is the spin-orbit coupling constant; and Δg_{ij} is defined in terms of the matrix elements of the orbital angular momentum operator $\underline{\hat{L}}$ by

$$\Delta g_{ij} = \sum_{n \neq 0} \frac{\langle 0|\hat{L}_i|n\rangle \langle n|\hat{L}_j|0\rangle}{\varepsilon_n - \varepsilon_o} \qquad (2\text{-}7)$$

where $|n\rangle$ corresponds to the unfilled molecular orbital Ψ_n of energy ε_n, and 0 refers to the lowest state containing the odd electron. The $\underline{D}$ tensor is given by

$$\underline{D}_{ij} = -\lambda^2 \Delta g_{ij} \qquad (2\text{-}8)$$

and is chiefly responsible for the fine structure of triplet state transition metal ions. When the species is a doublet state, this term is of no consequence. For organic triplet states, λ and $\Delta\underline{g}$ are usually small and the spin-spin effect becomes most important. Thus when a spin Hamiltonian contains only the $\underline{g}$ term, the resonance frequency ν is given by

$$\nu = \frac{\beta_e g^* \underline{H}}{h} \qquad (2\text{-}9)$$

where $\underline{H}$ is the magnitude of the magnetic field and g^* is given by

$$g^* = [g_{xx}^2 a^2 + g_{yy}^2 b^2 + g_{zz}^2 c^2]^{1/2} \qquad (2\text{-}10)$$

where g_{xx}, g_{yy}, and g_{zz} are the principal values of the $\underline{g}$ tensor and a, b, and c are the direction cosines of the applied magnetic field with respect to the principal fine-structure axes. In solution ESR spectra, the observed g-factors are the averages of the principal values. With powdered samples, the observed spectrum is the envelope of a spectrum from all possible orientations of the radical. For radicals with an axial symmetry, the spin Hamiltonian can be simplified as

$$\mathcal{H} = \beta_e g_{11} \cdot H_{11} \cdot S_{11} + \beta_e g_{\perp} H_{\perp} S_{\perp} \qquad (2\text{-}11)$$

where $g_{11} = g_{zz}$ and $g_{1} = g_{xx} = g_{yy}$. The resonance equation is thus given by

$$\nu h = g(\theta)\beta_e \underline{H} \qquad (2\text{-}12)$$

where $g(\theta) = (g_{11}^2 \cos^2\theta + g_{\perp}^2 \sin^2\theta)^{1/2}$ and θ is the angle between the molecular symmetry axis and the direction of the applied magnetic field.

For most organic radicals, the spin-orbit coupling constants λ associated with light atoms have small negative values and the term $(\varepsilon_n - \varepsilon_o)$ in (2-7) is usually large. Consequently the deviation of the g-factor for organic radicals from the free electron value is small and generally does not excees three or four units in the third decimal place. Introduction of an oxygen or sulfur atom into the radical increases the g-factor and may be used for identification of oxygen and sulfur radicals. Other detailed studies show that the g-factor depends on both the temperature and the nature of the substituents in the radical (21). Precise measurement of g-factors may therefore reveal more information about the nature of the chemical bonding in organic radicals. Thus small variations in g-factors have been found in a series of silver complexed cycloalkenyl radicals corresponding to the relative stability of the radicals associated with the δ-coordination (22). More general discussions on g-factors can be found elsewhere (23,24).

For paramagnetic diatomic molecules with net orbital magnetic moments, the g-factor depends on the complicated coupling between the orbital, spin, and rotational moments.

3. Electron-Nucleus Interaction: Hyperfine Structure

For radicals with magnetic nuclei, the hyperfine structure of ESR spectra is produced by the interaction of the electron magnetic moment with the nuclear spin of those nuclei covered by the molecular orbital of the unpaired electron. This interaction splits further the two spin levels in a magnetic field. The hyperfine coupling is often given by the Hamiltonian H_{SN}:

$$H_{SN} = a\,\underline{\hat{I}} \cdot \underline{\hat{S}} + \underline{\hat{I}} \cdot \underline{B} \cdot \underline{\hat{S}} \qquad (2\text{-}12)$$

Here the first scalar term is the isotropic splitting, and the second traceless tensor term is the anisotropic dipolar splitting. Therefore the complete spin Hamiltonian for a doublet state radical expressed in terms of spin variables only (thus it is for the "fictitious spin" of a system in which spin-orbit coupling is nonexistent) is

$$H_S = g_e\beta_e\underline{HS} - g_N\beta_N\underline{HI} + a\,\underline{\hat{I}}\cdot\underline{\hat{S}} + \underline{\hat{I}}\cdot B\cdot\underline{\hat{S}} \qquad (2\text{-}13)$$

The second term here is the nuclear Zeeman interaction where g_N is the nuclear g-factor, β_N the nuclear Bohr magneton, and $\underline{I}$ the nuclear spin.

The constant of the isotropic hyperfine splitting a is given by the Fermi (25) approximation

$$a = \frac{8\pi}{3} g_e\ \beta_e\ g_N\ \beta_N|\Psi(o)|^2 \qquad (2\text{-}14)$$

where $|\Psi(o)|^2$ is the probability distribution function for the electron, evaluated at the nucleus (o). Obviously $\underline{a}$ vanishes unless there is a finite probability density at the nucleus; since the p, d, f .... orbitals have nodes at the nucleus, the electron must have some s-orbital character. Experimentally one finds that an isotropic hyperfine splitting can be observed, even when the unpaired electron is in a pure π-level or pure p-orbital. The common explanation of this fact is the spin-polarization mechanism in which the filled s-orbitals are polarized by the half-filled p-orbital (occupied by the unpaired electron) so that the α-spin orbital has a different radial function from that of the β-spin orbital. The resulting imbalance of spin-density at the nucleus $\rho(o)$ produces an isotropic splitting. Thus $\rho(o)$ can be defined at the nucleus o as:

$$\rho(o) = P(o,\alpha) - P(o,\beta) \qquad (2\text{-}15)$$

where $P(o,\alpha)$ and $P(o,\beta)$ are probabilities of finding spins α, β at the nucleus. Clearly $\rho(o)$ can be positive, zero, or negative. The experimentally determined values of isotropic hyperfine splittings would normally provide information regarding only the magnitude of the spin density on the nucleus concerned but not the sign. For first- and second-row elements, spin density due to polarization is usually taken to be positive. That is, the polarized spin felt by the nucleus is parallel to the unpaired electron. This is, however, complicated by a further effect that spin density on adjoining atoms can contribute to the overall isotropic splitting from the nucleus under consideration. Let us consider the following two examples.

a. The Methyl Radical. The radical is nearly planar (2), and the protons thus lie in the nodal plane of the p-orbital of the unpaired electron. In spite of this fact, isotropic hyperfine splittings of both the carbon ($^{13}CH_3$) and the protons have been observed (26). This is explained by the spin polarization in the sp^2-bonding orbitals represented schematically in Figure 1. Here

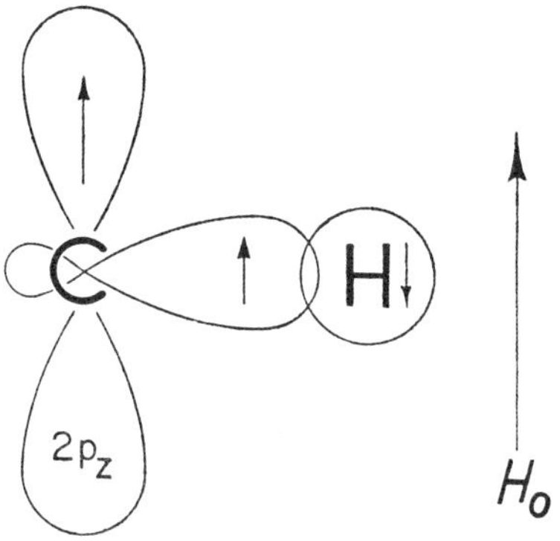

FIGURE 1. Schematic representation of spin polarization in the sp^2-bonds of a methyl radical.

the unpaired electron in the 2pz-orbital of the carbon polarized the C-H σ-bond electrons. In accordance with Hund's rule of maximum multiplicity, there will be a small net <u>positive</u> spin density on the carbon, and the Pauli principle requires that a net <u>negative</u> spin density be induced on the protons. The "signs" of the spin densities have been confirmed experimentally for a disubstituted methyl radical, $H^{13}\dot{C}(CO_2H)_2$ (27,28) and thus this "spin polarization" mechanism appears to be reasonable and is now universally accepted.

b. The Diatomic Radical NO. The unpaired electron in nitric oxide (NO) occupies a π-orbital built from 2p-levels on N and O atoms. Here the isotropic hyperfine splitting of

N, a_N will arise in part from a "positive" spin on nitrogen. However, the appreciable spin density on the adjoining oxygen atom will give a "negative" contribution to a_N and hence the net value of a_N may be either positive or negative, and this information is not readily available from electron spin resonance study.

We now return to eq. (2-13) and consider briefly the last term describing the anisotropic dipolar hyperfine interaction and its effect on ESR spectra. Let us consider an electron in a p-orbital of a magnetic nucleus in the presence of an external magnetic field. When the applied field is parallel to the symmetry axis of the orbital, the nuclear spin is quantized strictly along the direction of the applied field, and the additional field at the electron ΔH due to the nucleus is in the same direction as the applied field. When the external field is at an angle θ to this orbital axis, the nuclear spin should be quantized relative to this new axis. However, the isotropic component of the interaction due to spin polarization is usually larger than the aniso-tropic component, and the isotropic component is always parallel to the applied field. This allows the quantization of the nucleus to remain in the direction of the applied field, and the relatively small anisotropic splitting can be expressed by (29)

$$\Delta H = \mu_I \frac{3\cos^2\theta - 1}{r^3} \qquad (2\text{-}16)$$

where μ_I is the nuclear magnetic moment quantized along the direction of the applied field, r is the distance between the nuclear and the electron moments, and θ is the angle between the direction of the applied field and r. The expression (2-16) has to be averaged over the entire probability distribution of the unpaired electron and

$$\langle \Delta H \rangle = \mu_I \langle \frac{1}{r^3} \rangle \langle 3 \cos^2\theta - 1 \rangle \qquad (2\text{-}17)$$

where

$$\langle \frac{1}{r^3} \rangle = \int \Psi^*_{(r)}\ r^{-3}\ \Psi_{(r)}\ d\tau \qquad (2\text{-}18)$$

and

$$\langle 3 \cos^2\theta - 1 \rangle = \int \Psi_{(\theta)} (3 \cos^2\theta - 1) \Psi_{(\theta)} d\tau$$

$$(2\text{-}19)$$

If the electron cloud is spherical, $\langle \cos^2\theta \rangle = 1/3$ and therefore $\langle \Delta H \rangle = 0$. $\langle \Delta H \rangle$ will be nonzero only if the unpaired electron has p- or d-character. The scalar value of the anisotropic hyperfine splitting $\underline{B}$ can be expressed in the form

$$\underline{B} = \frac{2}{5h}\, g_e\ \beta_e\ g_N\ \beta_N \langle r^{-3} \rangle \qquad (2\text{-}20)$$

Experimentally the hyperfine interaction of an unpaired electron in a p- or sp-hybrid orbital is often found to be an axially symmetric tensor of the form a + 2$\underline{B}$ along the orbital direction, and a-$\underline{B}$ perpendicular to it. A detailed analysis of the hyperfine structure of ESR spectra therefore reveals two important characteristics of radicals: $\underline{a}$ the isotropic parameter associated with $\underline{s}$-character and $\underline{B}$ the anisotropic parameter associated with $\underline{p}$-character. Often values of $|\Psi(o)|^2_{ns}$ and $\langle r^{-3} \rangle_{np}$ are known for various common nuclei from s.c.f. wave functions (30,31). Thus it is possible to estimate the relative *s*- and *p*-characters of the unpaired spin orbital by comparing the observed hyperfine splittings *a* and *B* with the theoretical parameters a_{ns} and B_{np}. Molecular structural information can then be derived. For example, planarity of the methyl radical is substantiated by a detailed analysis of the ESR spectra of the radical oriented in a γ-irradiated single crystal of sodium acetate (32). Some computed atomic parameters and their corresponding a_{ns} and B_{np} hyperfine constants for various common nuclei are collected in Table 1.

TABLE I
Atomic Parameters and Nuclear Hyperfine Coupling Constants[a]

Nucleus	Nuclear Spin	$\Psi^2(o)_{ns}$ (a.u.)	a_o (gauss)	$\langle r^{-3}\rangle_{np}$ (a.u.)	B_o (gauss)
^{1}H	1/2		508		
^{2}D	1		78		
^{10}B	3		240		6.35
^{11}B	3/2	1.408	720	0.775	18.9
^{13}C	1/2	2.767	1110	1.692	32.5
^{14}N	1	4.770	550	3.101	17.0
^{15}N	1/2		771		23.9
^{17}O	5/2	7.638	1653	4.974	51.5
^{19}F	1/2	11.966	17110	7.546	541
^{29}Si	1/2	3.807	1208	2.041	31.0
^{31}P	1/2	5.625	3636	3.319	102
^{33}S	3/2	7.919	2715	4.814	27.9
^{35}Cl	3/2	10.643	1665	6.709	50.0
^{37}Cl	3/2		1385		41.8

[a]Taken from reference (95).

It is now well known that radicals introduced or generated in single crystals are usually oriented (33), and detailed analysis of their hyperfine interactions by ESR is relatively simple. The crystals may be rotated about an axis perpendicular to the applied field, or more satisfactorily, the magnet is rotated about the crystal. However, in photochemical studies it is not always possible to work with single crystals, and most photolyses are carried out in solution, in a polycrystalline environment, or in a low-temperature glass matrix. The transient radicals thus generated and trapped will be randomly oriented relative to the applied magnetic field. If the radicals are randomly trapped rigidly in the matrix, the anisotropic dipolar interaction produces smearing of the hyperfine structure (34) and/or variation of line-widths (35). However, if the radical as a whole tumbles in the matrix with a frequency greater than the width of the hyperfine lines, the dipolar broadening effect is eliminated as the anisotropic interaction is averaged to zero. It should be remembered that anisotropy in the g-tensor and electron spin-spin interaction also leads to line broadening. In general, radicals in liquid solution of low viscosity produce well-resolved isotropic hyperfine structure in their ESR spectra, and it is not uncommon that reasonably resolved isotropic splittings may be observed for hydrocarbon radicals trapped in polycrystalline solids with large spherical cavities, such as adamantane (36), or in low-temperature glasses.

We now turn to the last of the tree main magnetic interactions, the nuclear Zeeman term. The general approach (7) is to divide the spin Hamiltonian into two distinct parts:

$$\mathcal{H}_S = \mathcal{H}_o + \mathcal{H}_1$$

$$= (g_e \beta_e \underline{HS} - g_N \beta_N \underline{HI}) + \underline{a}\ \underline{\hat{I}} \cdot \underline{\hat{S}} \qquad (2\text{-}21)$$

in which $\mathcal{H}_o$, the Zeeman term, is the main part and $\mathcal{H}_1$, the contact term, is regarded as a small perturbation. It can be shown by perturbation theory that when $\underline{a}$ is considerably smaller than $g_e\ \beta_e \underline{H}$, and the applied field is in the z direction, the spin energy levels modified by hyperfine perturbation are defined by the quantum numbers M_S and M_I:

$$W_{M_I M_S} = g_e \beta_e \underline{H}\ M_S + \underline{a}\ M_I M_S \qquad (2\text{-}22)$$

Electron spin resonance transitions between these energy levels are governed by the selection rules $\Delta M_S = \pm 1$ and $\Delta M_I = 0$, and $\underline{a}$ determines the energy separation of the hyperfine lines. ESR spectra involving only allowed transitions are regarded as first-order. Organic radicals usually exhibit first-order spectra since in a field of about 3000 gauss, the hyperfine frequencies for most organic radicals are much smaller than the resonance frequency of ~9,000 Mc/s. At magnetic fields of 3000 gauss or higher, the electron and nuclear spins precess independently about the direction of the applied field. Thus the hyperfine splittings in hydrocarbon radicals with one, two, and three equivalent protons are shown in Figure 2. The transitions for n protons with spin I = 1/2 give rise to (2nI + 1) lines with intensities proportional to the binomial coefficient.

Second-order ESR spectra involving both the allowed ($\Delta M_S = \pm 1$, $\Delta M_I = 0$) and the "forbidden" ($\Delta M_S = 1$, $\Delta M_I = 1$) transition have also been observed for some organic radicals (37). When the hyperfine constant $\underline{a}$ is

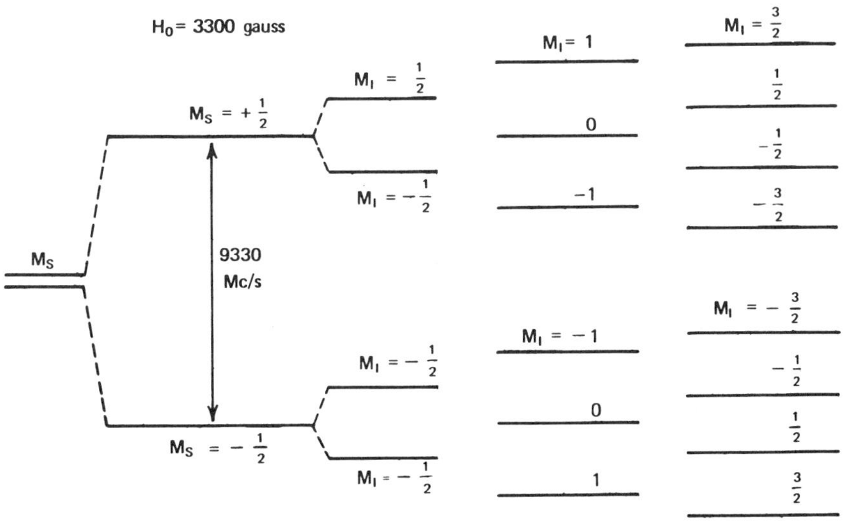

FIGURE 2. Hyperfine splitting in free radicals for 1, 2, and 3 protons.

relatively large or when the applied field is lowered, the M_S and M_I values are not sharply defined and the "forbidden" transitions become detectable. Detailed consideration of second-order ESR spectra are to be found elsewhere (38).

4. Electron Spin-Spin Interaction: Zero-Field Splitting

Similar to electron-nuclear hyperfine coupling, the interaction between the magnetic moments of two electrons involves the Fermi contact term $\mathcal{H}_F$ and the classical dipolar term $\mathcal{H}_{SS}$:

$$\mathcal{H}_F = g_e^2 \beta_e^2 \frac{8\pi}{3} \sum_{1,2} \delta(\underline{r}_{1,2})\, \hat{\underline{S}}_1 \cdot \hat{\underline{S}}_2 \qquad (2\text{-}23)$$

$$\mathcal{H}_{SS} = g_e^2 \beta_e^2 \sum_{1,2} \frac{1}{r_{1,2}^3} \left[\hat{\underline{S}}_1 \cdot \hat{\underline{S}}_2 - \frac{3(\hat{\underline{S}}_1 \cdot \underline{r}_1)(\hat{\underline{S}}_2 \cdot \underline{r}_2)}{r_{1,2}^2}\right] \qquad (2\text{-}24)$$

When the molecule is in a singlet state with all electrons paired, the classical dipolar interaction averages to zero and the contribution to the spin-spin energy from the contact term (2-23) depends on the δ-function. This latter contribution, however, has no consequence in electron spin resonance and forms part of the relativistic correction to calculation on two-electron systems. For two electrons in a triplet state, the contact interaction vanishes, but the classical dipolar coupling then lifts the degeneracy between the three spin functions. This phenomenon is often known as zero-field splitting (ZFS). The splitting depends on the orientation of the spins and on their spatial distribution. The energy levels are found by diagonalization of the matrix of $\mathcal{H}_{SS}$ with respect to the zeroth order spin basis functions. Experimentally spin-spin splitting in triplet states of large organic molecules can be observed by ESR spectroscopy.

The spin Hamiltonian for the interaction between electrons 1 and 2 and external field $\underline{H}$ is given by

$$\mathcal{H}_S = g\beta_e \underline{H}\cdot(\hat{\underline{S}}_1 + \hat{\underline{S}}_2) + \mathcal{H}_{SS} \qquad (2\text{-}25)$$

When the energy due to the first term is large compared to that due to $\mathcal{H}_{SS}$, the triplet functions are eigenfunctions of $\mathcal{H}_S$ in eq. (2-25), and the transitions are governed by $\Delta M_S = 1$ (Figure 3). If the $\Delta M_S = 1$ transitions are observed experimentally at high field, extrapolation of the resonance frequency to zero field may yield the zero-field splitting. This phenomenon was first observed by Hutchison and Mangum (39) for the photoexcited triplet state of naphthalene doped in a single crystal of durene. In general the experimental

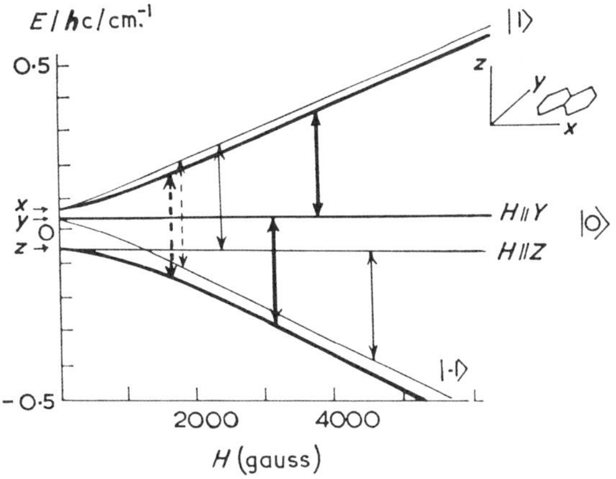

FIGURE 3. Fine structure of the triplet state of naphthalene as a function of the external field H_0. Heavy lines: H_0 parallel to y-axis; light lines: H_0 parallel to z-axis; dotted arrows: the $\Delta M_S = 2$ transition.

results on triplet aromatic molecules are often described by the zero-field splitting parameters $\underline{D}$ and $\underline{E}$ (40,41):

$$H_S = g\beta_e \cdot \underline{H} \cdot \underline{\hat{S}} + \underline{D}\hat{S}_z^{\,2} + \underline{E}(\hat{S}_x^{\,2} - \hat{S}_y^{\,2}) \qquad (2\text{-}26)$$

where g is the phenomenological g-factor, and $\hat{S}_s$, $\hat{S}_y$, and $\hat{S}_z$ are the components of the total spin $\underline{\hat{S}} = \underline{\hat{S}}_1 + \underline{\hat{S}}_2$; z is taken perpendicular to the molecular plane. The zero-field splitting (ZFS) of the spin levels can then be found from eq. (2-24) with $\underline{H} = 0$. For molecules with a threefold or sixfold symmetry axis, $\underline{E} = 0$ and the energy of the $M_S = 0$ state differs from that of the doubly degenerate $M_S = 0$ states:

$$e_{M_S \neq 0} = 1/3\underline{D} \quad \text{and} \quad e_{M_S = 0} = -2/3\underline{D} \qquad (2\text{-}27)$$

For molecules of lower symmetry, quantization of spins can occur along all of the three symmetry axes and this leads to additional ZFS between the $M_{S\neq 0}$ states of $2\underline{E}$. The three nondegenerate levels have eigenvalues:

$$e_1 = 1/3\underline{D} + \underline{E}, 2e_2 = 1/3\underline{D} - \underline{E}, 2e_3 = -2/3\underline{D} \tag{2-28}$$

Thus, as shown in Figure 3, two $\Delta M_{S=1}$ transitions can be observed. For $\underline{H}$ parallel to the z-axis, the separation between the two transitions is equal to $(c/g\beta_e)|2\underline{D}|$, and for $\underline{H}$ parallel to x- and y-axis, the separations are $(c/b\beta_e)[|\underline{D}| - |3\underline{E}|]$ and $(c/g\beta_e)[|D| + |3E|]$, respectively. One might therefore expect that ESR study of oriented triplet molecules will lead to information concerning the symmetry of the triplet states.

In the usual X-band ESR spectrometer, the applied field $\underline{H}$ is approximately 3000 gauss. The competition between the strong applied field and the molecular axes for the spin quantization leads to very large anisotropy of the $\Delta M_S=1$ splitting and makes it very difficult to observe these transitions for an assembly of randomly oriented triplet molecules. However, van der Waals and deGroot (42) have demonstrated that the so-called $\Delta M_S=2$ transition at intermediate field has an intensity about 1 to 2% of the high field $\Delta M_S=1$ transitions. The intermediate field transition is much less anisotropic (Figure 3) and can be readily observed. This discovery has made a great impact on photochemistry of triplet states as ESR can be used to study randomly oriented triplet organic molecules photoexcited in rigid glasses.

It can be shown that for conventional spectrometers

with the microwave field perpendicular to the applied field, the resonance position $\underline{H}_r$ at low field ($\Delta M_S=2$) is related to zero-field splitting parameters:

$$\underline{H}_r = (2g\beta)^{-1} [(h\nu)^2 - 4(\underline{D}^2 + 3\underline{E}^2)]^{1/2} \quad (2\text{-}29)$$

Unfortunately the parameters $\underline{D}$ and $\underline{E}$ cannot be determined separately from such an intermediate field measurement, and the parameter $(\underline{D}^2 + 3\underline{E}^2)^{1/2}$ is usually referred to as $\underline{D}^*$. However, $\underline{D}$ and $\underline{E}$ can be obtained directly from the $\Delta M_S=2$ transition by using spectrometers with the microwave field parallel to the applied field (41,42).

Let us briefly mention here that proton hyperfine splittings have been observed in the high-field ESR spectra of the naphthalene triplet (33,40) in single crystals. Recently Grivet (43) has succeeded in observing hyperfine splittings in the intermediate field ($\Delta M_S=2$) line of the triplet of partially deuterated naphthalene in rigid solution with a conventional spectrometer. Fischer and Hausser (44) have also observed hyperfine splittings in the $\Delta M_S=2$ line of some fluorohydrocarbons. These observations will undoubtedly have great significance in photochemical studies of excited triplet states since they allow the convenience of rigid glass technique to yield important information about the distribution of unpaired electrons in the excited state.

Although triplet state studies in rigid glasses are more convenient, it must be pointed out that they have yielded far less quantitative and precise data than have the single-crystal investigations (39,45). Initially the disadvantage of single-crystal studies is that suitable host crystals are difficult to find. However, with time the single-crystal method has been developed to have wide applicability with many suitable host crystals being found for a great variety of guest triplet states. Furthermore, extremely precise ZFS data and details about the dynamic properties of triplet states become attainable in single crystals using the "near zero field" ("low-field") method (46-48). The advantages of using the low-field method over the intermediate and high field studies have been summarized by Gerkin and Winer (47b). Electron-nuclear double resonance (ENDOR) techniques have also been employed to study organic triplet states in single crystals in conjugation with the ordinary ESR method (49,50).

Finally we will briefly consider the "traditional" and arbitrary distinction between a "diradical" and a triplet state molecule. Most of the stable organic diradicals studied by ESR consist of identical monoradical halves bonded together by a central linking group. The monoradical spatial wave functions are usually centered on a given half of the molecule, and the overlap of the monoradical functions is negligible. Thus a "diradical" is usually defined as a molecule having two "independent unpaired electrons" with negligible electron-electron dipolar coupling. For a symmetric diradical, the hyperfine coupling constant $\underline{a}_d$ is equal to the monoradical value $\underline{a}_m$. In the limit of strong exchange, $J >> \underline{a}_m$, the two monoradical moieties will behave as a triplet spin species and a new hyperfine spectrum appears with $\underline{a} = 1/2\underline{a}_m$. Two mechanisms for the electron exchange have been considered. One involves an indirect polarization through the central linking group; the other involves a direct coupling through space when the two monoradical moieties are sufficiently close. Numerous examples of nitroxide diradicals, phenoxy diradicals, hydrocarbon diradicals, and many others have been described in the ESR literature (51-53).

C. Spin-Density: The Semi-Empirical McConnell Relation

In Section II.B.3 we briefly mentioned the origin of hyperfine structure of free radicals. To a photochemist interested in ESR of transient radicals, the hyperfine coupling constant is perhaps the most informative single parameter. It is also useful diagnostically since characteristic multiplet splittings of most common magnetic nuclei are readily recognizable. A few elegant nonempirical calculations (54-57) of hyperfine coupling constants in small π-radicals have been carried out to test the theory, but the results are not very encouraging. In this section we shall mention briefly the semi-empirical approach to establishing a linear relationship between the hyperfine constant $\underline{a}$ and the π-electron spin-density.

The suggestion of spin polarization was first made by McConnell (58), Bersohn (59), and Weissman (60) to account for the surprising observation of proton hyperfine splitting in aromatic π-radicals, as the π-electron theory usually ignores δ-π interaction. The problem has been approached by approximate valence bond, molecular orbital, and unrestricted Hartree-Fock methods, all involving some degree of a breaking down of the δ-π separation hypothesis. All three methods as discussed thoroughly by McConnell and Chesnut (61) lead to a linear relationship

$$(\underline{a}_H)_i = Q\rho_i \qquad (2\text{-}30)$$

where $(\underline{a}_H)_i$ is the hyperfine coupling constant in gauss of the proton bonded to the *i*th carbon, ρ_i is the spin-density on the *i*th carbon, and Q is a quantity depending on atomic integrals and excitation energies which is the same for each carbon in a molecule. Calculated values of Q using different methods, however, give only the correct order-of-magnitude results, and Q is therefore usually taken as an empirically determined constant with an accepted value of ~-22.5 gauss. The negative sign of Q has been confirmed by Ghosh and Whiffen (62) experimentally. Equation (2-30) is usually known as the McConnell Relation. In recent years a variety of semi-empirical methods have been developed for obtaining more satisfactory molecular wave functions for evaluation of spin-density distribution in odd-electron systems (7). So far, the McConnell equation (2-30) bears up reasonably well under experimental test. Colpa and Bolton (63) have proposed some empirical modification to account for the experimental result that large coupling constants in aromatic cations tend to be larger than the corresponding constants in anions. Symons (64) has also suggested a modified equation including adjacent atom terms.

The usefulness of the McConnell relationship has been steadily increasing with application of its extensions to ^{13}C (65,66), ^{14}N (67,68), ^{33}S (69), and Si (70). Many attempts have also been made to derive a similar relationship for ^{19}F (71-74) but a recent assessment (75) of the current theory of fluorine hyperfine interaction indicated that more work is required before a satisfactory quantitative treatment is possible. Mackey and Wood (76) have considered the application of eq. (2-30) to a series of Group IV-B atom-substituted methyl radicals and concluded that the spin polarization parameter Q differs by not more than a factor of two when the α-atom changes from carbon to Pb.

In contrast to α-protons having negative spin-density in aromatic π-radicals, hyperfine coupling to β-protons in aliphatic radicals gives rise to positive spin-density at the proton (77). The theoretical interpretation of this phenomenon is somewhat less clear (78) but is generally attributed to "hyper-conjugation." The theory of hyperconjugation has been discussed elsewhere (79,80). In the analysis of ESR hyperfine coupling of the methyl protons in an ethyl radical, the basic concept of hyperconjugation is that there is overlap between the C-H δ-bonds of the methyl group and the p-orbital of the unpaired electron resulting in some redistribution of positive spin-density at the β-proton sites. For a rotating methyl group, the isotropic coupling is about +26 gauss.

For β-protons in methylene groups which are fixed in the radical (Figure 4), a wide range of values of the coupling is possible (33) and this has been interpreted

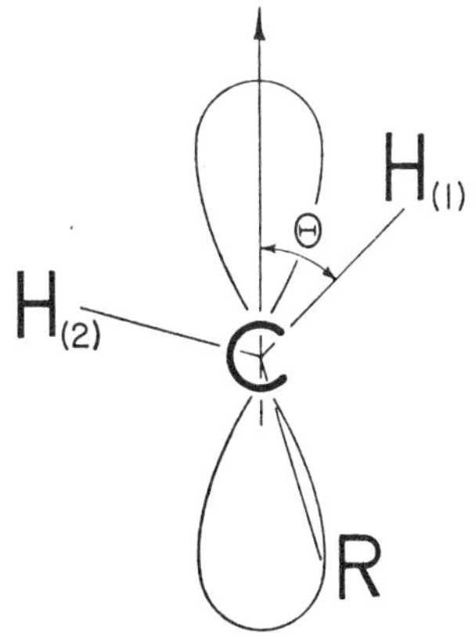

FIGURE 4. Isotropic hyperfine interaction with β-protons.

(28,81-82) in terms of the variation in angle θ between the methylene C-H δ-bonds and the p-orbital direction, projected perpendicular to the δ-bonds (Figure 4). The coupling thus follows a $\cos^2\theta$ law, and the two methylene β-protons can be distinguished by having different hyperfine splittings. For example, the ESR spectrum of the radical $(HO_2C)CHCH_2(CO_2H)$ trapped in a crystal showed that the β-proton $H_{(1)}$ has a larger interaction (~36 gauss) than the β-proton $H_{(2)}$ (~28 gauss) in the methylene group (83,84).

We shall mention briefly the hyperfine interaction with β-protons in substituted aromatic radicals, such as the toluene radical anion. In such cases, there are two ways in which unpaired spin-density can be induced at the methyl proton sites. The first is by hyperconjugation inducing positive spin-density and the second is by spin polarization causing negative spin-density. The resulting sign of the spin-density at the methyl proton sites will depend upon the relative contributions of these two opposing effects. In some cases, these two effects of the two mechanisms can be separated experimentally (85) by a systematic study of a series of cations and anions in which the effect of hyperconjugation is found to be more significant in cations than in anions.

Hyperfine interactions with γ-protons in aliphatic radicals are usually very small (86). However, significant long-range hyperfine coupling with γ-protons in the neopentyl (87) radical and in a variety of bicyclic and polycyclic alkyl radicals (88-91) has been observed.

III. SOME EXPERIMENTAL ASPECTS

A comprehensive and thorough discussion of all basic aspects concerning ESR spectrometers has been given by

Poole (92) and by Wilmshurst (93). Excellent summaries of techniques were recently given by Ingram (94) and Ayscough (95). In this section we intend to present only a simple abstract of the salient features concerning the performance of ESR spectrometers, and the basic problems associated with experimental analysis of ESR data.

A. Performance of ESR Spectrometers

In general the most desirable features in a spectrometer are high resolution and sensitivity. Resolution of an ESR spectrometer depends basically on the time stability of the magnetic field and the microwave frequency as well as the homogeneity of the magnetic field. The spectrometer sensitivity, expressed in terms of minimum detectable number of unpaired electrons N_m, is related to a variety of experimental parameters (94) grouped in the following equation:

$$N_m = \left(\frac{4k\ T_S}{g^2 \beta_e^{\ 2}}\right) \cdot \left(\frac{\Delta H}{H_o}\right) \cdot \left(\frac{1}{Q_o \pi \eta}\right) \cdot \left(\frac{Fk\ T_D\ \Delta\nu}{P_o}\right) \qquad (3\text{-}1)$$

The first group in eq. (3-1) concerns the g-value and the absolute temperature T_S of the sample being measured. The second term is related to the experimental conditon H_o, the resonance field, and the spectral property of the sample ΔH, the line-width measured in gauss. The third group involves the parameters of the resonance cavity; the factor Q_o measures the relative microwave energy stored in the cavity, and the filling-factor η measures the effective microwave magnetic field filled by the sample in the cavity. The last group of parameters is mainly associated with the detector and the maximum available microwave power P_o. The properties of the detector are the noise of the amplifier F, the temperature T_D and the band width $\Delta\nu$. For a typical X-band spectrometer, N_m is of the order of 5×10^{10} ΔH spins.

Thus the design of most commercial instruments now available is governed by resolution and sensitivity considerations. Highly homogeneous and stable magnetic fields can be achieved with low-impedance electromagnets, and recent development of solid-state sources may lead to a more stable microwave source than the klystron. In a conventional ESR spectrometer the sample is placed in a microwave cavity of constant frequency held in a homogeneous magnetic field, and the ESR spectrum is obtained by sweeping slowly through a range of magnetic field. The field is modulated at some audio- or radio-frequency (usually ~100 kHz), and this produces signals at the crystal detector whose output can be amplified at the modulation frequency. In general, sensitivity is higher at high-modulation frequencies because of the linear variation of excess crystal noise with the inverse of frequency. However, the recently developed Schottky diode makes it possible to obtain good sensitivity at relatively low frequency.

The most popular commercial ESR spectrometers employ microwave X-band components. At a frequency around 9,000 MHz, the resonance field for most organic radicals will be about 3300 gauss. A variety of high Q cavities with designs for specific purposes (e.g., optical transmission, large sample access) are also available with most commercial spectrometers. For higher sensitivity, Q-band spectrometers are also available since increase of the resonance frequency (hence the magnetic field) should give higher absolute sensitivities as suggested by the second term in eq. (3-1). The magnetic field associated with Q-band spectrometers is normally about 13,000 gauss and is still higher if very high-frequency millimetre spectrometers are used. However, sample size is often very limited in Q-band or higher-frequency spectrometers. Apart from the sensitivity consideration, it is often desirable to have both X-band and Q-band sources available for the purpose of separating and identifying radicals existing simultaneously. By changing the microwave frequency, radicals with different g-values will be separated in the ESR spectra, but their hyperfine splittings are unaffected.

Another practical factor which can affect the performance of the ESR spectrometer is the nature of the sample to be studied. If the sample is in the form of a liquid with high dielectric constant (e.g., water), a specially thin sample-holder must be employed to keep the liquid sample well out of the regions of microwave electric field strength. Nevertheless, the "filling-factor" will always be smaller with such a "lossy" sample, and the sensitivity of detection will be reduced according to the third term of eq. (3-1) as well as due to the reduced volume of the sample allowed by the thin sample cell. Richards and coworkers (96) have designed a low-frequency spectrometer operating at ~130 gauss which allows use of large volumes of aqueous solution and have discussed the advantage and disadvantage of such low-field measurements as applied to quantitative concentration studies. Lebedev and Taranukha (97) on the other hand, studied the use of a slow-wave spiral to increase the filling factor of small "lossy" samples in a cavity. Samples having large line-width caused by a number of broadening effects will reduce the detection sensitivity and resolution. Although sensitivity could be improved with increasing microwave power, care must be taken to avoid broadening of line-width by microwave saturation.

For better analysis of poorly resolved ESR spectra, Silsbee (98) has attempted to use Fourier transforms of experimental spectra as compared to the model ones.

During the last few years the introduction of the time-average method has revolutionized the standard techniques for detecting very weak signals buried in noise. The availability of commercial small digital computers at low cost has made possible this great time-averaging technique as a standard accessory for most conventional ESR spectrometers. Such a device is often called a computer of average transients, or CAT for short. A comprehensive discussion of the principles of signal-to-noise improvement by the time-averaging method has been given by Ingram (94,99). The basic idea of the method can be summarized in the following. If $\underline{n}$ repeated sweeps have been fed into the computer digital memory circuits for storage, the stored magnitude of the signal will add linearly and thus will be directly proportional to $\underline{n}$, whereas the random noise stored will be proportional to $\sqrt{\underline{n}}$ only. Thus the signal-to-noise ratio will grow as $\sqrt{\underline{n}}$, and for a large number of sweeps, the actual signal strength can be integrated out of the noise background. This effect can be dramatically shown in Figure 5, where the spectrum after a single sweep from a trace quantity of DPPH shows little, if any, of the ESR signal, but the spectrum after 4,000 sweeps recovers a distinct signal from the noise. Because of the large number of sweeps often required for this effect, it is desirable to have rapid field-scanning facilities synchronized to a fast digitizing circuit in the CAT. Digital computers with as low as one microsecond dwelling time are now available commercially. The severe limitation is, however, the speed at which the magnetic field can be swept linearly.

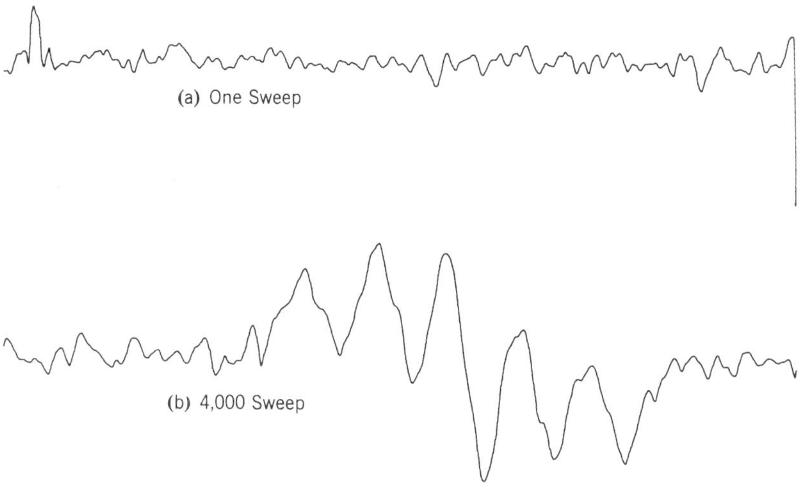

FIGURE 5. Spectra obtained from a weak solution of DPPH in benzene using a CAT: (a) One 20-ms sweep, (b) 4,000 20-ms sweeps.

Thus the method of time averaging is not only effective in enhancement of the static signal but also makes possible observation of rapidly changing ESR spectra due to short-lived transient radicals.

Finally, we must mention the excellent work of Carrington (100), who with his coworkers developed the Stark modulation cavity for electron resonance study in the gas phase.

B. Analysis of Experimental ESR Spectra

Detailed accounts of interpretation and analysis of complex ESR spectra have been given by many authors (95,101,102) but there are certain aspects which are worth repeating here. We shall be concerned in this section with only isotropic and powder spectra since these are the most likely spectra a photochemist will encounter in his study. Techniques and methods of analysis of single-crystal spectra can be found elsewhere (95,101).

1. Ambiguities of Multiradical Spectra

Often more than one kind of radical may be trapped in UV-irradiated samples at low temperature. Interpretation of the resultant complex ESR spectra requires special care to identify whether some of the ESR lines are part of a hyperfine component or whether they are due to another kind of radical. Such ambiguities may sometimes be resolved in the following ways.

a. By gradually changing the sample temperature. Assuming the chemical rates of decay are appreciably different for different kinds of radicals at relatively high temperature, it may be possible to distinguish the radicals by allowing the sample to warm up slowly, hoping that one kind of radical will disappear from the observed spectra more quickly than the others. For example, photolysis of a benzene solution of triphenylmethanethiol at 77°K leads to a complex ESR spectrum (103) due to both the sulfur radical $Ph_3C\dot{S}$ and the carbon radical $Ph_2(\dot{C}_6H_6)CSH$. When the irradiated sample was warmed to 153°K for several hours, the carbon radicals disappeared and the ESR spectrum of the sulfur radicals was clearly observed. However, it must be noted that higher temperature can, in some cases, lead to further interesting complications such as intramolecular radical isomerization (89,104) and secondary reactions producing new kinds of radicals (105).

Another interesting aspect one may encounter in low-temperature solid-state photolysis is the phenomenon of pairwise trapping of two identical radicals formed by the photodecomposition of a single molecule. For example, photolysis of solid azobisisobutyronitrile at 77°K leads to pairwise trapping of two $\dot{C}(CH_3)_2CN$ radicals separated by a distance of ~6 Å (106) and photolysis of tetraphenylhydrazine crystal at low temperature produces $\dot{N}(C_6H_5)_2$ radicals trapped in pairs with a separation of about 6 Å (107). These radical pairs behave magnetically like triplet states and will modify the $\Delta M_S=1$ spectra as well as exhibit new spectra of $\Delta M_S=2$. Thus the high-field spectra ($\Delta M_S=1$) may be composed of those radical pairs and the isolated radicals. For hycrocarbon radical pairs trapped in irradiated polymer, Iwasaki and coworkers (108) have shown that with increasing temperature the thermal decay of the radical pairs is faster than that of the isolated radicals. This is reasonable as recombination in radical pairs would be facilitated by the short distance. The resulting ESR spectra at higher temperature become characteristic of the isolated radicals.

b. By microwave power saturation. Radicals of different nature and structure usually have different relaxation times and therefore can sometimes be separated by broadening sufficiently the lines of one kind of radical at a certain microwave power level. In this respect it is worthwhile to point out that radicals with orbitally degenerate ground states usually have exceedingly short relaxation times, and their lines are difficult to saturate by high power (109-113).

One further point should be noted regarding interfering ESR absorption due to quartz sample tubes after prolonged exposure to UV-irradiation. This is often annoying and can be minimized by using "suprasil" instead of ordinary quartz. These "quartz signals" are normally due to paramagnetic color centers involving trapped electrons and can be readily saturated at high power.

c. By changing the microwave frequency. If the different kinds of radical have appreciable variation in their g-values, their components in an isotropic complex ESR spectrum can be separated by obtaining the spectrum at two different microwave frequencies (e.g., X- and Q-band). Their individual hyperfine lines are, however, independent of the microwave frequency.

d. By changing the solvent. Photolysis of organic compounds in a solution with an unfiltered light source often leads to formation of both solvent and solute radicals. In fact, we have demonstrated (113) that at low temperature, most of the common solvents including methanol, ethanol, ether, benzene, dioxane, isopentane, tetrahydrofuran, and others produce ESR spectra upon irradiation with an unfiltered high-intensity light source. Similarly, transient species have also been observed by optical absorption in the flash photolyses of most of these common solvents (114). If one suspects that the complex spectra contain components due to solvent radicals, a change in solvent or use of deuterated solvents may help to resolve the ambiguity.

2. Measurement of g-Values

The g-values of most organic free radicals are within 2% of the free electron value and thus do not provide a potentially powerful criterion for radical identification. However, Norman and Pritchett (115) have noted that substitution of hydroxy or alkoxy groups on the α-carbon in simple hydrocarbon radicals increases the g-value by about 0.001, while α-aldehydic or ketonic groups increase the value still higher, about 0.002. It has already been mentioned in Section II.B.2 that because of large spin-orbit coupling, most radicals with spin localized or heteroatoms, such as oxygen and sulfur, have g-values appreciably different from that of a free electron. These limited observations may serve as a guide for identification of some radicals.

Experimentally the measurement of the g-value in an isotropic, symmetrical ESR spectrum is straightforward by using the simple equation

$$h\nu = g\beta_e H_o \qquad (3\text{-}2)$$

$$\text{numerically, } g = \frac{\nu\,(\text{MHz})}{1.39969\ H_o\ (\text{gauss})}$$

where H_o is the magnetic field at the center of the spectrum. Both H_o and ν can be measured with high accuracy by using a fluxmeter and a frequency counter, and the resulting measurements lead to a maximum error in g of about $\pm$ 0.00012. If the costly fluxmeter and counter are not available, g-values can be determined with reasonable accuracy by measuring the microwave frequency with a wavemeter and calibrating the field against a standard of known g-value. The peroxylamine disulfonate ion $NO(SO_3)_2^{=}$ is about the best standard for this purpose; its g-value is equal to 2.00550 and its ^{14}N hyperfine splitting is 13.00 gauss. The radical ion can be made more stable for calibration by making up the aqueous solution in 0.05M sodium carbonate.

In an asymmetric spectrum, the center of the spectrum may not coincide with the "cross-over" point referred to as the base line. This uncertainty is removed when a hyperfine pattern is observed. In the absence of hyperfine structure, Blois et al. (15) have shown that it is necessary to run the sample in a series of tubes of different diameter, and the true g-value is then obtained by extrapolating to zero tube diameter.

With a second-order spectrum, the g-value cannot be directly measured from the center of the spectrum as the second-order lines are all shifted to low field and the true g-value will be less than that measured at the "center" of the second-order spectrum. A simple correction is given by Atkins and Symons (101).

In solution samples or in a solid matrix where the radicals can tumble rapidly, only the average g-value can be measured. For powder or polycrystalline samples the spectrum is the envelope of spectra arising from all possible orientations of the radical. It is often possible (95) to determine directly from the shape of the first derivative ESR line the values of either the axially symmetric (g_1, g_1, g_{11}) or the fully anisotropic (g_1, g_2, g_3) g-tensor. The major disadvantage is that the orientation of the radical in the medium cannot be determined, but this is less important in most photochemical studies.

3. Determination of Experimental Hyperfine Coupling Constants

In general there are various ways to improve the hyperfine resolution in a given spectrometer by optimization of a few experimental parameters. In practice this is often difficult to accomplish as there may be overriding requirements that the experiments be carried out under a given set of conditions. For example, it is often necessary to carry out photolysis at a very low temperature in order to obtain an observable steady-state concentration of the transient radical. It is clearly easier to deal with isotropic spectra, but these require that the radicals rotate rapidly in the medium (116). At low temperature in a solid matrix the restriction of rotation may contribute to a large anisotropic broadening, resulting in smearing of the hyperfine structure. On the other hand, reducing the temperature may increase the relaxation time and thus decrease the line-width of the hyperfine lines. Fortunately since the monumental work of Fessenden (86) the technique has been modified for in situ photolysis of liquids at low temperature using cyclopropane and ethane as inert liquid solvents. This allows tumbling of the radicals in the liquids, and well-resolved spectra with narrow lines can be obtained. Obviously other line-width broadening effects due to saturation by high-power and high-modulation amplitude must be avoided. It is also important to ensure that radical concentrations are low enough to avoid electron spin exchange which can cause either broadening or narrowing of the spectra, depending on the circumstances (7). Often it is necessary to exclude the presence of oxygen in the sample before being photolysed as small traces of oxygen can induce paramagnetic relaxation of the radicals and thus broaden the line-width.

In the analysis of a complex ESR spectrum, there are no sets of simple rules to follow. Through experience and perhaps some chemical intuition, careful examination of the spectrum often reveals trends or patterns which may help in the analysis. Some useful patterns of spectra from equivalent protons at various stages of resolution have been offered by Poole and Anderson (117). Obviously effort must be made to make certain that all lines expected or unexpected have been obtained. Lines in the wings of a spectrum are often most important since here the overlapping is usually less. However, the intensity of these outer lines can be very small relative to the centered lines and their observation can be difficult under low signal-to-noise ratio condition. For some systems with "equivalent nuclei," anomalous relaxation may give rise to alternation of hyperfine line-widths (118) and in the extreme case may lead to "disappearance" of certain hyperfine lines.

When complete resolution cannot be obtained even under the best experimental conditions, reasonable coupling constants can be estimated by comparison of the observed spectrum with a number of model spectra simulated by computation using a combination of suitably chosen splitting constants and line-widths. This is a relatively simple procedure for poorly resolved "isotropic" spectra, and standard computer programs for such purposes are readily available (e.g., from the Quantum Chemistry Program Exchange, Indiana University). With powder spectra having considerable structure which arises from the g and hyperfine tensors, the theory of the spectrum shape becomes much more complicated (119). Recently an extended computer simulation of powder spectra of "polycentric" π-radicals has been given by Owen and Vincow (120).

After analyzing the spectrum in terms of various hyperfine coupling constants, the next important step is to assign the constants to the appropriate magnetic nuclei. This is not always straightforward, and ambiguous and mistaken assignments are not unknown in the literature. Frequently for organic radicals, confirmation of assignments can be made by selectively replacing magnetic nuclei with isotopes of different nuclear spin and/or magnetic moments. For examples, hydrogen replaced by deuterium, ^{14}N by ^{15}N, etc. However, subtle complications may arise from deuterium substitution in degenerate π-radicals (121-124). Observation of splittings from the 1.1% of naturally occuring ^{13}C, when possible, also can help in the assignments of proton coupling constants (125,126). Despite these various experimental approaches, not all spectra can be assigned unambiguously. As a last resort, an experimentalist may then turn to simple molecular orbital calculations with the hope of providing some guidance for the assignment. For this purpose, the most popular methods are the approximate SCF theory of McLachlan (127) and the INDO calculations suggested by Pople and coworkers (128). The former method is particularly suitable for calculations involving large aromatic π-radicals. However, it must be recognized that such theoretical approaches to the prediction of hyperfine coupling constants (or spin-densities) are not infallible. Indeed, significant improvements of theory are usually made after it fails to predict the "unexpected" and the "unpredictable" experimental result.

While we are still on the subject of cautiousness in interpretation of ESR spectra, it is worthwhile to point out here a few notorious cases which merit special care. These are the elusive alkoxy radicals and the conspicuous nitroxide radicals.

Photolysis of hydrogen peroxide or organic peroxides in the presence of organic substrates has been a very popular technique in ESR studies. The $\dot{O}H$ or $\dot{R}O$ radicals generated can readily abstract hydrogen atoms from the substrate yielding the corresponding organic radicals. It is now realized that $\dot{O}H$ and $\dot{R}O$ radicals in liquid solution have an extremely short relaxation time due to various strong perturbations of the π-levels (129). Their spectra are therefore too broad for detection, and this fact should be kept in mind when analyzing a complex ESR spectrum produced by the peroxide photolysis technique. Liquid photolysis of organic amines and hydrazines and the reaction of NO_2 with liquid olefins are known to produce nitrogen radicals. However, there is often a distinct possibility that nitroxide radicals are also formed in these systems (130,131). Since the isotropic hyperfine coupling constants for a nitroxide radical are usually very similar to those of amino-type radicals, interpretation of ESR spectra in these systems would require extra caution.

C. Measurements of Spin Concentrations

Although quantitative measurements of spin concentrations are very important in kinetic studies, the determination of this absolute parameter by ESR remains most difficult. Nevertheless, this should not discourage a photochemist who is accustomed to similar problems, such as measurements of photoexcited triplet-state concentrations in flash photolysis.

A glance at the sensitivity eq. (3-1) should convince us to shy away from absolute spin measurements and compromise with the "calibration" technique using suitable primary or secondary standards. This becomes relatively straightforward if suitable primary standards with accurately known concentration can be found. The comparison is based on the fact that the area under the ESR absorption is directly proportional to the number of spins, provided the absorption is not saturated and all variable spectrometric parameters remain constant. In practice, two major groups of problems may develop.

The first group of problems involves the selection of standards and their method of comparison. A list of desirable properties for standards has been given by Alger (102) of which the most important is using a standard having the same line-width as that of the unknown. The more common standards employed are DPPH, $CuSO_4 \cdot 5H_2O$ and $MnSO_4 \cdot H_2O$. There are two general methods of comparing the standard with the unknown, the substitution and the superposition techniques. The former method, in which the standard and the unknown are measured one at a time, is less satisfactory than the latter method. A simple technique using a polycrystalline sample of CaO containing 0.01% $MnSO_4$ placed in a sealed capillary inserted into the unknown has been found to be convenient and reliable (132).

The other group of problems is associated with the actual measurement and the stability of the spectrometer. Since the number of spins is proportional to the area under the absorption curve and conventional spectrometers record the first derivative, the most reliable approach is to perform a double integration. A simple procedure using a desk calculator for numerical double integration of the first derivative curve has been given by Ayscough (95). Various mechanical, electrical analog devices and digital computers have been coupled to spectrometers for this purpose. The analog devices depend strongly on the stability of the spectrometer, and in practice when weak signals are to be measured at the limit of spectrometer sensitivity, the base-line drifts often render these analog devices impossible to use. This problem can be partially resolved by using a digital computer in which integration is performed after the spectrum is stored in the memory. Some careful base-line correction can be made before trial integration. The Fabri-tek 1070 instrument computers are hard-wired programmed for such operation, and a thorough examination performed in our laboratory indicated that a normal integration uncertainty of less than 1% can be obtained with this computer.

In some studies in which only the knowledge of the

relative spin concentration as a function of time is required, the measurement becomes greatly simplified. Thus the relative concentration is directly proportional to the "peak-to-peak" height of the derivative, provided the line is not saturated. In such cases, an accuracy of 2% can be expected. This is in marked contrast with the poor accuracy of "absolute spin concentration" measured by the comparison method, which can involve errors ranging from 20 to over 100%, depending on the system and the standard used.

In recent years ESR has proved to be an extremely powerful technique in the kinetic study of atoms in the gas phase (133). For gaseous systems Krongelb and Strandberg (134) have developed a method of measuring atom concentrations using a calibration procedure against molecular oxygen. Since most of such gas-phase studies involve a microwave discharge flow system at low pressures, and atoms such as N and H have rather long relaxation times,special care must be taken in the experimental design to ensure that all atoms reach their equilibrium magnetization before flowing through the resonance cavity for measurement. For example, Brown and Brennen (135) have dramatically shown that when nitrogen atoms are generated by discharge outside the cavity and flow toward the magnetic field, a change in flow or in the atom relaxation time could alter the ESR intensity by as much as a factor of seven, even though the atomic concentration is constant. In photochemical gas-phase reactions, this "hot atom" effect is less general but still must be considered.

IV. APPLICATION OF PHOTOCHEMISTRY TO ESR STUDIES

While the potential role of ESR in photochemistry has not yet been fully realized, there is little doubt that the photolysis method along with other radical-producing techniques has contributed to the success of many ESR studies. In this section we describe briefly some model photochemical systems which have been successfully examined by ESR. These studies in general did not completely concern "photochemistry" *per se*, but their success should point the way to a promising future for the use of ESR in photochemistry.

A. Methods for Producing Radicals

From the standpoint of ESR spectroscopy, the main objective is finding the best method to produce radicals yielding high-resolution ESR spectra. Obviously depending on the particular radical species, photolysis sometimes fails to produce the radicals or the required resolution, and employment of other available techniques is often necessary. In general, ESR spectral parameters of organic radicals are not strongly dependent on their method of production, and a good collection of confirmed spectral information is helpful for identifying poorly resolved ESR spectra in some photochemical systems. It is thus desirable to consider briefly the photolysis method and other standard techniques available.

Much of the earlier photolysis work was confined to the use of high-intensity steady illumination. Few gas-phase photolysis experiments have been successfully studied by ESR. In gaseous polyatomic radicals, only magnetic dipole transitions are accessible by ESR, and these spin transitions have probabilities about 10^4 times lower than the electric dipole transitions. This is one of the reasons why optical detection of gas-phase radicals is more sensitive than ESR observation. In addition, coupling of electronic orbital and spin momenta with molecular rotational momentum will lead to exceedingly complex ESR spectra for gas-phase polyatomic radicals. However, gas-phase photolysis or pyrolysis of organic substrates can be performed in a flow system in which the products including the radicals are trapped by condensation at low temperature. The limitation is the difficulty of controlling the gas-phase reactions so that only one elementary step can take place. Since the detection in this case is made on the condensed phase, it is virtually impossible to carry out reactivity studies of the trapped radicals in subsequent reactions. Gas-phase radiolysis will suffer the same limitation and could even be worse than photolysis because of the often lower yields of radical species. The more successful gas-phase ESR studies all involve microwave discharge in flow systems dealing with atoms and/or diatomic and triatomic species in which electric dipole transitions are observed. The discharge method is nonselective, and the most severe limitation at present appears to be the flow rate since all radicals are generated outside the cavity and some radicals may not live long enough to reach the cavity for detection. *In situ* flash photolysis is perhaps the most suitable technique for gas-phase ESR study.

Considerable success has been achieved in the study of gas-solid systems using a rotating cryostat developed by Thomas (136). This technique involves organic reactants continuously frozen on the surface of a rotating drum at low temperature and then subjected sequentially to bombarding molecules or atoms from the gas phase. Bennett, Thomas, and coworkers (137-138) have used this technique as a general method for preparing specific free radicals and radical ions. Other advantages of this method are its ability to control single-step reactions between specific radicals and molecular reactants and the convenience of being able to select the matrix independently of the radical precursor. Since the radicals are actually studied in a solid matrix, their ESR spectra resolution strongly depends upon the choice of the matrix. Thus Bennett and coworkers (138) have succeeded in preparing the elusive phenyl radical for high-resolution ESR study by the reaction of Na atom with phenyl iodide in both benzene and water matrix at 77°K. When the parent halide was used as the matrix, the observed spectra were poorly resolved.

Solid-state photolysis and radiolysis have always been popular as methods for producing radicals. The advantage of photolysis is that it is more selective than the high-energy radiolysis and the matrix can be selected independently of the radical precursor. Experimentally the solid-state radiolysis is, however, much simpler as polycrystalline samples can be used without "light-scattering" problems. In situ solid-state photolysis often gives a low steady-state radical concentration due to the cage effect. In photochemical homolyses, pairwise trapping of radicals is well known (129). In addition, further photolysis of the primary radicals may occur and leads to "secondary" radicals. Some of these drawbacks can be overcome by using the rotating cryostat (138) in conjunction with photolysis (Figure 6). Most solid-state photolyses were carried out with a low-temperature rigid glass; however, high-temperature photolysis of organic substrates (139,140) embedded in optically transparent solid polymer films or in a transparent solid solution of bulky hydrocarbon, such as adamantane (141), should be possible. The greatest advantage of in situ solid-state photolysis over all other methods is no doubt its applicability to ESR studies of phosphorescent triplet-state organic molecules.

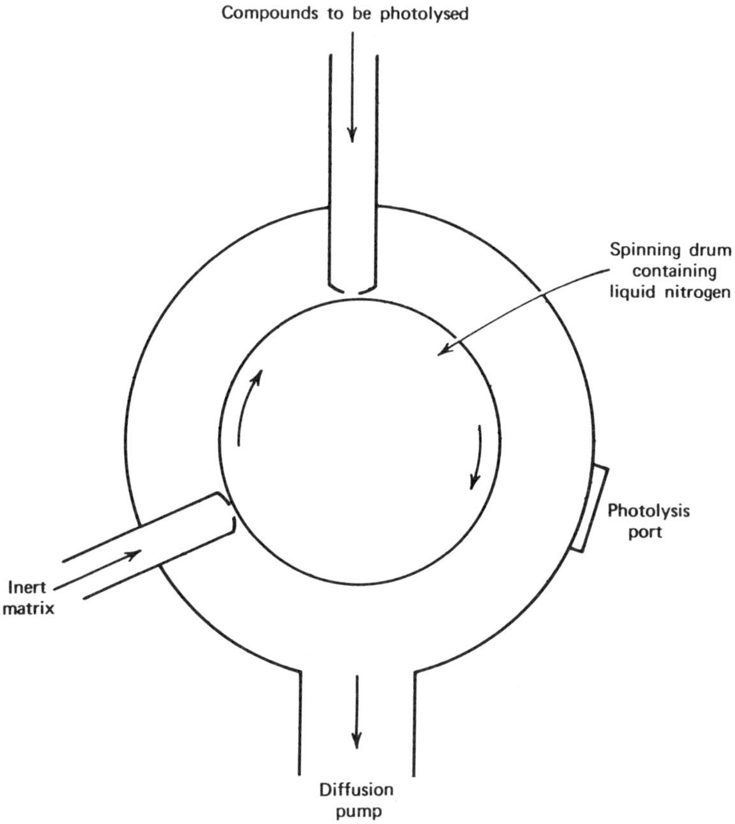

FIGURE 6. General arrangement of the rotating cryostat adapted for photolysis study.

Returning to solid-state radiolysis, this method is not only powerful in producing organic radicals and radical ions in both single crystals and polycrystalline matrix, but is also a convenient technique to generate paramagnetic color centers by radiation damage in inorganic solids. Because of the high cost involved, only a few laboratories are equipped to perform in situ radiolysis study. The samples are usually transferred to an ESR spectrometer after being irradiated, and the paramagnetic centers and radicals must live long enough in the solid to give rise to ESR spectra.

Another interesting method of producing radicals in the solid state is the application of mechanical stress to polymeric materials. For example, Campbell and Peterlin (142) have produced free radicals with well-defined spectra by stretching nylon under vacuum within the ESR cavity. Such an effect is fascinating and makes one wonder if similar results can be obtained by subjecting other kinds of solids to various types of mechanical stress or high pressure.

For liquid-phase ESR studies there are many interesting techniques available. The classical work of Fessenden and Schuler (86) demonstrated that well-resolved spectra of unstable radicals can be generated in static liquid systems by radiolysis at low temperatures. Subsequently Kochi and coworkers (143) have achieved the same goal by photolytic means. Both methods lead to much information of great theoretical and experimental interest concerning radical structure, but their applicability is limited to systems that can exist as a liquid phase at low temperatures. On the other hand, pyrolysis of static solution systems at high temperatures can also produce unstable species for ESR study. For example, McIntosh and Wan (144) have recorded a series of well-resolved spectra of various triphenylsiloxy-diphenylmethyl radicals (Figure 7) in a liquid solution of benzophenone and triphenylsilane at high temperatures. More stable radicals and radical-ions can also be readily generated in static liquid systems by appropriate redox reactions of alkali metals with organic substrates such as polycyclic aromatic hydrocarbons.

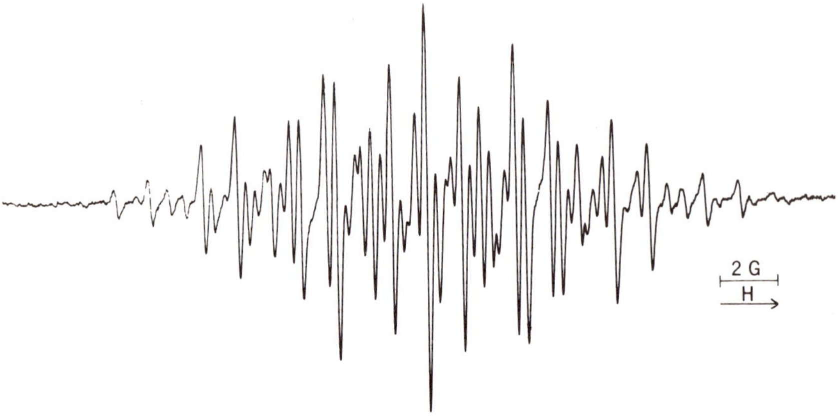

FIGURE 7. ESR spectrum of $Ph_2COSiPh_3$ radical produced by thermal reaction of benzophenone with triphenylsilane. Figure reproduced from reference (144).

Static systems, however, are not usually suitable for rate studies. Fessenden (145) was the first to realize this and modified the static system for "intermittent radical production" using pulsed radiolysis. With the advances in electronic digital equipment, Smaller and coworkers (146) have subsequently fully developed the pulse-radiolysis technique for ESR studies and have successfully detected radicals with lifetimes as short as two microseconds. Concurrent developments of such "intermittent radical production" concepts have also been accomplished in photolytic systems by using either a rotating sector technique (147,148) or the flash photolytic technique (6). At present the pulse-radiolysis technique enjoys the advantage of a short and intense pulse at a rapid repetition rate. Only flash photolytic systems using a pulsed laser can approach these desirable conditions. These techniques will no doubt be continuously improved, and their future in ESR study of transient radicals appears quite promising.

Study of nonrepetitive biological transient reactions in solution has led to the development of continuous-flow systems first pioneered by Piette (149), Chance (150), and their coworkers. This was subsequently extended by Stone and Waters (151) and Dixon and Norman (152) as a simple and convenient method for the formation of organic radicals and the study of radical reactivity. Some of the current activities of the flow technique in ESR studies have been reviewed (153,154).

Basically the flow method involves the mixing together of two solutions of reactants capable of producing radicals and flowing the mixture through the ESR cavity. Various designs of flow cells have been introduced. In some of them, observation of the radicals is possible within a millisecond after the mixing, and in others, facility is provided for adding a third reactant solution at a specific time before or after the mixing of the other two. Since there is an upper limit to the flow rate, the success of this method depends on the magnitudes of the rate constants of both the primary radical producing process, k_1, and the process of radical destruction, k_2. For a diffusion-controlled process, $k_2 \sim 10^{10}\ M^{-1}\ sec^{-1}$, and k_1 is required to be in the range of 10^2 to $10^6\ M^{-1}\ sec^{-1}$. The reduction of H_2O_2 with Ti^{+++} giving $\dot{O}H$ radicals is among a number of initiating reactions which fulfill this requirement, and investigation of the reactions of $\dot{O}H$ radicals with various organic substrates has been very successful. Apart from the practical limitation, the major disadvantage of this method is the consumption of large quantities of reacting solutions and the difficulty of removing the undesirable oxygen from them. Because of the dimensions and geometry of the flow cell within the ESR cavity, meaningful quantitative rate measurements are difficult to obtain. Nevertheless, Weissman and coworkers (155) have applied a stopped-flow technique for ESR study of the rate of H atom transfer between phenol and phenoxyl radical. The flow system can also be applied to photochemical systems to circumvent the difficulty of obtaining high enough transient radical concentration for better resolution, as demonstrated by Livingston and Zeldes (156).

Since the Ti^{+++}/H_2O_2 reduction is an important system for producing OH radicals and has been used for many studies involving alcohols, it is perhaps interesting to note some of the ESR properties of the Ti^{+++} ion. The aqueous Ti^{+++} ion is believed to be coordinated as $Ti(H_2O)_6^{+++}$ and has no observable ESR spectrum because of the very short relaxation time. However, if the symmetry is reduced to tetragonal or lower, the orbital momentum should be completely quenched and a narrow ESR line is expected. This phenomenon has been observed in water-alcohol solutions (157). Recently Bolton and coworkers (158) have further observed some proton hyperfine structure from the water ligands of the aqueous Ti^{+++} complex in a 20% butyl alcohol-water solution. There has been a standing controversy in the interpretation of the spectra detected in the Ti(III), Ti(IV)-H_2O_2 system (159-162). More recent careful studies (163) appear to have resolved the problem, and assignment of at least part of the spectra as due to the $[TiOO]^{+++}$ complex seems reasonable.

The development of the electrochemical technique for generation of radicals and radical ions directly in the resonance cavity added a new dimension to the applications of ESR spectroscopy. Following some early electrolysis experiments (164,165), Maki and Geske (166-169) developed the elegant *in situ* generation technique which paved the way to a full realization of the value of the electrochemical method in ESR studies. Extension of this *in situ* technique was made by Fraenkel and coworkers (170) to employ external generation of the radical species and to flow them through the cavity by a variety of pumping methods for ESR detection. The electrochemical method has obvious advantages over redox production of radicals and radical ions by chemical reagents. For example, the choice of solvent is less restricted in the electrochemical technique with a larger range of controlled redox potentials available. In the case of radical anions, their interaction with the bulky tetraalkylammonium cations used as the supporting electrolytes is much smaller than with the metal cations formed when metallic reductants are employed. Some of the disadvantages of such electrochemical methods include the difficulty of correlating quantitative ESR interpretations in terms of electrochemical parameters; often precise polarographic measurements are required to establish the conditions for carrying out the electrolytic reduction and for optimum ESR observations. Dohrmann and Vetter (171) have studied the dependence of minimum detectable life time of radicals on current density, electrolytic conductivity, and the dimensions and nature of the electrode. An overall assessment of the electrochemical technique has been given by Adams (172). In recent years Dessy and coworkers (173) have extended the electrochemical method to prepare organometallic radicals for ESR study, and Hirasawa and coworkers (174-175) have coupled the pulse-electrolysis technique with ESR to study kinetics of fast electrode processes. Because of the versatility in arrangements of the electrolysis cell within the resonance cavity, combined *in situ* photo-electrochemical study appears possible.

B. Photochemical Generation of Paramagnetic Species

Having briefly summarized most of the common methods available for ESR studies, we now turn to some illustrations of those fruitful areas where photochemical technique has made a significant contribution.

1. Inorganic Radicals

A recent review by Symons (176) dealt with ESR studies of radiation damage in inorganic solids. It is well known that trapped electrons or F centers may be produced by photolysis of inorganic compounds. Recently Froben and Willard (177) reported the photoionization of metal atoms, such as Li, K, Na, Mg, and Cd, and observed by ESR the electrons trapped in 3-methylpentane at 77°K.

The hydrogen atom, the simplest inorganic radical, continues to be the favorite subject of many ESR studies. Jen, Foner, and coworkers (178) have extensively studied trapped hydrogen atoms in rare gas matrices by the photolysis method. Delbecq and coworkers (179) produced interstitial H atoms by UV-irradiation of alkali halide crystals containing U-centers. Photolysis of KCl or KBr doped with KSH or KSeH (180) also led to trapping of H atoms at 20°K. Upon annealing to 110°K, ESR spectra of S^-, H_2S^- and Se^- were observed (180). H atoms have been produced in good yield by photolysis of Co(II) in frozen acidified aqueous solutions (181). Recently Weiss and coworkers (182) have investigated the effect of dissolved oxygen on the ESR intensity of trapped H atoms. Wong and Wan (183) have also studied the effect of paramagnetic metal ions in enhancement of ESR absorption by H or D atoms formed by γ- or UV-irradiation of frozen acids. An ESR spectrum attributed to trapped P atoms has been observed by photolysis of phosphine in a rare gas matrix (184). Nitrogen and oxygen atoms have not been observed in a photolysed matrix; the former is difficult to produce by irradiation and the latter would probably give lines too broad for detection. A similar reason can also explain the failure to observe halogen atoms in a matrix. However, I atoms have been extensively studied by steady-state photolysis of molecular iodine vapor (185-187). Photolysis of rare gas matrices containing small amounts of Cd, Cr and Mn metal atoms together with some HI yielded paramagnetic species of the metal monocations (188a). Apparently the presence of HI serves to scavenge the electrons photoejected from the metal atom. Similar paramagnetic metal monocations have been observed in the photoreduction of corresponding M^{++} ions (188b).

A number of diatomic and polyatomic inorganic radicals have been produced in photochemical systems. Some of these radicals are listed in Table 2. Recently Martin and Swift (189) observed, during the photolysis of ceric perchlorate acid glass, ESR spectra similar to those of methyl radicals and intuitively assigned them to $H_3\dot{O}$ radicals. Given that these spectra have been correctly analyzed, it is difficult to understand how their ESR parameters could relate to the δ radical $H_3\dot{O}$. Confirmation of the assignment must await further detailed study involving ^{17}O isotopic substitution.

Inorganic molecular radical ions have also been produced by the photolysis method. Pairwise trapping of SO_4^- radicals separated by 13-19 Å has been observed in UV-irradiated potassium persulphate crystals (190). Synthetic sodalites containing sulfur and displaying

TABLE II
Some Inorganic Radicals Produced by Photolysis Method[a]

Radical	Phase	g-Value	Hyperfine Splitting (gauss)
$\dot{C}N$	Ar matrix		$a_N^{iso} = 4.6$
$\dot{C}lO$	Rigid H_2SO_4 solution	$g_{av} = 1.9927$	$a_{Cl}^{iso} = -5.7$
$\dot{N}H_2$	Ar matrix	$g_{av} = 2.0048$	$a_N^{iso} = 10.3$;
			$a_H^{iso} = 23.9$
$\dot{N}D_2$	Ar matrix		$a_N^{iso} = 12.1$;
			$a_D^{iso} = 3.68$
$\dot{P}H_2$	Kr matrix	$g_{av} = 2.0087$	$a_P^{iso} = 80$;
			$a_H^{iso} = 18$
$\dot{P}Cl_2$	Solid	$g_{\perp} = 2.021$	$a_P^{\perp} = 28$; $a_{Cl}^{\perp} \simeq 0$
		$g_{11} = 2.001$	$a_P^{11} = 269$; $a_{Cl}^{11} = 15$
$\dot{P}Cl$	Solid	$g_{av} = 2.013$	$a_P^{iso} = 1206$;
			$a_{Cl}^{iso} = 62.7$
$\dot{S}iH_3$	Liquid (-153°C)		$a_{^{29}Si}^{iso} = 266$;
			$a_H^{iso} = 7.96$

[a]Compiled from references (34), (95), (101), (198c), and (258).

photochromic activity have been studied by ESR (191). The color center responsible for the ESR spectrum was assigned to an electron trapped at a chlorine vacancy and interacting with four equivalent ^{23}Na atoms. The radical ion S_2^- was believed to be the origin of this photochromic electron transfer process. The radical ion N_2^- has been observed in UV-irradiated alkali azides (192) and photolysis of alkali cyanides produced a radical believed to be $C_2N_2^-$ (193). Dihalide radical ions, I_2^-, Br_2^-, and Cl_2^- have been prepared by photolysis of a metaphosphate glass containing the corresponding halide, but only Cl_2^- showed an ill-defined ESR spectrum (194).

2. Organic π-Radicals

The photolysis method of producing radicals in low-temperature matrices usually leads to less-well-resolved ESR spectra and has been applied to relatively few systems. The development of the liquid photolysis of alkyl peroxides in the presence of organic substrates at low temperatures (143) marked the beginning of an exciting chapter for application of photochemistry to high-resolution organic radical ESR studies. High-intensity photolysis of hydrogen peroxide or alkyl peroxide with organic compounds in a flow system at room temperatures (156,195) has also been fruitful in ESR investigations. To illustrate the potential of these methods in producing high-resolution ESR spectra, Figure 8a shows the spectrum of ethyl radicals with partially resolved second-order splittings obtained at -114°C by Krusic and Kochi (196). Numerous ESR spectra suitable for framing have been given by these authors for a large variety of radicals. Figure 8b gives the ethyl alcohol radical spectrum at 30°C obtained by Livingston and Zeldes (197) in a photochemical flow system. These authors have extensively studied alcohol and nitrile radicals. We will briefly summarize most of the interesting data into the following groups of organic π-radicals.

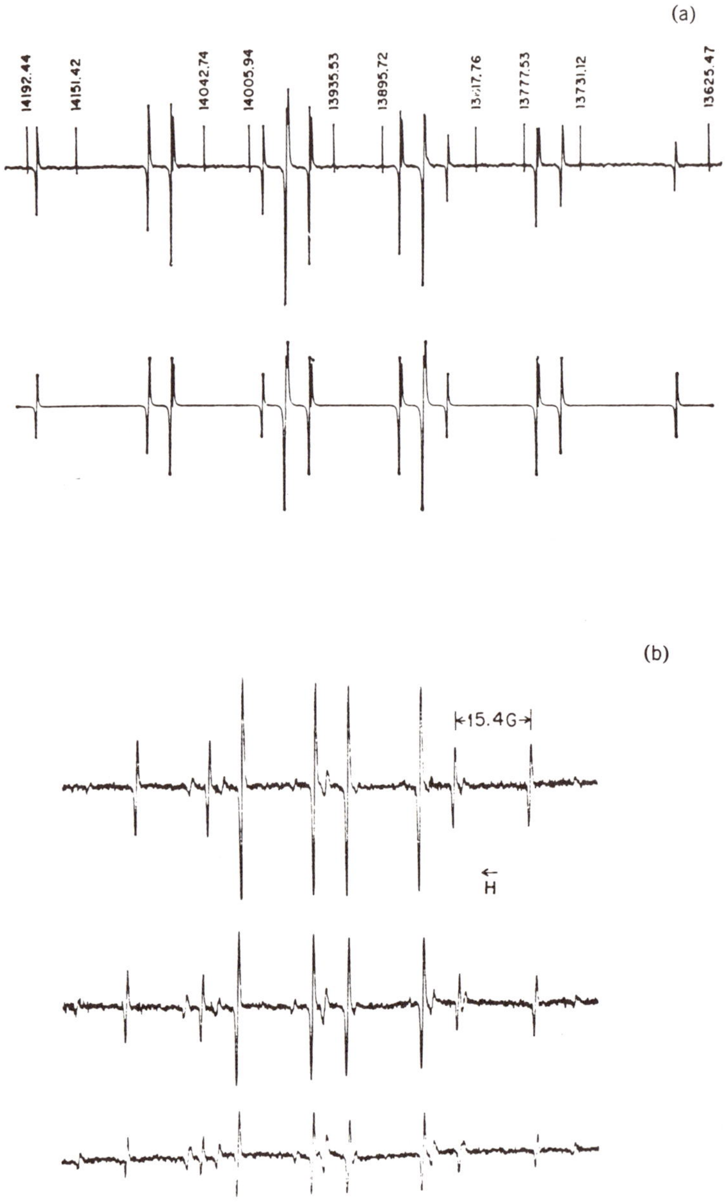

FIGURE 8. High-resolution ESR spectra of radicals produced by liquid photolysis: (a) The ethyl radical, reproduced from the work of Krusic and Kochi (196); (b) The ethyl alcohol radicals, reproduced from the work of Livingston and Zeldes (197).

a. Aliphatic Hydrocarbon π-Radicals. Some of the aliphatic radicals studied by Kochi and Krusic (143,196,198) are listed in Table 3. Their monumental work greatly strengthens our knowledge of radical structure. Those who are interested in this aspect should consult the original references. The simplest aliphatic radical $\dot{C}H_3$ first escaped detection in the liquid phase but was later observed by Riveros and Shih (199) in an aqueous photolysed solution containing 2% H_2O_2 and 10% CH_3CHO. These authors also studied the isotopic effect in the hyperfine splitting on the basis of vibrational consideration at different temperatures. Kochi and Krusic (196) also produced a high yield of methyl and other alkyl radicals at -110°C by reacting trimethylboron (and other organoboron, -aluminum, and -gallium compounds) with photochemically generated t-butoxy radicals. While there is no doubt that the methyl radicals produced in these systems are planar in structure, observation of a novel nonplanar type of methyl radicals trapped in a UV-irradiated porous vycor glass containing CH_3I at 77°K has been claimed (200). This "nonplanar" methyl radical has a smaller splitting of 19.3 gauss, shows an inverse line-width alternation, and possesses a different thermal stability from that of the "normal" planar radical. Further study including isotopic substitution is necessary to confirm this analysis.

We will also mention here the Group IV organometallic alkyl radicals studied by Krusic and Kochi (201a) and their extremely interesting investigation of the $\dot{C}_3H_3$ radical and its analogs (198b). These authors also studied the unsubstituted and substituted cyclopropylcarbinyl and allylcarbinyl radicals in solution (201b,c) and found that these radicals display appreciable γ-H

TABLE III
Hyperfine Splittings (gauss) for Aliphatic π-Radicals in Photochemical Systems[a]

Radical	a_{α}^{H}	a_{β}^{H}	a_{γ}^{H}	a_{δ}^{H}
$\dot{C}H_3$	22.8	--	--	--
$CH_3\dot{C}H_2$	22.3	26.8	--	--
$CH_3CH_2\dot{C}H_2$	22.1	30.3	0.27	--
$(CH_3)_2\dot{C}H$	22.1	24.7	--	--
$(CH_3)_3C\dot{C}H_2$	21.8	--	1.0	--
$CH_3\dot{C}HCH_2CH_3$	21.9	24.6 (CH_3) 27.2 (CH_2)	0.39	--
$CH_3CH_2\dot{C}HCH_2CH_3$	21.7	27.6	0.44	--
$CH_3\dot{C}HCH_2CH_2CH_3$	21.8	24.8 (CH_3) 24.8 (CH_2)	--	--
$(CH_3)_2\dot{C}CH_2CH_3$	--	18.1 (CH_2) 22.8 (CH_3)	--	--
$H_2C{=}C{=}\dot{C}H$	12.6 (CH)	--	18.9 (CH_2)	--
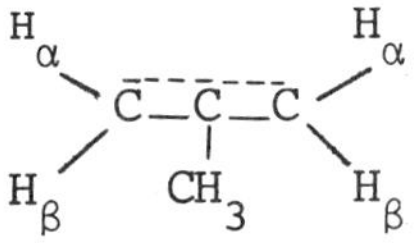	3.2 (CH_3)	14.6 ($CH_{2\alpha}$)	13.8 ($CH_{2\beta}$)	--
$CH_2{=}CH\text{-}\dot{C}H_2CH_2$	22.1	28.5	0.61	0.35
cyclo-$C_5H_9\dot{C}H_2$	21.3	21.3	--	--
$C_6H_5\dot{C}H_2$	16.3	5.15 (H_o)	1.77 (H_m) 6.18 (H_p)	--
$C_6H_5CH_2\dot{C}H_2$	22.0	29.2	--	--
$CH_3\text{-}CH{=}C{=}\dot{C}H$	11.7	--	18.7 (CH)	19.6 (CH_3)
$CH_3\text{-}C{\equiv}C\text{-}\dot{C}H_2$	18.5	--	--	12.3

[a]Taken from references (143), (196), and (198).

splittings. Some rationalization of the mechanisms of the radical-producing processes based on the ESR structural information appeared possible in these cases. Finally Davies and Roberts (202a) have claimed a general method for preparing specific alkyl radicals in solution by the photolysis of di-t-butyl peroxide in the presence of alkylboranes, and Hudson et al. (202b) have also studied methyl radicals containing heteroatoms such as S, Si, Ge, Sn, B, and P by similar methods.

b. Cyclic and Bicyclic Hydrocarbon Radicals. the γ-irradiation method has been particularly successful in producing well-resolved ESR spectra of cyclic and polycyclic radicals in solutions (86) and in solids (88,89,91,203,204). However, a few interesting cyclic and bicyclic radicals have also been observed in photochemical systems and these are given in Table 4. As mentioned earlier, appreciable γ-H splittings were observed in cycloalkyl radicals. Other cycloalkyl radicals have been listed by Ayscough (95), who also pointed out the peculiar feature of the near-constant α-proton splittings in the C_4-C_7 cycloalkyl radicals.

c. Aromatic Radical Ions. Aromatic radical ions have always been a popular subject for ESR studies, and in recent years these studies have played an important role in elucidation of biological problems. A great variety of methods have been employed in the oxidation or reduction of aromatic molecules to form radical ions, notably chemical methods and the electrolytic technique. Aromatic radical ions have also been produced photochemically. For example, Vincow and coworkers (205) have prepared radical cations of benzene, hexamethylbenzene, perylene, naphthalene, etc., by photoionization of the molecules in a rigid glass, while Hulme and Symons (206) have

TABLE IV
Cyclic and Bicyclic Hydrocarbon π-Radicals
Prepared by Photolytic Method[a]

Radical	a_H^{α}	Hyperfine Coupling Constants (gauss) a_H^{β}	a_H^{γ}	Other
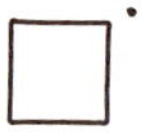	21.3	36.7	1.09	--
	6.04	--	--	--
cyclo-C_6H_6F	--	--	--	a_F^{iso} = 112
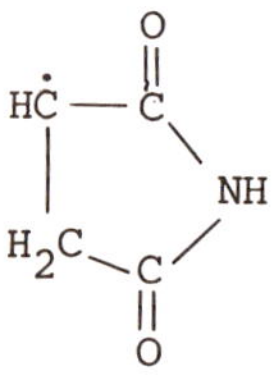	20.58	29.57	--	a_N = 0.40 a(NH) = 1.43
cyclo-C_7H_7	3.91	--	--	--
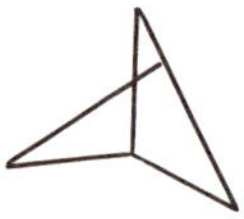	12.6	4.4	7.85	--

[a]Taken from references (90), (143), (217), and (269).

produced hexamethylbenzene and durene cation radicals in photolysed liquid solutions. Ayscough and Sargent (207a), Russell and Geele (207b), and Hudson et al. (75,208) have also generated anion radicals of nitrobenzene and benzophenone in solution by photolysis. Detailed accounts of radical ions have recently been given (109,209).

d. Halogen Substituted Hydrocarbon Radicals. McDowell and coworkers (210-211) have examined in great detail the $\dot{C}F_3$ radical in a UV-irradiated low temperature argon matrix containing CF_3I. Hudson and Hussain (212) have studied $\dot{C}Cl_3$ in the liquid state by the photolysis method. In contrast with the planar $\dot{C}H_3$ radical, both $\dot{C}F_3$ and $\dot{C}Cl_3$ are nonplanar as revealed by their ESR spectra. This phenomenon has been elegantly explained by Pauling (213) in terms of the different electronegativities of C, H, Cl, and F atoms.

Hudson and coworkers (75,214) have studied a number of fluorobenzyl radicals prepared by photolysis of solutions containing the corresponding fluorotoluene and di-t-butyl peroxide at -70°C. These workers (215) also investigated other chlorine and fluorine substituted alkyl radicals in photochemical liquid systems at -60°C. Their extensive studies revealed a number of interesting features concerning the halogen hyperfine splittings and the conformation of the radicals. The ESR parameters of some of these halogen substituted radicals are given in Table 5.

e. Radicals Derived from Alcohols, Ketones, Ethers, Nitriles, and Sulfides. A large number of radicals derived from alcohols and ethers have been prepared photochemically for ESR studies. Livingston and Zeldes have extensively studied alcohol radicals (19b,156,197,216), amide radicals (217), aliphatic nitroanions (218), and nitriles (219) by photolysis of solutions containing the corresponding molecules and hydrogen peroxide. Their illuminating work reveals many features of these radicals. Most of the alcohol radicals have hyperfine coupling constants slightly dependent on both temperatures and solvents. Ayscough and McClung (220) have also studied in detail the effect of temperature variation on proton hyperfine splitting of the biacetyl semidione radical in the range 250-400 K. Livingston and Zeldes (156) observed

TABLE V
Halogen-Substituted Hydrocarbon π-Radicals Generated in Photochemical Systems[a]

Radical	Hyperfine Coupling Constants (gauss)
$\dot{C}F_3$	a_F = 140.5 (Xe matrix), 141.8 (Kr matrix)
$\dot{C}Cl_3$	$a_{^{35}Cl}$ = 6.25, $a_{^{37}Cl}$ = 5.20
$\dot{C}HCl_2$	$a_{^{35}Cl}$ = 3.5, a_H = 20.5
$\dot{C}FCl_2$	a_F = 84.6, a_{Cl} = 10.5
$CF_3\dot{C}H_2$	a_H = 23.9, a_F = 29.9
$CF_3\dot{C}HCl$	a_H = 22.2, a_F = 23.4, a_{Cl} = 3.7
$CH_2F\dot{C}H_2$	a_H^{α} = 22.3, a_H^{β} = 26.9, a_F = 47.1
$CH_2Cl\dot{C}H_2$	a_H^{α} = 21.5, a_H^{β} = 11.5, a_{Cl} = 17.4
$CH_2Cl\dot{C}HCH_2Cl$	a_H^{α} = 21.3, a_H^{β} = 11.4, a_{Cl} = 14.2
$CH_2ClC{\equiv}C\dot{C}H_2$	a_H^{α} = 17.6, a_H^{δ} = 9.7, a_{Cl} = 4.9
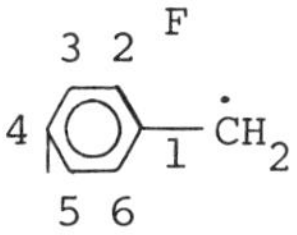	$a^{\alpha}_{(CH_2)}$ = 16.45, a_F = 8.17, a_H^3 = 1.80, a_H^4 = 6.48, a_H^5 = 1.80, a_H^6 = 5.36
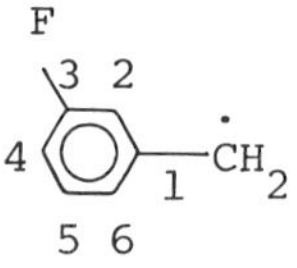	$a^{\alpha}_{(CH_2)}$ = 16.54, a_F = 4.87, a_H^2 = 5.31, a_H^4 = 6.36, a_H^5 = 1.85, a_H^6 = 5.06
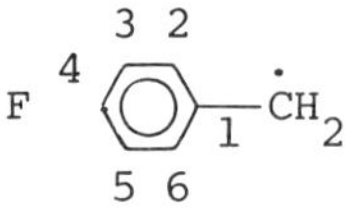	$a^{\alpha}_{(CH_2)}$ = 16.89, a_F = 14.53, a_H^2 = 5.47, a_H^3 = 1.74, a_H^5 = 1.74, a_H^6 = 5.47
$C_6F_5\dot{C}H_2$	$a^{\alpha}_{(CH_2)}$ = 16.80, a_F^2 = 9.5, a_F^3 = 4.9, a_F^4 = 17.1, a_F^5 = 4.9, a_F^6 = 9.5

[a]Taken from references (75), (212), (214), and (215).

a spectrum of 109 resolved lines for the $C_6H_5\dot{C}HOH$ radical, and in the case of alkyl alcohol, two stereoisomer radicals are observed.

Alkyl radicals derived from trialkoxyboranes have been generated photochemically by Krusic and Kochi (196). Hudson and Root (221) have studied aliphatic and cyclic ether radicals and substituted phenoxymethyl radicals in liquid solutions. The phenoxymethyl radical has also been observed in a photochemical system by Baum and Norman (222). Adams (223) has examined radicals generated by reactions of alkoxy radicals with alcohols, ethers, sulfides, and disulfides in photolytic flow systems at room temperature. Ketyl radicals have been extensively investigated by Wilson and coworkers (224). The aromatic ketyls exhibit hindered rotation in protic solvents, and the proton hyperfine splittings are found to vary with both temperature and solvent. Figure 9 shows that ESR spectra of acetophenone ketyl are appreciably different in methanol and in ethanol at room temperature (224). Because of the controversial nature of the photoreduction of aromatic ketones in amine solvents, the careful work of Davidson and Wilson (224d) on the photoreduction of fluorenone in triethylamine has given support to the postulate of an electron transfer mechanism for the N atom to the photoexcited carbonyl group. The impact of ESR study on mechanistic photochemistry will be considered in the next section.

Some of the radicals derived from alcohols, ethers, ketones, nitriles, and sulfides are given in Tables 6, 7, and 8.

f. Organo Silyl Radicals. Although hydrocarbon radicals have been extensively studied for many years, only recently has attention been given to the ESR study of

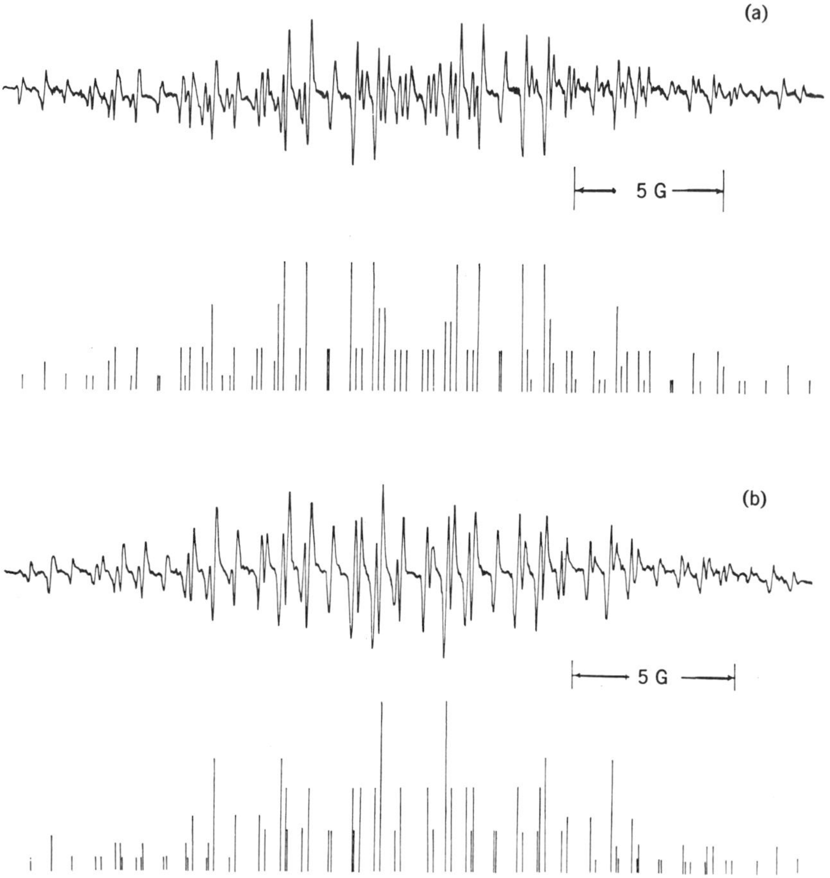

FIGURE 9. ESR spectra of acetophenone ketyl at 25°C in (a) methanol and (b) ethanol. Reproduced from the work of Wilson (224a).

their analogous silyl radicals (198c,225-226). Unlike the alkyl radicals, silyl radicals are nonplanar. There are indications that among the series R_nSiH_{3-n} the radicals become more nonplanar as R is replaced by H. This again could be attributed to differences in electronegativities of Si, C, and H atoms as suggested by Pauling (213). A number of "aliphatic" silyl radicals

TABLE VI
π-Radicals Derived from Alcohols and Carbonyl Compounds in Photochemical Systems[a]

Radical	Temperature (°C)	Hyperfine Coupling (gauss) a_H^{α}	a_H^{β}	a_{OH}	Others
$\dot{C}H_2OH$	26°	17.38	--	1.15	--
$CH_3\dot{C}HOH$	26°	15.4	22.2	1.13(-67°)	--
$\dot{C}H_2CH_2OH$	-71°	31.67	22.0	--	--
$(CH_3)_2\dot{C}OH$	26°	--	19.66	0.70	--
$\dot{C}HOHCH_2OH$	26°	17.54	9.94	1.04 0.34	--
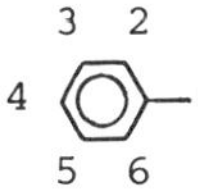 $\dot{C}HOH$	28°	14.97	--	0.58	$(a^1 = 1.56)$
		$(a^2 = 1.67,\ a^3 = 4.58,\ a^4 = 5.16,\ a^5 = 5.93)$			
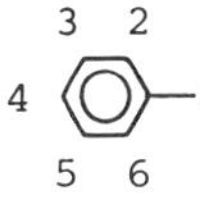 $\dot{C}OHPh'$	28° (cyclohexane)	$(a_{OH} = 2.28,\ a^2 = a^{2'} = a^6 = a^{6'} = 3.27;$ $a^3 = a^{3'} = a^5 = a^{5'} = 1.25;\ a^4 = a^{4'} = 3.69)$			
	22° (Et_3N)	$(a_{OH} = 5.33;\ a^2 = 3.21,\ a^3 = 1.22,\ a^4 = 3.66)$			
$CH_2(OH)\dot{C}HCl$	-60°	20.7	18.0	--	$a_{Cl} = 2.8$
$CH_2(OH)\dot{C}Cl_2$	-60°	--	13.4	--	$a_{Cl} = 4.1$
$\dot{C}H_2COCH_3$	25°	19.75	--	--	$a_{CH_3} < 1$
$CH_2ClCO\dot{C}H_2$	-60°	19.75	--	--	$a_{Cl} = 0.42$
$CH_3CO\dot{C}(OH)CH_3$	32°	--	13.41	2.07	$a^{\delta}_{CH_3} = 2.58$
$NH_2CO\dot{C}H_2$	32°	21.34	$(a_N = 1.72;\ a_{NH} = 2.04$ and $2.43)$		

[a]Taken from references (156), (216), (217), (197), (224), and (215).

TABLE VII
π-Alkyl Radicals Derived from Ethers in Photochemical Systems[a]

Radicals	Temperature (°C)	Hyperfine Coupling (gauss) a_H^{α}	a_H^{β}	a_H^{γ}	Others
$CH^{3}O\dot{C}H^{2}$	25°	16.6	--	--	--
$CH_3CH_2O\dot{C}HCH_3$	29°	13.9	21.62	1.5	--
$(CH_3CH_2O)_3\dot{C}$	-40°	--	--	0.37	--
$CH_3OCF_2\dot{C}HCl$	-60°	22.2	--	--	a_F = 26.9 a_{Cl} = 3.5
$CH_3OCF_2\dot{C}Cl_2$	-60°	--	--	--	a_F = 21.1 a_{Cl} = 4.0
$(CH_3O)_2BO\dot{C}H_2$	-90°	20.14	--	--	a_B = 0.35
$(CH_3CH_2O)_2BO\dot{C}HCH_3$	-90°	18.55	23.52	--	a_B = 0.66
$(C_2H_5O)_2Si(H)O\dot{C}HCH_3$	-97°	17.76	23.35	--	a_H^{Si} = 0.21
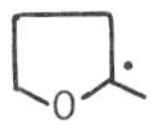	-40°	13.60	28.06	0.76 1.90	--
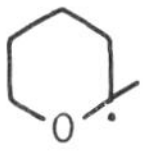	-40°	16.07	24.41	0.69 0.76	--
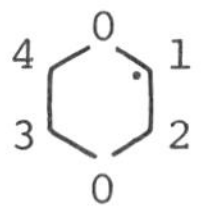	29°	16.92	(α + β) 46.92	= 1.06	--
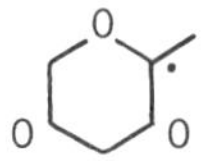	-40°	0.25	--	--	--
Anisole	-40°	17.77	(a^o = 0.54, a^m = 0.27; a^p = 0.49)		
Thioanisole	-40°	16.47	(a^o = 0.79; a^m = 0.32; a^p = 0.69)		

[a] Taken from references (216), (215), (221), (196), and (223).

TABLE VIII
π-Alkyl Radicals Derived from Nitriles and Sulfides in Photochemical Systems[a]

Radicals	Temperature (°C)	Hyperfine Coupling (gauss) a_H^{α}	a_H^{β}	a_H^{γ}	a_N	a_{OH}
$\dot{C}H_2CN$	-8°	20.98	--	--	3.51	--
$CH_3\dot{C}HCN$	40°	20.33	22.99	--	3.41	--
$(CH_3)_2\dot{C}CN$	36°	--	20.72	--	3.35	--
$CNCH_2\dot{C}HCN$	33°	20.82	23.85	--	3.49	--
$HO\dot{C}HCN$	32°	18.19	--	--	3.40	2.02
$HOCH_2\dot{C}HCN$	43°	21.34	20.34	--	3.47	0.68
$CN\dot{C}HCOOH$	30°	19.90	--	--	3.10	--
$CH_3CH_2S\dot{C}HCH_3$	25°	16.8	19.8	1.6	--	--
$CH_3CH_2SS\dot{C}HCH_3$	25°	16.8	20.6	--	--	--

[a]Taken from references (219) and (223).

have been prepared by Hudson and coworkers (225), and by Krusic and Kochi (198c) by the steady-state photolysis of di-t-butyl peroxide in the parent silane. Some ESR data of these silyl radicals are compiled in Table 9. However, it appears that ESR spectra of aromatic substituted silyl radicals have not yet been reported in photochemical systems. Because of the controversial question of d-orbital participation in π-bonding, radicals such as triphenylsilyl or benzylsilyl would be extremely interesting for ESR study. Geoffroy and Lucken (226) have recently claimed that the triphenylsilyl radicals are produced from X-irradiated single crystals of triphenylsilane at room temperature, but the analysis of the not-well-resolved

TABLE IX
Hyperfine Coupling Constants (gauss) of Some Silyl Radicals Generated by Photolysis[a]

Radical	Temperature (°C)	$a_{^{29}Si}$	a_H^{α}	a_H^{β}	a_H^{γ}
$\dot{S}iH_3$	-153°	266	7.96	--	--
$CH_3\dot{S}iH_2$	-121° -70°	--	11.82 12.11	7.98 8.21	--
$(CH_3)_2\dot{S}iH$	-123° -70°	183.05 --	16.99 17.29	7.19 7.30	--
$(CH_3)_3\dot{S}i$	-120° -70°	181.14 183	-- --	6.28 6.34	--
$(CH_3CH_2)_3\dot{S}i$	-70°	--	--	5.73	--
$(CH_3(_3Si\dot{S}i(CH_3)_2$	-70°	137	--	8.21	0.47
$(CH_3)_3Si\dot{S}i(CH_3)H$	-70°	--	16.30	8.15	0.30
$[(CH_3)_3Si]_3\dot{S}i$	-70°	--	--	--	0.44
$(CH_3)_3SiCH_2\dot{S}i(CH_3)_2$	-70°	--	--	5.56(CH_2) 6.43(CH_3)	-- --

[a]Taken from references (198c) and (225).

spectra raised many unanswered questions. Further studies of silyl radicals are very desirable.

g. Organic Oxygen and Sulfur Radicals. Stable substituted phenoxy radicals can be generated by photolysis of the corresponding phenols in liquid inert solvents (227). Although peroxide and alkoxy radicals are well known in liquid photolysis sytems and have been, in fact, the most popular and successful reactants in producing organic radicals, their detection by ESR has been most elusive in the liquid phase. We have mentioned before some of the arguments for the failure to observe $\dot{O}H$ and $R\dot{O}$ radicals in solution. The t-Butylperoxy radical $RO\dot{O}$ has been correctly identified by Morton and Ingold (228) with the aid of ^{17}O isotopes; the hyperfine splittings for the inner and the terminal ^{17}O nuclei are 16.4 and 21.8 gauss, respectively. Symons (129b) has also interpreted the singlet ESR signal observed during photolysis of liquid di-t-butylperoxide in the presence of some hydroperoxide, as the ozonide radical $(CH_3)_3COO\dot{O}$. This radical has an averaged g-value of 2.004 and is analogous to the $F_3COO\dot{O}$ radical observed by Fessenden (229). The oxygen frame in both ozonide radicals is expected to be strongly bent, and the g-tensor is therefore insensitive to the R group. Another peroxy radical $F_3CO\dot{O}$ has been reportedly observed during low-temperature liquid photolysis of CF_3OF and CF_3OOCF_3 (230), but the analysis of the fluorine hyperfine coupling has been questioned (129a).

Organic nitroxides are considered as "oxygen radicals" here. Since ^{16}O has no nuclear spin, nitroxides normally exhibit nitrogen hyperfine splittings. The nitroxides are also much more stable than the peroxy radicals. Thus various "spin trappers" have been developed to "trap" transient radicals into forming stable nitroxides for ESR studies. Janzen and Blackburn (231) have used phenyl-t-butylnitrone as trapper for radicals generated during photolysis of organometallic compounds:

$$R + C_6H_5CH{=}\overset{\overset{O^-}{|}}{\underset{+}{N}}{-}C(CH_3)_3 \rightarrow C_6H_5\underset{\underset{R}{|}}{C}H{-}\overset{\overset{\dot{O}}{|}}{N}{-}C(CH_3)_3 \quad (4\text{-}1)$$

Similarly, Lagercrantz and coworkers (232) have used organic nitroso compounds as "spin traps," and Leaver and coworkers (233) studied nitroxides formed during photolyses of indoles, phenols, thiols, and disulfides using tertiary nitrosobutane as radical trapper. It is interesting to note here that organic thiyl radicals $R\dot{S}$ have not yet been conveniently detected in liquid photolytic systems. Thus during the photolysis of butane-1-thiol or dibutyl disulphide in the presence of t-nitrosobutane (233) the formation of transient $R\dot{S}$ radicals is inferred from the nitroxide formed which has a considerably higher g-value of 2.00714 and a larger nitrogen splitting of 18.52 gauss. Interestingly enough, this proposed S-nitroxide radical is very unstalbe and rapidly disappears when the photolytic source is removed. On the other hand, Mackor and coworkers (234) have observed alkoxy-alkyl nitroxides formed during the photolysis of alkyl nitriles in hydrocarbon solvents and reported a much larger nitrogen splitting of 27-30 gauss. These alkoxy-alkyl nitroxides are more stable than the S-nitroxides. It can be safely predicted that these "spin trapping" methods will become powerful techniques for mechanistic studies of photochemical reactions.

Photolyses of nitroso compounds (235,236) and organic nitro compounds (237,238) can also lead to formation of nitroxide radicals. In the photolysis of nitrobenzene in dimethylformamide, Wong and Wan (237b) showed that in addition to the stable nitroxide radicals, some short-lived primary radicals can be observed using the rapid-scan ESR techniques. UV-irradiation of chloronitrobenzene in ether solution produced phenylalkoxy-nitroxides (239), and nitroxide radicals have been observed in the photolysis of caryophyllene nitrosite both in liquid and in solid phase (240). Finally we mention here the observation (241) for the first time of a very interesting neutral t-butylferrocenylnitroxide radical, even though this ferrocenylnitroxide was not prepared photochemically.

Similar to alkoxy radicals, few organic thiyl radicals have been observed in photochemical liquid systems. However, asymmetric spectra attributed to $R\dot{S}$ radicals with anisotropic g-tensors, have been observed in UV-irradiated thiols, sulphides, and disulphides in a solid matrix at 77°K (103,242-245). A typical "sulfur pattern" is shown in Figure 10, and the typical values

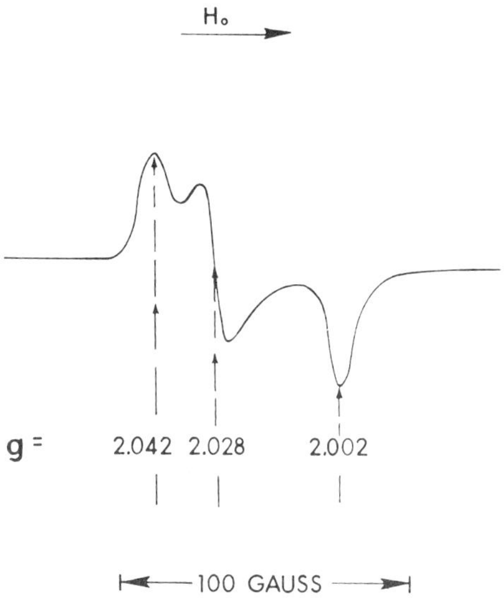

FIGURE 10. The ESR "sulfur pattern" of sulfur radicals trapped in a low-temperature matrix. Taken from reference (359b).

g_1, g_2, and g_3 are used as "fingerprints" for sulfur radicals in the solid matrix. $R\dot{O}$ and $R\dot{S}$ radicals both have the unpaired spin mainly localized on the oxygen and sulfur atoms, and generally neither show appreciable hyperfine interaction with the alkyl protons. Observation of ^{33}S hyperfine structure in $R^{33}\dot{S}$ radicals confirmed the identification of these sulfur radicals (246). A few $R\dot{S}e$ radicals have also been observed (243) and they are characterized by an axially symmetrical g-tensor.

h. Organic nitrogen and Phosphorous Radicals. UV-irradiation of solutions of aromatic amines containing alkylperoxides gives rise to nitroxides, presumably via intermediate radicals of the diarylnitrogen type (130,247):

$$ROOR + h\nu \rightarrow 2\ R\dot{O} \qquad (4\text{-}2)$$

$$R\dot{O} + Ph_2NH \rightarrow PH_2\dot{N} + ROOH \qquad (4\text{-}3)$$

$$Ph_2NH + ROOH \rightarrow R\dot{O} + Ph_2\dot{N} + H_2O \qquad (4\text{-}4)$$

$$Ph_2\dot{N} + ROOH \rightarrow Ph_2NOH + R\dot{O} \qquad (4\text{-}5)$$

$$Ph_2NOH + R\dot{O}(\text{or } \dot{O}H) \rightarrow Ph_2N\dot{O} + ROH(\text{or } H_2) \qquad (4\text{-}6)$$

Kavun and Buchachenko (247) have found that the "primary" amino radicals derived from thioamines in the presence of hydroperoxide are very stable and are not readily converted into "secondary" nitroxide radicals. Kommandeur and Wiersma (107) studied the photodecomposition of tetraphenylhydrazine in rigid solution at 77°K, and the observed ESR spectrum was assigned to the dimer of the diphenyl amino radical (248). Shida et al. (249) have also studied the photodimer of tetraphenylhydrazine and the photochromic dimer of triphenylimidazolyl at low temperature by optical and ESR techniques. These authors noted that photolysis of these compounds induces the homolysis while γ-irradiation leads to heterolytic dissociation. Blinder et al. (250) reported the ESR spectrum of the monomer tetraphenylpyrryl radical in solution produced from the photodissociation of the dimer, octaphenylbipyrrole. The ESR spectrum of the triphenylamine cation radical has been observed during UV-irradiation of triphenylamine solutions in the presence of silica gel (251). Photolysis of m-dinitrobenzene leads to formation of nitrogen radicals with a C-N-H fragment (252), and the UV-irradiation of azobenzene in cumene solutions produces a stable hydrazyl type of nitrogen radicals (253). Johnson and Gutowsky (254) have studied the ESR spectra of photochemically generated radicals from viologen and bipyridyl in aqueous alcohol solutions.

Recently dialkylamino radicals have been generated by direct photolysis of the corresponding tetrazenes in liquid cyclopropane solution at low temperatures (255):

$$R_2N\text{-}N{=}N\text{-}NR_2 \xrightarrow[-90°C]{h\nu} 2R_2\dot{N} + N_2 \qquad (4\text{-}7)$$

Wood and coworkers (141) have observed alkylimino radicals in an adamantane matrix by an interesting photochemical conversion of the aminoalkyl radicals first formed from γ-irradiation of alkylamines:

$$RR'\dot{C} - NH_2 \xrightarrow{h\nu} RR'C{=}\dot{N} + H_2 \qquad (4\text{-}8)$$

The unsubstituted methylene imino radical, $CH_2{=}\dot{N}$, was first observed by Cochran and coworkers (256) by UV-irradiation at 4.2°K of a solid deposit containing 1% HI, 9% HCN, and 9% Ar. This imino radical was also produced recently by direct photolysis of formaldazine (257), an interesting analog of butadiene:

$$CH_2{=}N\text{-}N{=}CH_2 \xrightarrow{h\nu} 2CH_2{=}\dot{N} \qquad (4\text{-}9)$$

The electronic structures of these imino radicals are particularly interesting as they are π-radicals as compared to the δ-iminoxy radicals $RR'C{=}N\text{-}\dot{O}$. From the well-resolved ESR spectra of the alkylimino radicals, Wood et al. (141) were able to estimate that the unpaired spin-density is mainly localized in the p-orbital of the nitrogen, while the nitrogen lone pair of electrons is largely in the 2s orbital. The ESR hyperfine coupling constants of some alkylamino and alkylimino radicals are given in Table 10.

TABLE X
Hyperfine Coupling Constants (gauss) of Some Alkylamino and Imino π-Radicals Generated by Photolysis[a]

Radical	Temperature (°C)	a_N	a_H^{β}	a_H^{γ}
$(CH_3)_2\dot{N}$	-90°	14.78	27.36	--
$(C_2H_5)_2\dot{N}$	-90°	14.27	36.90	not resolved
$[(CH_3)_2CH]_2\dot{N}$	-90°	14.31	14.31	0.66
$C_2H_5CH{=}\dot{N}$	25°	9.5	78.5(CH)	2.77(CH_2)
$C_3H_7CH{=}\dot{N}$	25°	9.5	79.5(CH)	2.84(CH_2)
$(CH_3)_2C{=}\dot{N}$	25°	9.7	--	1.33(CH_3)
$(C_2H_5)_2C{=}N$	25°	9.5	--	1.25(CH_2)
$(CH_3)C(C_2H_5){=}\dot{N}$	25°	9.6	--	1.25
$H_2C{=}\dot{N}$	-277°	~11	87.4(CH_2)	--

[a]Taken from references (141), (255), and (256).

Only a few ESR studies of photochemically generated organic phosphorus-centered radicals have been reported, although divalent and tetravalent organophosphinyl radicals are often postulated as intermediates in many photochemical or organophosphorus reactions. Kochi and

Krusic (198e) have produced the radical $(CH_3)_3\dot{P}OC(CH_3)_3$ in a liquid photochemical system, and they assigned a trigonal bipyramidal radical structure as suggested by the large phosphorus hyperfine splitting (a_p = 618 gauss). Kokoszka and Brinckman (258) photolysed CH_3PCl_2 at -196°C and assigned the observed ESR spectrum to the radical $CH_3\dot{P}Cl_3$ with a large phosphorus splitting of 1077 gauss. The diphenylphosphinyl radical has been observed in the photolysis of tetraphenyldiphosphine at low temperature (259,260).

3. Organic δ-Radicals

We have briefly considered some of the voluminous ESR studies of organic π-radicals and will now turn to the much fewer investigations of organic δ-radicals. In general the δ-radical has the unpaired electron mainly in an orbital with nonvanishing amplitude at the nucleus of the radical center. Thus ESR hyperfine coupling constants can be used to assign whether the radical is δ- or π-type.

One of the factors that may contribute to fewer studies of δ-radicals is their higher reactivity. For example, the hydrocarbon δ-radicals are more reactive than the π-radicals, and the phenyl radical has never been observed by ESR in the liquid phase. Similarly, iminoxy radicals are much less stable than nitroxide π-radicals, and Norman, Gilbert, and coworkers have studied extensively some of the δ-iminoxy radicals by the use of flow techniques (261-263). This difference in reactivity was explained (138) by the fact that in δ-radicals the unpaired electron is located in a directional sp-orbital which projects away from the neighboring atoms, resulting in reduced repulsion between the radical and the reacting molecule in the activated complex.

Among the first few δ-radicals reported is the formyl

radical, $H\dot{C}O$, identified by Adrian, Cochran, and Bowers (264) in a photolysed solid CO matrix containing HI. Subsequently the formyl radical containing enriched ^{13}C isotope (265) and the analogous $F\dot{C}O$ (266) and $F^{13}\dot{C}O$ (265) radicals have also been observed. The latter fluorinated radicals were prepared by the photolysis of OF_2 in a solid CO matrix. Careful analysis of the experimental ESR spectra shows that these radicals are sharply bent; the angles are 125° and 110° for $H\dot{C}O$ and $F\dot{C}O$, respectively. In addition, these authors have estimated the spin-density in the unpaired δ-orbital: For $H\dot{C}O$ the spin-density is 0.268 in the hydrogen ls orbital and about 0.45 in the carbon Cy orbital, while for $F\dot{C}O$ the spin-densities in the F 2s and sp orbitals are 0.02 and 0.17, respectively, and about 0.53 in the carbon.

The reactive vinyl δ-radical $H\dot{C}{=}CH_2$ has been observed by Cochran, Adrian, and Bowers (267) in a photolysed argon matrix containing HI and acetylene. These authors concluded that the β-proton trans to the orbital of the unpaired electron gives the larger splitting of about 68 gauss while the other β-proton is less strongly coupled with a splitting of about 34 gauss. The β-protons of the vinyl radical generated in liquid-phase radiolysis (86) are, however, magnetically equivalent on the time scale of the ESR measurement. Vinyl radicals prepared by pyrolysis and trapped in a neon matrix (268) exhibit some preferred orientation. Recently the trans isomer of $\dot{C}D{=}CDF$ (269) was prepared by photolysis of an argon matrix containing F_2 and deuterated ethylene; it is also possible that secondary photolysis of the $\dot{C}D{=}CDF$ radical leads to formation of the ethynyl radical $\dot{C}{\equiv}CD$ (267). Kasai and coworkers (270) have prepared the vinylog of the vinyl radical, the butatrienyl $H\dot{C}{=}C{=}C{=}CH_2$ by the photolysis of an argon matrix containing HI and diacetylene. The butatrienyl radical was found to be bent and planar. The hyperfine coupling constants of some of these δ-radicals are included in Table 11.

TABLE XI
Hyperfine Splitting (gauss) for Some Organic δ-Radicals in Photochemical Systems[a]

Radical	Temperature (°C)	a_H^{α}	a_H^{β}	Other
$H\dot{C}O$	-269°	136	--	$a_{^{13}C}$ = 134
$F\dot{C}O$	-269°	--	--	a_F = 322.1 $a_{^{13}C}$ = 285.6
$\dot{C}H=CH_2$	-269°	+16	68(H_1) 34(H_2)	--
trans-$\dot{C}H=CHF$	-269°	-38	(cis-β_H)	a_F^{iso} = 69
$H\dot{C}=C=C=CH_2$	-269°	13.7(CH)	43.4(CH_2)	--
$\dot{C}_6H_5$	-269°	a_{ortho} = +17.4; a_{meta} = +5.9; a_{para} = +1.9		
$C_6H_5-\dot{C}=O$	- 86°	a_{meta} = 1.16	--	$a_{^{13}C}$ = 128.2
o-$C_6H_4F\dot{C}=O$	- 88°	$a_{meta\ H}$ = 0.85	--	a_F = 1.46
m-$C_6H_4F\dot{C}=O$	- 85°	$a_{meta\ H}$ = 1.25	--	a_F = 3.43
p-$C_6H_4F\dot{C}=O$	- 50°	$a_{meta\ H}$ = 1.14	--	a_F ≅ 0.3
$H_2N-\dot{C}=O$	32°	--	1.27(H_1) 30.41(H_2)	a_N = 21.69

[a]Taken from references (217), (265), (267), (269), (270), (271), and (277).

The long-awaited unambiguous observation of the elusive phenyl radical in photolysed solid matrix by Bennett (138), Kasai (271), and their coworkers is a considerable achievement in ESR spectroscopy. This observation resolved the uncertainty as to whether the unpaired electron was localized in the broken bond (δ-radical) or delocalized in the π-orbital system of the ring (π-radical). The accurate assessment (271) of both the isotropic and anisotropic parts of the hyperfine coupling constants of all the protons in the phenyl radical confirms the optical evidence obtained by Porter and Ward (272) that the phenyl is a δ-radical. Kasai's experimental result (271) (Table 11) also settled the large discrepancies among various theoretical calculations (273-275) on the phenyl radicals and showed that the SCF-INDO method (273) gives a much better overall agreement with the experimental data.

Although the $H\dot{C}O$ radical has been identified by ESR in a photochemical matrix, few analogous aliphatic $R\dot{C}O$ radicals have been reported in photolysis systems. Bennett and coworkers (138) have prepared the acetyl radical $CH_3\dot{C}O$ by the reaction between Na and acetyl chloride and have obtained an isotropic value of 5.1 gauss for the proton splitting. An ESR spectrum attributed to $CH_3\dot{C}O$ with $a_H \simeq 16$ gauss was earlier reported during photolysis of biacetyl (276) in a solid matrix. Recently a series of benzoyl δ-radicals (277) have been observed in the liquid photolysis of the corresponding benzaldehyde in a cyclopropane solution containing di-t-butyl peroxide. For the first time, it is possible to study in these radicals the delocalization of the unpaired electron from the acyl δ-system into the adjacent phenyl π-system. The significant conclusion from this study is the assignment of the largest hyperfine splitting to the *meta* proton (or substituent) followed by the *ortho* and *para* protons (or substituents), respectively (Table 11). This is in marked contrast to the familiar trend, *para* $\geq$ *ortho* > *meta*, observed with benzylic π-radicals, but is in good agreement with the theoretical prediction from an INDO calculation. The δ-nature of these benzoyl radicals is also confirmed by the observed large ^{13}C splitting at the radical center. Other acyl radicals derived from formamide (217,277) have been observed. These δ-radicals are characterized by g-values which are less than the free spin value of 2.00232.

C. Photochemical Generation of Organic Triplet-State Molecules

Great advances in the elucidation of electronic structure and the dynamics of optical spin polarization in organic triplet-state molecules have been made by ESR spectroscopy since the first successful experiment of Hutchison and Mangum (39) in 1958. Most of the triplet ESR studies can be grouped into two sections: the photo-excited phosphorescent triplet states and the photo-chemically prepared ground triplet-state intermediates. We shall be mainly concerned here with the zero-field splitting parameters of the organic triplet states; the dynamics and mechanistic aspects will be considered in a later section.

1. Phosphorescent Organic Molecules

The list of triplet states of aromatic molecules studied by ESR has been steadily growing in recent years. In aromatic hydrocarbons the lowest triplet state is the π,π^* type characterized by a relatively long radiative lifetime. Substitution of deuterium in the molecule often increases both the ESR intensity and the observed decay time. Aromatic molecules having heteroatoms with lone pairs can have either the π,π^* type or the n,π^* type as their lowest triplet state. In general the nature of the lowest triplet state depends not only on the molecule but also on the environment. For example, the lowest triplet state of acetophenone in a nonpolar solvent is the n,π^* type but is converted to π,π^* state in a strongly polar solvent (278). At present with conventional ESR spectrometers, only π,π^* triplet states have been detected because of their relatively longer phosphorescent lifetimes (> 0.1 sec) compared to those of the n,π^* states (<0.01 sec). However, the technique of optical detection of ESR (279-282) in short-lived triplet states has recently been developed to overcome this difficulty.

In conventional ESR detection, the signal-to-noise ratio depends on the population of the triplet state and hence is proportional to the triplet lifetime. With optical detection, the ESR transitions between the Zeeman levels of the triplet states are monitored by the radiative emission from those levels, which is roughly independent of the triplet lifetime. Thus the ESR zero-field parameters of the elusive benzophenone n,π^* triplet state have been measured by Sharnoff (279a) using the optical detection technique, and the spectrum showing the fine and superfine structure of $\Delta M_S = 1$ ESR transitions is reproduced in Figure 11. The high-resolution spectrum of the naphthalene triplet state in a dilute single crystal, showing the fine structure ($\Delta M_S = 1$) and the hyperfine structure, is reproduced in Figure 12. The "classic" spectrum (283) of randomly oriented naphthalene triplets in a rigid glass is given in Figure 13. Hyperfine structure has recently been observed also in the $\Delta M_S = 2$ transitions of a number of aromatic molecules in

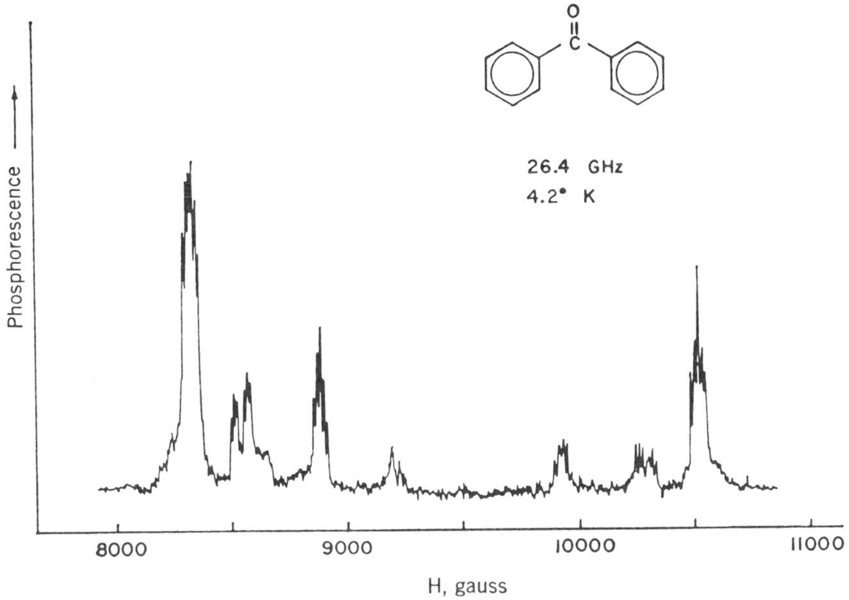

FIGURE 11. Optical detection of ESR of the triplet state of benzophenone showing the fine and superfine structure of $\Delta M_S = 1$ transitions. Reproduced from the work of Sharnoff (279a).

rigid glasses (43), and the spectrum due to phenazine is reproduced in Figure 14.

The zero-field splitting parameters of some aromatic hydrocarbons (9), aromatic carbonyl compounds (278,279a, 284) and aromatic nitrogen-containing compounds (9,285,286), in their phosphorescent triplet states, are compiled in Tables 12, 13, and 14, respectively. In the tables, the parameters D and E are obtained from the observed $\Delta M_S = 1$ transitions and from "low-field" single-crystal studies, the values of D^* are derived from the so-called $\Delta M_S = 2$ transition. Correlation of the zero-field splitting and the triplet-state energy of a large number of aromatic hydrocarbons reveals a good linear relationship (287).

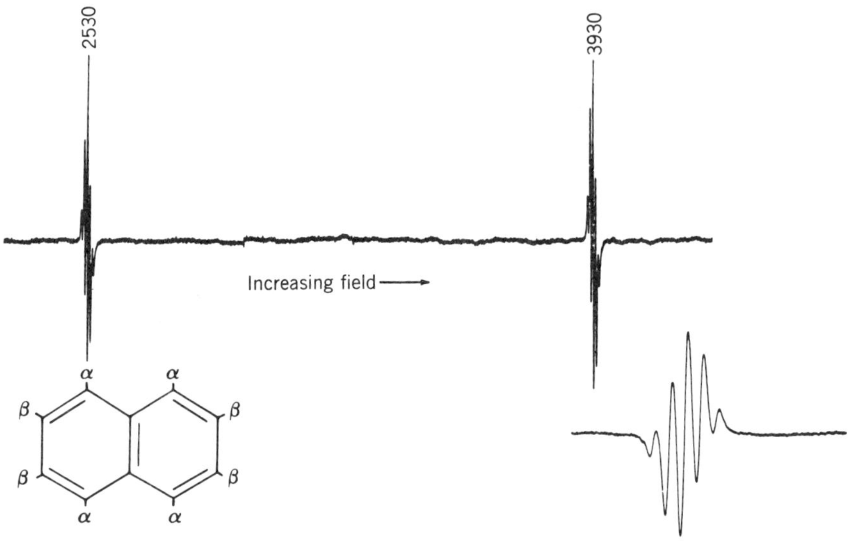

FIGURE 12. ESR spectrum of the triplet state of naphthalene showing the fine structure and the hyperfine splittings. From Varian Associates EPR at Work Series, No. 26.

Smaller et al. (288) have examined the correlation between the zero-field splittings and chemical properties of a large number of substituted aromatic compounds. In general, substituent effects are small and can be attributed to redistribution of spin-density. A good number of theoretical values of the zero-field splittings of nitrogen-containing aromatic compounds have been compared with the experimental values (285). It is not surprising that only a few aromatic carbonyl triplets have been studied as the majority of ketones and aldehydes possess lowest triplet states of the n,π^* type. A relatively large calculated zero-field splitting ($D/hc = +0.326\ cm^{-1}$; $E/hc = -0.037\ cm^{-1}$) for the n,π^* triplet state of formaldehyde has been predicted by Dixon (289).

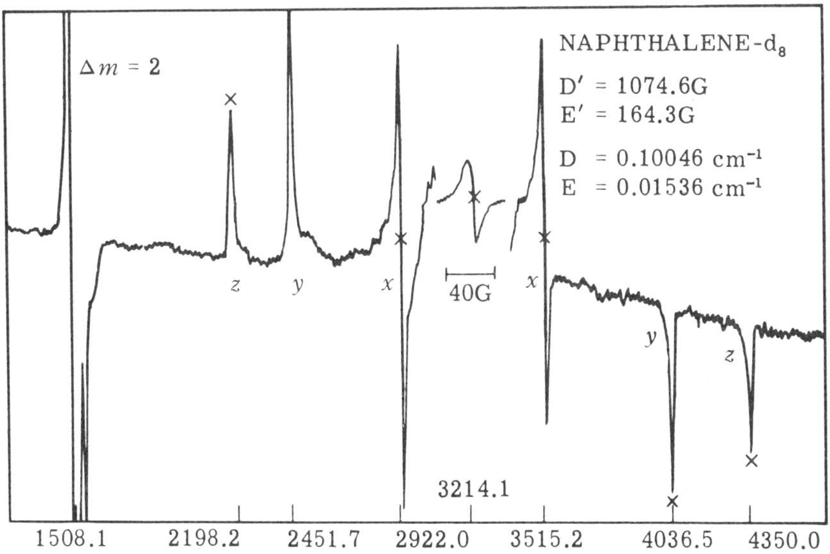

FIGURE 13. ESR spectrum of randomly oriented photoexcited naphthalene triplets showing the $\Delta M_S = 1$ transitions. Taken from reference (283).

However, for n,π^* triplet aromatic ketones and aldehydes, the increased spin delocalization onto the rings should reduce the zero-field splitting, and the failure of their ESR detection may be attributed mainly to their short lifetimes. Hayashi and coworkers (290) have examined the ESR fine-structure parameters of a number of molecular complexes in their phosphorescent state and have derived the dependence of the fine-structure values upon the degree of charge-transfer character in the lowest triplet state of the molecular complexes. Some of their results are collected in Table 15.

2. Ground-State Triplet Molecules

The existence of ground-state triplet carbenes, nitrenes, and their derivatives as intermediates (291) in

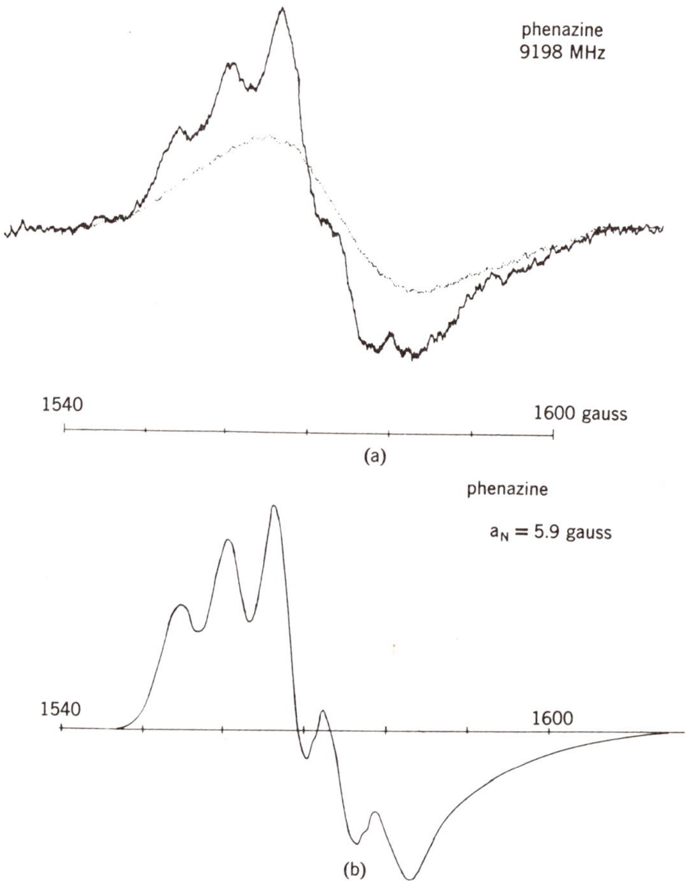

FIGURE 14. (a) ESR spectrum of randomly oriented phenazine triplets showing the ΔM_S = 2 transition and its hyperfine structure (b) Computer simulated spectrum. Taken from reference (43).

photochemical and organic reactions had been postulated long before their detection by ESR spectroscopy. Considerable information on molecular geometry and on the distribution of unpaired electrons in these triplet molecules can be obtained by detailed ESR studies, and these structural data are of great importance in the

TABLE XII
Observed Zero-Field Splitting Parameters of Some Aromatic Hydrocarbons in their Phosphorescent Triplet State[a]

Triplet Molecule in Rigid Glasses	D/hc (cm^{-1})	E/hc (cm^{-1})	D^*/hc (cm^{-1})
Benzene	0.156	0	0.156
Naphthalene	--	--	0.1048
Naphthalene-d_8	0.10046	-0.01536	0.1008
Anthracene	--	--	0.077
Anthracene-d_{10}	0.0724	0.0081	0.0737
Phenanthrene	0.100	0.047	0.129
Phenanthrene-d_{10}	0.1050	0.046	0.1321
Pyrene	--	--	0.093
Chrysene	--	--	0.1052
Coronene	0.096	--	0.096
Biphenyl	0.1092	0.0036	0.1094
Fluorene	0.1075	0.0033	0.1092
Toluene	--	--	0.171
1,3,5-Triphenylbenzene	0.111	0	0.111
Triphenylene	0.1353	0	0.1353
1,2-Benzanthracene	0.079	10.0141	0.083
1,2-Benzopyrene	0.090	10.0231	0.098
3,4-Benzopyrene	--	--	0.0758
1,12-Benzoperylene	--	--	0.0718
Triplet Molecule in Single Crystals			
Naphthalene (in Durene)	+0.10119	-0.01411	--
Naphthalene (in Biphenyl)	+0.09921	-0.01548	--
Anthracene (in Phenazine)	±0.07055	±0.00791	--
Phenanthrene (in Biphenyl)	±0.10043	±0.04658	--

TABLE XII
(Cont.)

Triplet Molecule in Single Crystals	D/hc (cm^{-1})	E/hc (cm^{-1})	D^*/hc (cm^{-1})
Pyrene (in Fluorene)	±0.0810	±0.0182	--
Chrysene-d_{12} (in p-terphenyl)	0.0896	0.0268	--
Triphenylene (in dodecahydrotriphenylene)	0.133526	0.000564	--
1,2,5,6-Dibenzanthracene (in 2,2'-Binaphthyl)	0.08838	0.02538	--
2-Methylnaphthane (in Naphthalene)	±0.09708	±0.01450	--
1-Fluoronaphthalene (in Durene)	0.10036	-0.01564	--
2-Fluoronaphthalene (in Durene)	0.10072	-0.01334	--

[a]Compiled from references (9), (39), (43), (45), (46), (47), (48), and (49).

TABLE XIII
Zero-Field Splitting Parameters of Some Aromatic Carbonyl Compounds in their Phosphorescent Triplet State[a]

Triplet	Phase	D/hc (cm^{-1})	E/hc (cm^{-1})	D^*/hc (cm^{-1})
Benzophenone (n,π^*)	Crystal[b]	± 0.152	± 0.021	--
Acetophenone (π,π^*)	Water-glycol rigid glass	10.12711	10.00411	--
Methyl-2-naphthyl-ketone (π,π^*)	EPA glass	0.096	0.027	0.105
2-Naphthylphenyl-ketone (π,π^*)	EPA glass	0.095	0.026	0.105
Naphthalene-2-carbaldehyde (π,π^*)	EPA glass	0.094	0.029	0.012
Formaldehyde (n,π^*)	Theoretical prediction	+0.326	-0.037	--
Chromone	Alcohol glass	--	--	0.098
3,4-Dimethyl-acetophenone	Alcohol glass	--	--	0.121

[a]Taken from references (279a), (278), (284), and (289).

[b]Optically detected.

TABLE XIV
Experimental Zero-Field Splittings of Some Nitrogen-Containing Aromatic Compounds in Their Phosphorescent Triplet State[a]

Triplet Molecule	D/hc (cm^{-1})	E/hc (cm^{-1})
1,5-Naphthyridine (in durene crystal)	+0.1030	-0.0167
Quinoxaline	$\|0.101\|$	$\|0.018\|$
1,10-Phenanthroline	$\|0.104\|$	$\|0.046\|$
4-Phenylpyridine	$\|0.111\|$	$\|0.005\|$
2,2-Bipyridyl	$\|0.112\|$	$\|0.010\|$
Phenoxazine	$\|0.1247\|$	$\|0.0119\|$
Aniline	$\|0.125\|$	$\|0.021\|$
N,N-Dimethylaniline	$\|0.120\|$	$\|0.017\|$
Diphenylamine	$\|0.093\|$	$\|0.020\|$
Benzonitrile	$\|0.135\|$	$\|0.019\|$
Tolunitrile	$\|0.134\|$	$\|0.022\|$
4,4-Dimethylamino-benzonitrile	$\|0.101\|$	$\|0.022\|$

[a]Taken from references (9), (285), and (286).

TABLE XV
Fine-Structure Constants (D^*) and Degree of Charge-Transfer Characters (% CT) Obtained with Photoexcited Triplet States of Molecular Complexes[a]

Triplet	D^*/hc (cm^{-1})	% CT
1,2,4,5-Tetracyano-benzene (TCNB)	0.1317	0
TCNB-Hexamethylbenzene	0.0231	95
TCNB-Pentamethylbenzene	0.0452	77
TCNB-Mesitylene	0.0844	42
TCNB-*p*,Xylene	0.0880	39
TCNB-*o*,Xylene	0.0918	36
TCNB-*m*,Xylene	0.0950	33
TCNB-Toluene	0.1189	11
TCNB-Benzene	0.1238	7

[a]Taken from reference (290b).

understanding of their reactivity. Many of these triplet intermediates can be generated readily by photolysis of their percursors in a low-temperature rigid glass.

Aromatic methylenes have been extensively studied by Wasserman, Yager, and coworkers (292-295), and a collection of these results is tabulated in a review by Trozzolo (296). These authors also studied the perfluoroalkylmethylenes (297). A number of triplet alternate methylenes of the propargylene homolog have also been investigated by Bernheim, Skell, and coworkers (298). In general, the zero-field splittings are usually larger in these methylenes than in the phosphorescent triplet aromatic hydrocarbons. The triplet diphenylmethylene is found to be almost linear but with mutually perpendicular phenyl rings. Linear geometry is also predominant in the triplet alternate methylenes. Higuchi (299) has derived a relationship between the zero-field parameters and the bond angle of the divalent carbon.

The long-awaited ESR detection of the parent triplet methylene, CH_2, was recently accomplished by Bernheim, Skell et al. (300) and independently by Wasserman, Yager, and coworkers (301). These authors observed the triplet CH_2 in the photolysis of either diazomethane or its cyclic isomer diazirine at 4°K. The latter investigation (301) also included CHD and CD_2 and after careful analysis of the zero-field parameters, the authors concluded that the triplet CH_2 is substantially bent with the most probable HCH angle being 136 $\pm$8°. This is consistent with the reinterpretation of the optical spectra of the triplet methylene in the gas phase by Herzberg and Johns (3 that the molecule is bent. The experimental evidence of a bent triplet CH_2 also agrees with the theoretical prediction given by Harrison and Allen (303). A tabulatior of some of the ESR results of various methylenes is given in Table 16.

Ground-state triplet dicarbenes in photochemical systems have also been extensively studied by ESR. Dowd and coworkers (304) examined the interesting, frequently postulated ground triplet trimethylenemethane obtained by photolysis of either 4-methylene-Δ'-pyrazoline or a single crystal of 3-methylene-cyclobutanone. This triplet molecule is axially symmetrical and the proton hyperfine splittings of 8.9 gauss observed in single-crystal experiments indicate that all protons in the molecule are equivalent when the axis perpendicular to the plane

TABLE XVI
Zero-Field Parameters of Some Intermediate Ground-State Triplet Carbenes Produced by Photolysis[a]

Triplet Carbenes	D/hc (cm^{-1})	E/hc (cm^{-1})
H–C̈–H	0.6881	0.00346
H–C̈–D	0.7443	0.00640
D–C̈–D	0.7563	0.00443
Ph–C̈–H	0.518	0.024
Ph–C̈–Ph	0.401	0.018
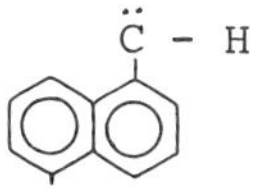	0.4555	0.0202
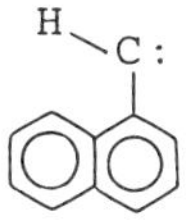	0.4347	0.0208
F_3C–C̈–H	0.712	0.021
C_3F_7–C̈–H	0.723	0.027
F_3C–C̈–CF_3	0.7444	0.0437
H-C≡C-C̈-H	0.6276	0.000
HC≡C-C̈-CH_3	0.6263	0.000
Ph-C≡C-C̈-H	0.5413	0.0035
CH_3-C≡C-C≡C-C̈-H	0.6087	0.000
Ph-C≡C-C≡C-C̈-H	0.5530	0.000
N≡C-C̈-H	0.8629	0.000
NC-C̈-CN	1.002	0.002

[a]Taken from references (296), (297), (298), and (301).

of the molecule is oriented parallel to the magnetic field. This finding is in good agreement with a theoretical prediction (305). Triplet diradicals produced by photolysis of 3H-indazoles (306) and spiro [fluorene-9,3'-indazoles] (307) have been examined. Such dicarbene intermediates can be "trapped" in solution by reaction with olefin giving a new triplet diradical which has been identified by ESR (308):

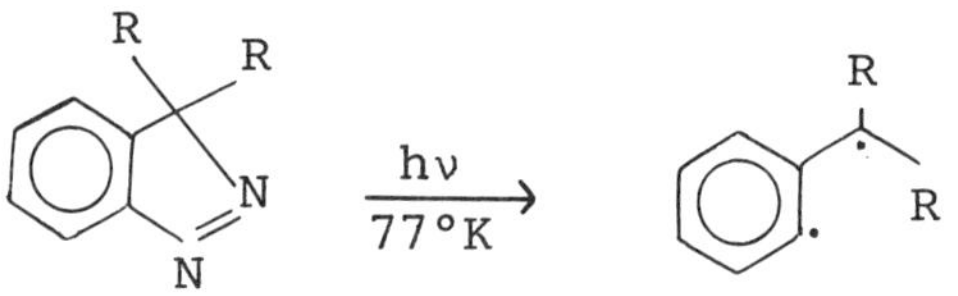

(4-10)

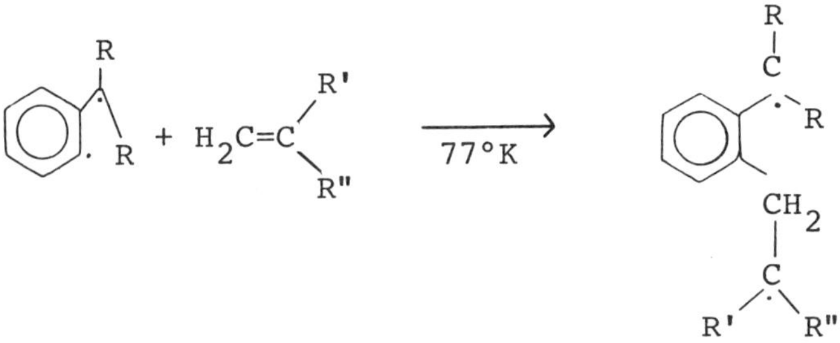

(4-11)

Similarly, 1,3-diradicals have also been identified by ESR in the photolysis of bicyclic azo compounds (309):

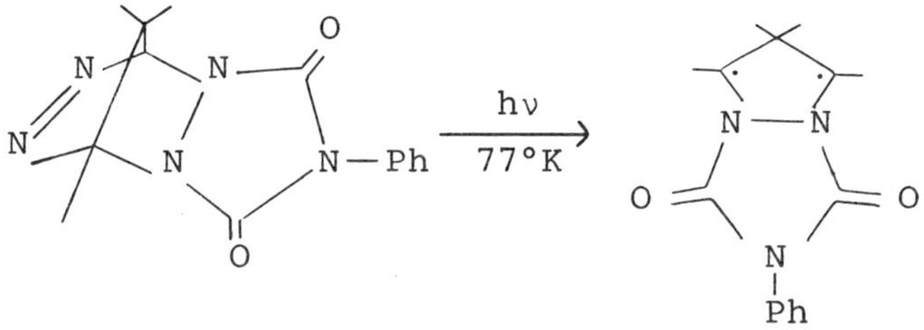

(4-12)

Such ESR characterization of 1,3-diradical intermediates is of considerable interest to mechanistic chemical studies. Some of the ESR results are given in Table 17.

The ground-state triplet nitrene and its derivatives

TABLE XVII
Zero-Field Parameters of Some Intermediate Ground-State Triplet Diradicals Produced in Photochemical Systems[a]

Triplet Diradicals		D/hc (cm^{-1})	E/hc (cm^{-1})
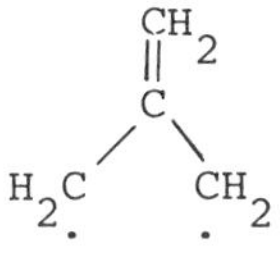		0.024	<0.001
		(a_H = 8.9 gauss for 6 protons)	
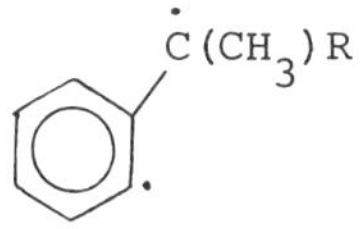	($R=CN$)	0.1069	0.0058
	($R=OCH_3$)	0.1217	0.0066
	($R=COOCH_3$)	0.1110	0.0069
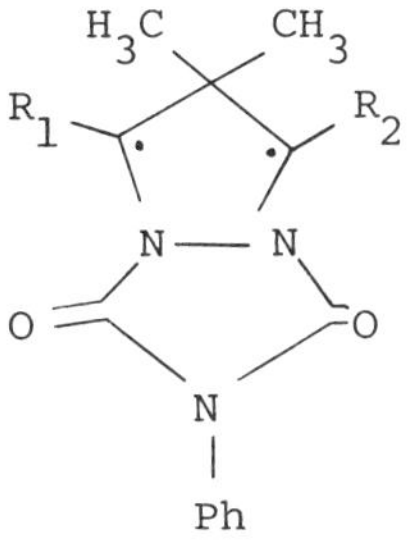	($R_1=R_2=Ph$)	0.048	<0.001
	($R_1=Ph$, $R_2=CH_3$)	0.055	0.0020
	($R_1=R_2=CH_3$)	0.068	0.0030
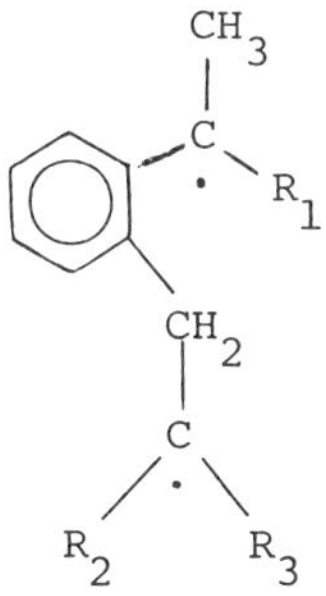	($R_1=CN$; $R_2=R_3=Ph$)	0.0156	<0.0008
	($R_1=CN$; $R_2=H$; $R_3=Ph$)	0.0181	<0.001
	($R_1=CN$; $R_2=H$; R_3=Vinyl)	0.0151	<0.001
	($R_1=OCH_3$, $R_2=R_3=C_6H_5$)	0.0177	<0.001

[a]Taken from references (304), (306), (308), and (309).

have been extensively studied by ESR. Dixon (310) first reported the zero-field parameters for $\dot{N}H$, and Hameka et al. (311) have subsequently shown theoretically that spin-orbit coupling accounts for about 25% of the experimental splitting. Presumably this is due to triplet nitrenes containing nitrogen atoms which are "heavier" than carbon. Alky nitrenes and cycloalkyl nitrenes have been examined by Wasserman et al. (312) and observation and correlation of zero-field parameters with bond angles and π-electron spin-densities in a series of aromatic nitrenes have been given by McDowell (313) and Wasserman (314) and their coworkers. Some of their results are given in Table 18.

Finally we mention here a number of interesting ESR observations of diradicals in miscellaneous photochemical systems. Strausz and coworkers (245) have studied the apparent diradicals of the types $\dot{C}H_2-(CH_2)_n-\dot{S}$, and $\dot{S}-(CH_2)_n-\underset{\dot{S}}{CH}-(CH_2)_4COOH$ by photolyses of the corresponding cyclic sulfides and cyclic disulfides. Photolysis of the 1:1 acridine-acridan molecular complex (315) in the charge-transfer optical band gives rise to carbon, nitrogen-centered radicals trapped in two kinds of radical pairs. Photodissociation of solid β-tetrachloro-α-ketonaphthalene leads to the observation of a triplet ESR spectrum (316) which was interpreted as being due to pairwise trapping of the primary naphthoxy radical and the chlorine atom.

V. APPLICATION OF ESR TO PHOTOCHEMISTRY

A complete understanding of the photochemistry of any given system is a monumental task necessarily encompassing many multidisciplinary domains. Naturally the photochemist is forever looking to new frontiers of

TABLE XVIII
Zero-Field Parameters of Some Intermediate Ground-State Triplet Nitrenes Produced by Photolysis[a]

Triplet Nitrene	$\lvert D/hc \rvert$ (cm^{-1})	$\lvert E/hc \rvert$ (cm^{-1})
$\ddot{N}H$	1.86	0
$\ddot{N}$-CN	1.544	<0.002
$Ph\ddot{N}$	1.000	0
Ph—SO_2—$\ddot{N}$	1.45	0
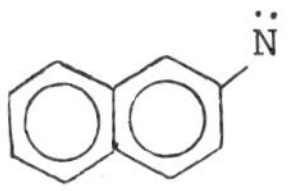	0.89	0
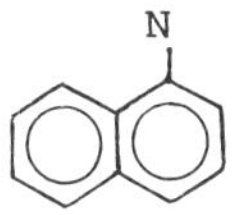	0.78	0
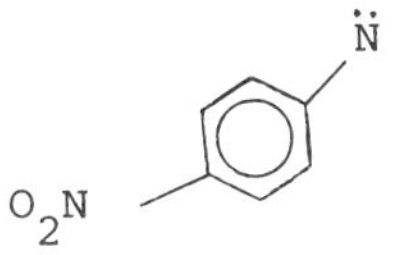	0.98	0
C_3H_7-$\ddot{N}$	1.607	0.0034
cyclo C_6H_{11}-$\ddot{N}$	1.599	<0.002
cyclo C_5H_9-$\ddot{N}$	1.575	<0.002
t-butyl nitrene	1.625	<0.002

[a]Taken from references (310), (312), and (313).

developing techniques which may be applied to photochemical systems and reveal additional information. ESR spectroscopy, when properly applied to photochemical systems, can shed some light on the nature and rate of elementary processes. It is self-evident from some of the ESR studies illustrated in Section IV that the ESR technique is of considerable importance in photochemical studies. We shall illustrate in the following pages some of the applications of ESR to solving complex problems concerning elementary photophysical and photochemical processes.

A. Photophysical Processes

Photophysical processes involving only diamagnetic singlet-state organic molecules cannot be studied by ESR. However, paramagnetic singlet molecules, such as molecular oxygen $^1\Delta g$, have been extensively investigated by ESR (13). In this section we shall concern ourselves mainly with ESR studies of the dynamics and properties of photoexcited triplet states, including the process of energy transfer from these states. An elegant evaluation of the current theory on radiationless transition in photochemistry has been reviewed by Jortner, Rice, and Hochstrasser (317), and a summary of both the radiative and nonradiative properties of excited triplet states has been given by El-Sayed (318a). Recently the effect of spin-orbit coupling on the phosphorescence lifetimes of aromatic molecules has been considered by Veeman and van der Waals (318b).

In most photochemical systems, the triplet state is populated via a radiationless transition from the lowest excited singlet state following initial excitation. This transition is also known as "intersystem crossing" in photochemistry. Very little is known experimentally about the exact nature of this process, but the rate or efficiency of such a transition in a given system can sometimes be estimated by chemical (319) or optical methods (320). Quantitative measurement by ESR of the triplet quantum yield is extremely difficult. Nevertheless, Guéron, Eisinger and Shulman (321) have developed a method to determine the triplet yield and the intersystem crossing efficiency for triplet molecules with lifetimes greater than 10 msec. The method employs extensive use of a computer program to evaluate the intensity of the observed $\Delta M_S = 2$ transition as a function of line-width and zero-field splittings. The calculated intensity related to the number of triplets in a given system is then compared with a standard having a known intersystem crossing yield. Clearly the ESR method of quantitative measurement of triplet yield is much less convenient and reliable than the chemical method suggested by Lamola and Hammond (319). However, in the case of ground triplet 1,3-diradical intermediates, the ESR technique may be the only reasonably reliable method available for quantitative estimation.

Kottis and Lefebvre (322) have suggested that if polarized light is used to excite randomly oriented molecules to the triplet state, observation of the changes in the $\Delta M_S = \pm 1$ ESR spectrum can reveal the correlation of the polarization properties of the excitation with the principal axis system of the triplet zero-field tensor. Such photoselection experiments have been carried out successfully by Lhotse and coworkers (323) and El-Sayed and Siegel (324) on a number of aromatic systems. Piette and collaborators (325) have studied the effect of metal complexation on the zero-field parameters and lifetimes of the phosphorescent triplet of aromatic-metal complexes with similar photoselection technique. The changes in zero-field parameters are attributed to changes in geometry. Some of their results are collected in Table 19.

Clarke (326) has studied the optical electron spin polarization in triplet anthracene and has observed ESR emission at 1.5°K which was attributed to a non-Boltzman distribution over the triplet spin levels at low temperature. The dynamics of optical spin polarization in triplet naphthalene at 1.6°K was also reported by Sixl and Schwoerer (327a) and van der Waals et al. (327b) have used a general method to study dynamics of populating and depopulating triplet spin levels by microwave-induced delayed phosphorescence. These experiments enable measurements of the lifetimes of each triplet spin state and thus can provide important information about intramolecular decay processes and intermolecular triplet energy transfer.

It is also worthwhile to mention here the relatively new and important method using phosphorescence microwave double resonance (PMDR) techniques in zero-field for the determination of intersystem crossing relative rates to the different zero-field levels of the lowest triplet state. A number of methods have been developed; these include the pulsed-excitation PMDR technique used by El-Sayed and coworkers (328a,b); the fast passage microwave population inversion technique proposed by Harris (328c); the microwave-induced delayed phosphorescence method used by van der Waals and coworkers (328d); and the microwave continuous-wave saturation technique earlier developed by El-Sayed (328e). In addition, the microwave pulse recovery method has been used by Wolfe (328f) to measure directly the spin-lattice relaxation in crystals.

The mean lifetime of phosphorescence has been correlated with the triplet-state energy in aromatic hydro-

TABLE XIX
Effects of Metal Ions on ESR Parameters of Triplet-State Molecules[a]

Molecule	D/hc (cm^{-1})	E/hc (cm^{-1})	D^*/hc (cm^{-1})	Half-life (sec)
o-Phenanthroline	0.1071	0.0386	0.1350	1.05
o-Phenanthroline with Zn^{++}	0.1078	0.0382	0.1367	2.09
2,2'-Bipyridyl	0.1132	0.0134	0.1132	0.64
2,2'-Bipyridyl with Zn^{+}	0.1108	0.0062	0.1110	0.80
2,2'-Bipyridyl with Mg^{++}	--	--	--	1.60
2,2'-Biquinoline	--	--	0.1158	0.56
2,2'-Biquinoline with Zn^{++}	--	--	0.0880	0.56
2,2'-Biquinoline with Al^{+++}	--	--	--	0.15

[a]Taken from reference (325).

carbons (329,330) and the latter has been correlated with the zero-field parameter (287). The reason for such a three-way relationship is not understood. On the experimental side, Brinen and coworkers (331a) have made careful quantitative phosphorescence studies by simultaneous ESR and optical measurements. They have also considered the effect of triplet-triplet reabsorption and concluded (331b) that optical measurements of phosphorescence decay are very susceptible to error due to reabsorption but that ESR determination is independent of triplet-triplet reabsorption. Excellent ESR lifetime studies have also been carried out in single crystals (9).

Intramolecular transfer of excitation is of considerable importance in photochemistry. Leermakers, Byers, Lamola, and Hammond (332) have demonstrated by optical emission the occurrence of intramolecular electronic energy transfer in 4-(1-naphthylalkyl)benzophenone. deGroot and van der Waals (333) have examined the temperature-dependent ESR spectra of the phosphorescent states of benzene, toluene, triptycene, and tribenzotriptycene. The latter two molecules consist of three benzene or naphthalene systems joined together, and their ESR spectra reveal the intramolecular excitation transfer between the benzene or naphthalene subsystems. At 20°K, the excitation is mainly localized in one of the subsystems, but at such higher temperatures as 77°K, the triplet excitation travels from one aromatic ring to another at a rate $>10^{10}$ sec^{-1}, resulting in ESR line shape corresponding to D_{6h} symmetry. Similarly, a dynamical equilibrium between possible conformational isomers of benzene leads to an interconversion rate of the order of 10^9 to 10^{10} sec^{-1}. These demonstrate the exclusive powerful ESR method in studies of this type of intramolecular excitation transfer process.

Intermolecular triplet-state energy transfer is well known in photochemistry. McDowell et al. (334) first demonstrated by ESR the triplet energy transfer from benzophenone to naphthalene in a rigid glass. The excitation transfer is found to be independent of orientation (335a), while the probability of excitation transfer per lifetime is dependent on concentration (335b). Triplet energy transfer from nitrobenzene (n,π^* excitation) to perfluoronaphthalene in rigid glass has also been observed (336). The effect of viscosity on triplet transfer process in glass has been investigated in detail (337). With mixed crystals, energy transfer effect can be elegantly studied. The pioneering work of Brandon, Gerkin, and Hutchison (338) on the triplet energy transfer from phenanthrene to naphthalene in biphenyl is the class experiment on energy transfer in single crystals. Subsequently numerous mixed-crystal systems have been investigated. For example, Hirota and Hutchison (339a) studied the decay rate of triplet biphenyl in durene, Ashford and Hutchison (45) measured the decay of 2-methylnaphthalene in naphthalene and interpreted the results in terms of a model of triplet energy transfer in these crystals. Harrigan and Hirota (339b) have studied the temperature-dependent triplet energy transfer processes from guest to host and guest to guest in single-crystal systems with small guest-host triplet energy separation. Energy transfer processes in crystal have also been studied by McGlynn and coworkers (340) using the delayed phosphorescence techniques.

The important energy transfer process from the triplet state of aromatic molecules to molecular oxygen giving singlet molecular oxygen ($^1\Delta g$) has been elegantly demonstrated in the gas-phase ESR studied by Kearns and

Maki (341) and Wasserman, Yager, and their coworkers (342). Production of singlet oxygen in the benzene-oxygen photochemical system was earlier reported by Snelling (343) in an optical study. Because of the importance of singlet molecular oxygen in photooxidation reactions (344) and in environmental studies (345), Kearns and coworkers (346) have examined theoretically the quenching of triplet-state molecules by oxygen and suggested that the process is usually 100-1000 times faster than quenching by enhanced intersystem crossing in the donor molecule. Experimentally Wasserman et al. (342) found that even at low oxygen pressure (0.2-1.0 mm), effectively all excited octafluoronaphthalene molecules transferred energy to oxygen before deactivation, and as high as 70% conversion of the total O_2 $(^3\Sigma)$ to $O_2(^1\Delta)$ was observed. Apparently the efficiencies of the transfer are independent of the triplet energy levels and the phosphorescent lifetimes for a series of donor molecules, but they are dependent on the chemical reactivity of the donor molecules toward oxygen. Some of the results are given in Table 20.

B. Photochemical Processes

The ESR method is useful in two aspects for studies of photochemical processes involving paramagnetic species: the qualitative mechanistic aspect and the quantitative rate measurement. For the mechanistic approach we have seen from Section IV how transient radicals can be observed during photolysis. The major disadvantages are: (a) It is not practical to determine accurately the quantum efficiency of radical production. Thus the radical process evidenced by ESR can well have been a very minor photochemical process, due to the high sensitivity of the modern ESR spectrometer. (b) In static

TABLE XX
Efficiencies of Energy Transfer from Excited Donor Molecules to Oxygen in Gas Phase[a]

Molecule	% O_2 ($^3\Sigma$) converted to O_2 ($^1\Delta$)	Triplet-State[b] Energy Level (ev)	Phosphorescence[b] Lifetime (sec) 77°K
Benzene	<0.7	3.66	--
Naphthalene	3 to 6	2.64	2.3
Octafluoronaphthalene	70	--	--
Anthracene	0.3	1.82	0.05
Biphenyl	2.4	2.84	3.1
Phenanthrene	0.4	2.68	3.5
Quinoxaline	3.5	--	--
Anthraquinone	0.1	2.72	--
Phenanthridine	0.1	--	--
1-Naphthaldehyde	2.5	2.44	--
1-Chloronaphthalene	0.4	2.57	0.29

[a]Taken from reference (342).

[b]Taken from J. G. Calvert and J. N. Pitts, Jr., Photochemistry, Wiley-Interscience, N.Y. (1966).

photolysis systems, it is often difficult to eliminate secondary radical-radical reactions as well as secondary photochemical processes. Perhaps it should also be emphasized here that failure to observe the desired radical intermediate does not exclude its existence in the reaction. Despite the difficulties involved in spin concentration determination, ESR has been extensively used in quantitative kinetic studies of radical reactions. The method constitutes a major advance in gas-phase kinetic studies involving H, N, and O atoms and $O_2(^1\Delta)$ molecules. Westenberg (347) and coworkers have recently improved the technique for in situ homogeneous radical reactions. Obviously ESR measurements are today the most convenient and reliable for monitoring concentrations of these gaseous atomic species. While most gas-phase studies involve discharge systems, photolysis can be applied to all phases, and kinetic studies of photochemical reactions in the condensed phase by ESR have also received widespread attention.

1. Qualitative Mechanistic Photochemistry

The ESR method is popular in photochemical studies involving triplet states. Photoreduction of aromatic ketones via an n,π^* triplet state is well known, although direct ESR observation of the short-lived n,π^* triplet in photochemical systems has not been successful. However, Pitts et al. (348) were among the first to report ESR detection of π,π^* triplet states in those substituted aromatic ketones which lack photoreactivity. Subsequently aromatic ketyl radicals produced in photoreduction via n,π^* triplet states have been observed by Wilson (224) and by Hammond and coworkers (349). Davidson and Wilson (224d) have also concluded from their ESR study of photoreduction of fluorenone in amine solvents that the reduction probably occurs by electron transfer from the N atom to the excited carbonyl group followed by a proton transfer from the amine radical cations. Biphotonic reactions involving triplet-triplet excitation have been extensively investigated by Russian workers, and the early work has been summarized by Terein (350). These higher excited triplet aromatic molecules are capable of ionization as well as sensitized decomposition of the solvent. Similar biphotonic processes in aromatic molecules have been investigated by Porter et al. (351) and Tanaka and coworkers (352) in rigid low-temperature glasses. Recently Santus and coworkers (188b) studied the photoionization of aromatic amino acids in the presence of divalent metal ions and attributed the ionization to a biphotonic triplet-triplet excitation process. Significant contribution to reaction mechanism involving ground-state triplet methylenes, nitrenes, and 1,3- and 1,5-diradicals by ESR has been revealed in Section IV.C.2.

The ESR study of solids at low temperatures can yield evidence for primary photochemical processes involving radical production. Svejda and Volman (353) have obtained ESR evidence for two primary photochemical processes of CH_3CN; one involves a C-C bond scission, the other involves a C-H bond split. They further substantiated the C-H split mechanism by observing the ethylidenimino radical $(CH_3)HC{=}\dot{N}$, resulting from the addition of the H atom to CH_3CN. The photochemistry of CH_3CN appears rather simple compared to the radiolysis of CH_3CN in which methyl radicals are produced by photochemical reactions in the γ-irradiated sample (354). Recently Sprague and Williams (355) have demonstrated that methyl radicals produced from CH_3CN in a low-temperature glass can abstract hydrogen from CH_3CN. From the ESR rate data, they reported an apparent activation energy for the abstraction process of 1.4 kcal $mole^{-1}$, which is very different from the gas-phase value of 10 kcal $mole^{-1}$ estimated by Wijnen (356). The reason for such a low activation energy is not fully understood, but the authors (355) suggested that H-atom abstraction may be a general and important process in low-temperature solid systems. Photolysis of formaldazine (257) presumably involves production of two isolated methylene imino radicals, while diphenylamino radical dimers are formed in the photodissociation of tetraphenylhydrazine (107). Photolysis of methanol at 77°K has been studied in some detail (357), and the ESR data are complicated by secondary radical reactions and secondary photochemical decomposition of radicals. ESR study of the photodecomposition of allylic alcohols (358) in a rigid glass indicates primary cleavage of the carbon-oxygen bond, giving allylic hydrocarbon radicals. On the other hand, the primary photochemical process in organic thiols, including triphenylmethanethiol (103), invariably involves an S-H split (245). The photolysis of inorganic sulfur compounds has also been studied by ESR (359). Photolysis of tetraphenyldiphosphine (260) apparently involves a primary cleavage of the P-P bond, and the photodecomposition of phenyl iodide (271) leads to a C-I split. The photolysis of biacetyl (276) at 77°K presumably leads to formation of acetyl radicals followed by secondary processes giving methyl radicals. ESR has also been used for photolysis studies of polyolefins and polystyrene at low temperature (360). Photoisomerization of o-nitrobenzaldehyde has been studied (361), and the radical process is probably of minor importance. Photoinduced radical isomerization in the photolysis of some aliphatic ketones has been observed (362). Low-temperature solid photochemical systems involving reactions of H and F atoms with unsaturated hydrocarbons have been extensively studied by ESR (see Section IV). Long and Willard (363) have studied the mechanism of hydrogen atom trapping and decay in a UV-irradiated 3-methylpentane-d_{14} matrix containing HI.

Direct observation of transient radicals in liquid photolysis has added a new dimension to mechanistic organic photochemistry. While the nature of the primary process in the photolysis of gaseous acetone is well known (364), the mechanism of the photochemical reaction in pure liquid acetone or in hydrocarbon solvent has not been established in detail. Zeldes and Livingston (216) have observed $(CH_3)_2\dot{C}OH$, $\dot{C}H_2COCH_3$, and some $\dot{C}H_3$ radicals when pure liquid acetone is photolysed and have suggested the following mechanism:

$$CH_3COCH_3 + CH_3COCH_3 \xrightarrow{h\nu} (CH_3)_2\dot{C}OH + \dot{C}H_2COCH_3 \quad (5\text{-}1)$$

$$\dot{C}H_2COCH_3 \rightarrow H_2C{=}C{=}O + \dot{C}H_3 \quad (5\text{-}2)$$

In the presence of an alcohol or ether the excited acetone abstracts an H atom from the carbon alpha to the O atom of the alcohol or ether. Similar photoreduction of perinaphthenone (365) has been observed. However, the nature of the excited state of acetone for liquid photoreduction in these studies has not been established. Free radicals produced from photoreduction of acetaldehyde, biacetyl and acetoin in the presence of good H-atom donors have been observed by Zeldes and Livingston (216). These authors also studied the photolysis of oxalic acid and esters (366).

Photolysis of acyl peroxides apparently leads to formation of alkyl radicals and carbon dioxide as observed by ESR (196):

$$\overset{O}{\underset{\|}{RC}}\text{-O-O}\overset{O}{\underset{\|}{C}}R + h\nu \rightarrow 2\dot{R} + 2CO_2 \qquad (5\text{-}3)$$

A further detailed ESR study of the photochemistry of aliphatic diacyl peroxides by Sheldon and Kochi (367) established that the quantum yield for CO_2 is 2 at 2537 Å and is independent of the alkyl groups. The disproportionation and combination of the alkyl radical pairs produced in eq. (5-3) have been studied in various solvents by direct ESR observation. In the case of cyclopropylacetyl peroxides, the primary cyclopropylmethyl radicals are observed at -150°C, whereas at -100°C or higher, only the isomer radicals, allylcarbinyl, are seen. However, since the dimeric products retain the propylmethyl moiety at temperatures as high as 30°C, and since the isomerization in eq. (5-4) is known to be very fast ($\sim 10^8$ sec^{-1}),

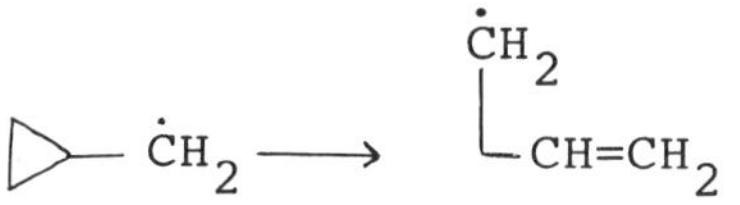
(5-4)

the authors concluded that cage combination is very fast and that only those alkyl radicals which diffuse from the solvent cage can be observed. They further suggested that cage combination of alkyl radicals is not affected by the structure and stability of the radical.

ESR evidence indicates that one primary process in the photolysis of liquid aliphatic tetrazenes (255) is the formation of two amino radicals and molecular nitrogen. Photochemical dissociation of Ph_3CN_3 (368) into $Ph_3\dot{C}$ and $\dot{N}_3$ radicals is found to be a minor process. Formation of nitroxide radicals in photolysis of aromatic nitro and nitroso compounds (235-238) is probably also a minor but concerted process. Nitroanions (218) have been observed during photolysis of basic aqueous or alcoholic solutions containing the corresponding nitroalkanes. Isaacs and Paxton (369) have studied the nature of some photolytic donor-acceptor systems, and photoinduced paramagnetism in trinitrobenzene-anion δ-complexes (370) has also been investigated.

Finally, as we mentioned in Section IV, indirect ESR evidence via "spin trapping" of intermediate radicals giving stable nitroxides has been applied in many organic mechanistic studies.

2. Quantitative Kinetic Studies

There have been relatively few quantitative kinetic ESR studies of photochemical systems. Using conventional steady-state decay measurements, the rate of photoreduction of DPPH in hydrocarbon solvents has been studied (371). Photolytic generation and decay rate of cumylperoxy radicals have been investigated, and a second-order decay with $2k = 4.4 \times 10^4\ M^{-1}\ sec^{-1}$ for the combination of cumylperoxy radicals was reported (372). The combination of tri-tert-butylbicycloheptatrienyl radicals produced by the photolysis of hexa-tert-butylbicycloheptatrienyl in hydrocarbon solvent has been examined (373) in the temperature range -112°C to + 22°C. An activation energy of 1.7 $\pm$ 0.3 kcal $mole^{-1}$ was reported for this second-order combination process. A lower activation energy has also been estimated by Sprague and Williams (355) for the methyl radical abstraction of an H atom in a solid CH_3CN matrix.

Hammond and coworkers (147,349,374) have developed a rotating sector method to measure by ESR the bimolecular decay of radicals produced by photolysis. Calvin and coworkers (148) have also studied the kinetics of ESR signal decay by intermittent UV-irradiation of charge-transfer complexes. A CAT was used by both groups to store the multiple decay curves monitored at a constant magnetic field. Thus Hammond and Weiner (374) were able to obtain the combination rate constant for benzophenone ketyl radicals which is in excellent agreement with the result of a flash kinetic spectrophotometric study. They have also investigated the effect of substituents on the reactivity of ketyl radicals. The authors discuss the caution required in the interpretation of rate data by means of commonly accepted models such as the Hammett relation. The hyperfine splittings of a number of novel ketyl radicals were also reported in their studies (349). The rates of recombination of cyanoalkyl radicals (147) and of hydrocarbon radicals (374a) produced by photolysis of corresponding azo compounds have also been reported. Ingold and coworkers (375) have recently initiated a series of ESR kinetic studies involving self-reactions of photochemically generated nitroxide radicals.

Flash photolysis coupled with ESR technique (6) has been used for kinetic studies. Bennett and coworkers (6b) used a multichannel sampling technique to obtain the full transient ESR spectrum by slowly scanning the magnetic field during flashes. Atkins et al. (6c) have modified the flash technique by using a pulsed laser. Bolton (6d) has coupled the flash photolysis to rapid scan of magnetic field for the observation and concentration determination of the transient radical. A function generator was used to drive a pair of external sweep coils and to trigger the CAT. Fabes (6e) has modified the commercial Varian 'Fieldial' control to obtain "external" field scan of about 150 gauss in 20 msec repetitively. The sweep was controlled by the computer while the flash was triggered at a predetermined point of the sweep. The kinetics of photochemical reactions of quinones in alcoholic solvent was reported by Bolton (6d), and the kinetics of the decay of the chloranil radical anion in dioxane was examined by Fabes (6e).

C. Photobiological Studies

The application of ESR to biological investigations has been steadily growing in recent years, particularly in the field of radiation damage and in spin label studies of macromolecules. Ingram (94) has written a monograph on ESR applications to biological and biochemical studies, and a book of current reviews on biological applications of ESR spectroscopy has been edited by Bolton, Borg, and Swartz (376).

Numerous ESR studies of the photodamage effect on biologically important molecules have been reported in recent years. Brustad (377) studied the UV-irradiation of the enzyme trypsin and found a linear correlation of the loss of enzyme activity with concentrations of the photoproduced radical. Leterrier and Douzou (378) have applied ESR techniques to study the mechanism of the photoreduction of flabin mononucleotide. A number of investigations deal with photodamage of aliphatic amino acids, glycine peptides, dipeptides, and alanine peptides (379-383). John et al. (383) have suggested that the primary photochemical reaction of solid alanine peptides is the decarboxylation followed by the thermal conversion of the primary radical via intermolecular H-atom transfer. Radicals produced by interaction of electrons with peptides have been investigated (384), and Lion et al. (385) reported that the reaction of purine bases and their derivatives with photoproduced electrons leads to organic radicals formed by H-atom addition. Photolysis of cysteine hydrochloride (386) at 77°K produces sulfur radicals which are converted by irradiation at longer wavelength to carbon radicals. Steen (387) studied the photochemical reaction of tryptophan in rigid ethylene glycol-water glass and concluded that photoionization of the molecule occurs via a biphotonic process. Reaction between the molecular ion and the solvent leads to formation of the ethyl radical. Tu and Wang (388) investigated the molecular mechanism in the hematoporphyrin-photosensitized phosphorylation of imidazole in aqueous solution.

The role of the triplet state in biological and biochemical systems continues to receive wide attention. Photoexcited triplet states of prophins and derivatives (389), prophyrins (390), aromatic amino acids (390), aromatic amines (391), and monoanionic thymine (392) have been observed at 77°K. Shiga and Piette (393) confirmed the interpretation of ESR data on triplet-state excitation in proteins by a simultaneous phosphorescence study. Shiga et al. (394) have followed up with various studies of the effects of solvent, oxygen paramagnetic ions, and other metal atoms on triplet excitation in protein molecules. Triplet energy transfer in dinucleotides has been investigated (395), which probably involves an excited singlet transfer mechanism. Some models of excitation energy transfer in biological systems have been reviewed by Terein (350). Photosensitized generation of singlet molecular oxygen has been unequivocally demonstrated by ESR (341-342). Because of the recent interest in the biological activity of singlet oxygen, the extreme sensitivity and specificity of ESR would no doubt be appreciated in such biological studies.

Finally we mention here some contributions of ESR studies to the important field of photosynthesis. A critical review of the developments of the knowledge in photosynthesis has been given by Clayton (396). Following the initial ESR studies in this field by Townsend et al.

(397) and Calvin and coworkers (398), the mechanism and kinetics of photosynthetic systems have been extensively investigated by ESR. For example, ESR signals have been observed in photosynthesizing algae and bacteria (399) on chromatophores and chloroplasts (400), in solid or dissolved chlorophyll (401) and in plant materials (402). The quantum yield of the photoproduced ESR signal in chromatophores from Rhodospirillum rubrum has been estimated (403) to be 0.8 at 8800 Å. From a systematic ESR study of chlorophyll spectra, a new model for the primary process in photosynthesis was proposed by Morin et al. (404). They suggested that the primary photobiological reaction of a chlorophyll-water micelle is the formation of chlorophyll radical cation and anion which provide the necessary reducing and oxidizing power for subsequent reactions.

The role of quinones in photobiological reactions involving chlorophyll has also been investigated (405,406). Despite the great effort and the multidisciplinary approach, progress in this field is slow because of the enormous complexity involved in photosynthetic systems. Hopefully the recent advances in experimental ESR technique, including the coupling of rapid scan ESR to flash photolysis, will help to elucidate the nature of the physical and chemical processes in photosynthesis.

ACKNOWLEDGMENT

The author wishes to thank Professors K. E. Russell and A. M. Winer and members of his own laboratories for helpful discussions.

REFERENCES

1. G. Porter, Science, 160, 1299 (1968).

2. G. Herzberg, Proc. Roy. Soc., 262A, 291 (1961).

3a. R. T. Meyer, J. Chem. Phys., 46, 967 (1967).

3b. O. P. Strausz, W. K. Duholke, and H. E. Gunning, J. Am. Chem. Soc., 92, 4128 (1970).

4. H. Masuhara, M. Shimada, and N. Mataga, Bull. Chem. Soc. Japan, 43, 3316 (1970).

5. T. Ogawa, G. A. Carlson, and G. C. Pimentel, J. Phys. Chem., 74, 2090 (1970), and references therein.

6a. T. J. Bennett, R. C. Smith, and T. H. Wilmshurst, Chem. Commun., 513 (1967); J. Sci. Instr., 2, 393 (1969).

6b. T. H. Wilmshurst, T. J. Bennett, and R. C. Smith, Proc. Roy. Soc., A302, 305 (1968).

6c. P. W. Atkins, K. A. McLauchlan, and A. F. Simpson, Chem. Commun., 179 (1968); Nature, 219, 927 (1968); J. Sci. Instr., 3, 547 (1970).

6d. B. J. Hales and J. R. Bolton, Photochem. Photobiol., 12, 239 (1970).

6e. L. Fabes, Ph.D. Thesis, Queen's University, Kingston, Ontario, Canada (1971).

7a. A. Abragam, The Principles of Nuclear Magnetism, Oxford (1961).

7b. G. E. Pake, Paramagnetic Resonance, Benjamin, New York (1962).

7c. S. A. Al'tschuller and B. M. Kozyrev, Electron Paramagnetic Resonance, Moscow (1961).

7d. C. P. Slichter, Principles of Magnetic Resonance, Harper and Row, New York (1963).

7e. A. Carrington and A. D. McLachlan, Introduction to Magnetic Resonance, Harper and Row, New York (1967).

7f. J. E. Wertz and J. R. Bolton, Electron Spin Resonance: Elementary Theory and Applications, McGraw-Hill, New York (1971)

8. D. J. E. Ingram, Free Radicals by ESR, Butterworths, London (1958).

9. C. Thomson, Quarterly Rev., 22, 45 (1968).

10. R. E. Jessee, P. Biloen, R. Prins, J. D. W. van Voorst, and G. J. Hoijtink, Mol. Phys., 6, 633 (1963).

11. G. S. Hammond and N. J. Turro, Science, 142, 1541 (1963).

12. W. S. Gleason, I. Rosenthal, and J. N. Pitts, Jr., J. Am. Chem. Soc., 92, 7042 (1970), and references therein.

13. R. P. Wayne, Adv. Photochem., 7, 311 (1969).

14. A. J. Stone, Proc. Roy. Soc., A271, 424 (1963); Mol. Phys. 6, 509 (1963); 7, 311 (1964).

15. M. S. Blois, Jr., H. W. Brown, and J. E. Maling, in Free Radicals in Biological Systems, Academic Press, New York (1961); 117.

16a. A. Hudson and G. R. Luckhurst, Chem. Rev., 69, 191 (1969).

16b. G. K. Fraenkel, J. Phys. Chem., 71, 139 (1967).

17a. C. C. Lin, Phys. Rev., 116, 903 (1959).

17b. R. F. Curl and J. L. Kinsey, J. Chem. Phys., 35, 1758 (1961).

18. A. Carrington, A. R. Fabris, and N. J. D. Lucas, J. Chem. Phys., 49, 5545 (1968); Mol, Phys., 16, 195 (1969), and many other references.

19a. P. W. Atkins, I. C. Buchanan, R. C. Grid, K. A. McLauchlan, and A. F. Simpson, Chem. Commun., 513 (1970); Chem. Phys. Lett., 8, 55 (1971).

19b. R. Livingston and H. Zeldes, J. Chem. Phys., 53, 1406 (1970).

19c. M. Schwoerer and H. C. Wolf, Proc. Colloque Ampère, 14, 87 (1966).

20. S. K. Lower and M. A. El-Sayed, Chem. Rev., 66, 199 (1966).

21. M. Blois, H. Brown, and J. Maling, Arch. Sci., 13, 243 (1960).

22. D. R. Gee and J. K. S. Wan, Chem. Commun., 641 (1970).

23. H. M. McConnell and R. E. Robertson, J. Phys. Chem., 61, 1018 (1957).

24. J. E. Wertz, Chem. Rev., 55, 829 (1955).

25. K. Fermi, Z. Physik, 60, 320 (1930).

26a. J. Turkevich and Y. Fujita, Science, 152, 1619 (1966).

26b. T. Cole, H. O. Pritchard, N. R. Davidson, and H. M. McConnell, Mol. Phys., 1, 406 (1958).

27. H. M. McConnell and R. W. Fessenden, J. Chem. Phys., 31 1688 (1959).

28. H. M. McConnell, C. Heller, T. Cole, and R. W. Fessenden, J. Am. Chem. Soc., 82, 766 (1960).

29. H. Zeldes, G. T. Trammell, and R. Livingston, J. Chem. Phys., 32, 618 (1960).

30. E. Clementi, C. C. J. Roothaan, and M. Yoshimine, Phys. Rev., 127, 1618 (1962).

31. R. E. Watson and A. J. Freeman, Phys. Rev., 123, 521 (1961); 124, 1117 (1961).

32. M. T. Rogers and L. D. Kispert, J. Chem. Phys., 46, 221 (1967).

33. J. R. Morton, Chem. Rev., 64, 453 (1964).

34. C. K. Jen, in Formation and Trapping of Free Radicals, A. M. Bass and H. P. Broide, Eds., Academic Press, New York and London (1960), Chapter 7.

35. P. D. Sullivan and J. R. Bolton, Adv. Mag. Reson., 4, 39 (1970).

36. D. E. Wood and R. V. Lloyd, J. Chem. Phys., 53, 3932 (1970).

37. E. de Boer and E. L. Makcor, Mol. Phys., 5, 493 (1962).

38. R. W. Fessenden, J. Chem. Phys., 37, 747 (1962).

39. C. A. Hutchison and B. W. Mangum, J. Chem. Phys., 29, 952 (1958); 34, 908 (1961).

40. C. A. Hutchison, Rec. Chem. Progr., 24, 105 (1963).

41. J. H. van der Waals and M. S. deGroot, J. Chem. Phys., 1643 (1964).

42. J. H. van der Waals and M. S. deGroot, Mol. Phys., 2, 333 (1959); 3, 190 (1960).

43. J.-PH. Grivet, Mol. Phys., 19, 389 (1970); J. Mispelter, J.-PH. Grivet, and J. M. Lhoste, Mol. Phys., 21, 999 (1971).

44. P. H. H. Fischer and K. H. Hausser, Chem. Phys. Letters, 1, 665 (1968).

45a. C. A. Hutchison, Rec. Chem. Progr., 24, 105 (1963).

45b. N. A. Ashford and C. A. Hutchison, J. Chem. Phys., 54, 1165 (1971).

46. R. W. Brandon, R. E. Gerkin, and C. A. Hutchison, J. Chem. Phys., 37, 447 (1962); 41, 3717 (1964).

47a. R. E. Gerkin and A. M. Winer, J. Chem. Phys., 50, 3114 (1969),

47b. R. E. Gerkin and A. M. Winer, J. Chem. Phys., 56, 1359 (1972).

48. R. E. Gerkin and P. Szerenyi, J. Chem. Phys., 50, 4095 (1969); 55, 2575 (1971).

49a. C. A. Hutchison, in The Triplet State, A. B. Zahlan, Ed., Cambridge University Press (1967), 63.

49b. A. M. Ponte Goncalves and C. A. Hutchison, J. Chem. Phys., 49, 4235 (1968).

49c. R. H. Clark and C. A. Hutchison, J. Chem. Phys., 54, 2962 (1971).

50. P. Ehret and H. C. Wolf, Z. Naturforsch, 23a, 1740 (1968).

51a. P. W. Kopf, K. Morokuma, and R. W. Kreilick, J. Chem. Phys., 54, 105 (1971).

51b. P. W. Kopf and R. W. Kreilick, J. Am. Chem. Soc., 91, 6569 (1969).

52a. H. Lemaire, A. Rassat, P. Rey, and G. R. Luckhurst, Mol. Phys., 14, 441 (1968).

52b. R. M. Dupeyre, H. Lemaire, and A. Rassat, J. Am. Chem. Soc., 87, 3771 (1965).

52c. R. Briere, R. M. Dupeyre, H. Lemaire, C. Morat, A. Rassat, and P. Rey, Bull. Soc. Chem. France, 11, 3290 (1965).

52d. H. R. Falle, G. R. Luckhurst, H. Lemaire, Y. Marechal, A. Rassat, and P. Rey, Mol. Phys., 11, 49 (1966).

53a. S. H. Glarurn and J. H. Marshall, J. Chem. Phys., 47, 1374 (1967).

53b. G. R. Luckhurst and G. F. Pedulli, Mol. Phys., 18, 425 (1970).

53c. W. J. van der Hart and L. J. Oosterhoff, Mol. Phys., 18, 281 (1970).

54. R. McWeeny and B. T. Sutcliffe, Mol. Phys., 6, 493 (1963).

55. B. T. Sutcliffe, J. Chem. Phys., 39, 3322 (1963).

56. K. Kayama, J. Chem. Phys., 39, 1507 (1963).

57. M. Yamazaki, M. Sakamoto, K. Hijikata, and C. C. Lin, J. Chem. Phys., 34, 1926 (1961).

58. H. M. McConnell, J. Chem. Phys., 24, 632, 764 (1956).

59. R. Bersohn, J. Chem. Phys., 24, 1066 (1956).

60. S. I. Weissman, J. Chem. Phys., 25, 890 (1956).

61. H. M. McConnell and D. B. Chesnut, J. Chem. Phys., 28, 107 (1958)

62. D. K. Ghosh and D. H. Whiffen, Mol. Phys., 2, 285 (1959).

63. J. P. Colpa and J. R. Bolton, Mol. Phys., 6, 273 (1963).

64. T. F. Hunter and M. C. R. Symons, J. Chem. Soc. (A), 1770 (1967).

65. M. Karplus and G. K. Fraenkel, J. Chem. Phys., 35, 1312 (1961).

66. A. D. McLachlan, H. H. Dearman, and R. Lefebvre, J. Chem. Phys., 33, 65 (1960).

67. J. C. M. Henning, J. Chem. Phys., 44, 2139 (1966).

68. A. Carrington and J. Dos Santos-Veiga, Mol. Phys., 5, 21 (1962).

69a. P. D. Sullivan, J. Am. Chem. Soc., 90, 3618 (1968).

69b. I. Degani, L. Lunazzi, G. F. Pedulli, C. Vincenzi, and A. Mangini, Mol. Phys., 18, 613 (1970).

70. F. Gerson, J. Heinzer, and H. Bock, Mol. Phys., 18, 461 (1970).

71. M. Kaplan, J. R. Bolton, and G. K. Fraenkel, J. Chem. Phys., 42, 955 (1965).

72. J. Sinclair and D. Kivelson, J. Am. Chem. Soc., 90, 5074 (1968).

73. W. E. Geiger and W. M. Gulick, J. Am. Chem. Soc., 91, 4657 (1969).

74. A. Hinchliffe and J. N. Murrell, Mol. Phys., 14, 147 (1968).

75. A. Hudson and J. W. E. Lewis, Mol. Phys., 19, 241 (1970).

76. J. M. Mackey and D. E. Wood, Mol. Phys., 18, 783 (1970).

77. M. C. R. Symons, Advan. Phys. Org. Chem., 1, 284 (1963).

78. D. B. Chesnut, J. Chem. Phys., 29, 43 (1958).

79. C. A. Coulson, Valence, Oxford University Press, Fairlawn, New Jersey (1961).

80. M. J. S. Dewar, Hyperconjugation, Ronald Press Co., New York (1962).

81. A. D. McLachlan, Mol. Phys., 1, 233 (1958).

82. M. C. R. Symons, J. Chem. Soc., 277 (1959).

83. C. Heller and H. M. McConnell, J. Chem. Phys., 32, 1535 (1960).

84. D. Pooley and D. H. Whiffen, Mol. Phys., 4, 81 (1961).

85. J. P. Colpa and E. deBoer, Mol. Phys., 7, 333 (1964).

86. R. W. Fessenden and R. H. Schuler, J. Chem. Phys., 39, 2147 (1963); 38, 773 (1963).

87. J. Lin and F. Williams, J. Phys. Chem., 72, 3707 (1968).

88. R. Marx and L. Bonazzola, Mol. Phys., 19, 899 (1970).

89. L. Bonazzola and R. Marx, Mol. Phys., 19, 405 (1970).

90. P. J. Krusic and J. K. Kochi, J. Am. Chem. Soc., 91, 4566 (1969).

91. J. Gloux, M. Guglielmi, and H. Lemaire, Mol. Phys., 17, 425 (1969).

92. C. P. Poole, Electron Spin Resonance, Interscience, New York (1967).

93. T. H. Wilmshurst, Electron Spin Resonance Spectrometers, Adam Hilger, London (1966).

94. D. J. E. Ingram, Biological and Biochemical Applications of ESR, Adam Hilger, London (1969).

95. P. B. Ayscough, ESR in Chemistry, Methuen, London (1967).

96. J. G. Kenworthy, G. F. Longster, and R. E. Richards, Trans. Faraday Soc., 62, 534 (1966).

97. A. S. Lebedev and O. M. Taranukha, Theoret. Exp. Chem. U.S.S.R. (Eng. Transl.), 1, 171 (1965).

98. R. H. Silsbee, J. Chem. Phys., 45, 1710 (1966).

99. E. W. Firth and D. J. E. Ingram, J. Sci. Instr., 44, 821 (1967)

100a. A. Carrington, D. H. Levy, and T. A. Miller, Rev. Sci. Instr., 38, 1183 (1967).

100b. A. Carrington, Proc. Roy. Soc., A302, 291 (1968).

101. P. W. Atkins and M. C. R. Symons, The Structure of Inorganic Radicals, Elsevier, Amsterdam (1967).

102. R. S. Alger, EPR Techniques and Applications, Interscience, New York (1968).

103. J. K. S. Wan, Chem. Commun., 429 (1967).

104. P. B. Ayscough and H. E. Evans, J. Phys. Chem., 68, 3066 (1964)

105a. P. B. Ayscough, J. H. O'Donnell, and K. J. Ivin, Trans. Faraday Soc., 61, 1110 (1965).

105b. J. K. S. Wan, J. R. Morton, and H. J. Bernstein, Can. J. Chem., 44, 1957 (1966).

106. M. C. R. Symons, Nature, 213, 1226 (1967).

107. D. A. Wiersma and J. Kommandeur, Mol. Phys., 13, 241 (1967).

108. M. Iwasaki, T. Ichikawa, and T. Ohmori, J. Chem. Phys., 50, 1984 (1969).

109. M. T. Jones, in Radical Ions, E. T. Kaiser and L. Kevan, Eds., Interscience, New York (1968), 256.

110. M. J. Stephen and G. K. Fraenkel, J. Chem. Phys., 32, 1435 (1960).

111. M. T. Jones, J. Chem. Phys., 42, 4054 (1965).

112. G. Bignol and W. Muller-Warmeith, Phys. Letters, 11, 292 (1964)

113. S. K. Wong and J. K. S. Wan, unpublished results.

114a. R. W. Yip, personal communication.

114b. S. K. Wong, W. Sytnyk, and J. K. S. Wan, Can. J. Chem., 49, 994 (1971).

115. R. O. C. Norman and R. J. Pritchett, Chem. Ind., 2040 (1965).

116. S. I. Weissman, J. Chem. Phys., 22, 1378 (1954).

117. C. P. Poole and R. S. Anderson, J. Chem. Phys., 31, 346 (1959).

118. J. H. Freed and G. K. Fraenkel, J. Chem. Phys., 39, 326 (1963).

119. R. Lefebvre and J. Maurani, J. Chem. Phys., 42, 1480 (1970).

120. G. S. Owen and G. Vincow, J. Chem. Phys., 54, 368 (1970).

121. R. G. Lawler and G. K. Fraenkel, J. Chem. Phys., 49, 1126 (1968).

122. W. V. Volland and G. Vincow, J. Chem. Phys., 48, 5589 (1968).

123. G. Vincow, M. L. Morrell, F. R. Hunter, and H. J. Dauben, J. Chem. Phys., 48, 2876 (1968).

124. A. Carrington, H. C. Longuet-Higgins, R. E. Moss, and P. F. Flood, Mol. Phys., 9, 187 (1965).

125. H. L. Strauss and G. K. Fraenkel, J. Chem. Phys., 35, 1738 (1961).

126. J. R. Bolton and G. K. Fraenkel, J. Chem. Phys., 40, 3307 (1964); 41, 944 (1964).

127. A. D. McLachlan, Mol. Phys., 3, 233 (1960).

128. J. A. Pople, D. L. Beveridge, and P. A. Dobosh, J. Chem. Phys., 47, 2026 (1967); J. Am. Chem. Soc., 90, 4201 (1968).

129. M. C. R. Symons, Ann. Rev. of Phys. Chem., 20, 219 (1969).

129b. M. C. R. Symons, J. Am. Chem. Soc., 91, 5924 (1969).

130. J. Pannell, Mol. Phys., 5, 291 (1962); 7, 317 (1963).

131. M. C. R. Symons, J. Phys. Chem., 74, 3834 (1970).

132. D. R. Gee, K. E. Russell, and J. K. S. Wan, Can. J. Chem., 48, 2740 (1970).

133. A. A. Westenberg, Science, 164, 381 (1969).

134. S. Krongelb and M. W. P. Strandberg, J. Chem. Phys., 31, 1196 (1959).

135. R. L. Brown and W. Brennen, J. Chem. Phys., 47, 4697 (1967).

136. A. Thomas, Trans. Faraday Soc., 57, 1979 (1961).

137. J. E. Bennett and A. Thomas, Proc. Roy. Soc., A280, 123 (1964).

138. J. E. Bennett, B. Mile, A. Thomas, and B. Ward, Adv. Phys. Org. Chem., 8, 2 (1970).

139. C. Thomsom, J. Chem. Phys., 41, 1 (1964).

140. N. Trublin, R. Santus, and M. Ptak, Comp. Rend., 260, 1134 (1965).

141. D. E. Wood, R. V. Lloyd, and D. W. Pratt, J. Am. Chem. Soc., 92, 4115 (1970).

142. D. Campbell and A. Peterlin, Polymer Letters, 6, 481 (1968).

143. J. K. Kochi and P. J. Krusic, J. Am. Chem. Soc., 90, 7155 (1968).

144. A. R. McIntosh and J. K. S. Wan, Can. J. Chem., 49, 812 (1971); Mol. Phys., 22, 183 (1971).

145. R. W. Fessenden, J. Phys. Chem., 68, 1508 (1964).

146. B. Smaller, J. R. Remko, and E. C. Avery, J. Chem. Phys., 48, 5174 (1968).

147. S. Weiner and G. S. Hammond, J. Am. Chem. Soc., 90, 1659 (1968); 91, 2182 (1969).

148. D. F. Ilten, M. E. Kronenberg, and M. Calvin, Photochem. Photobiol., 7, 331 (1968).

149. I. Yamazaki, H. S. Mason, and L. Piette, J. Biol. Chem., 235, 2444 (1960).

150. B. Chance, L. Bicking, and V. Legallais, in Free Radicals in Biological Systems, Academic Press, New York (1966), 101-11

151. T. J. Stone and W. A. Waters, Proc. Chem. Soc., 253 (1962).

152. W. T. Dixon and R. O. C. Norman. J. Chem. Soc., 3119 (1963).

153. R. O. C. Norman and B. C. Gilbert, Advan. Phys. Org. Chem., 5, 53 (1967).

154. A. L. J. Beckwith, Intra-Science Chem. Report, 4, 127 (1970).

155. M. R. Arick and S. I. Weissman, J. Am. Chem. Soc., 90, 1654 (1968).

156. R. Livingston and H. Zeldes, J. Chem. Phys., 44, 1245 (1966).

157a. E. L. Waters and A. H. Maki, Phys. Rev., 125, 233 (1962).

157b. S. Fujiwara and M. Codell, Bull. Chem. Soc. Japan, 37, 49 (1964).

158. R. Johnson, P. W. Murchison, and J. R. Bolton, J. Am. Chem. Soc., 92, 6354 (1970).

159. H. Fischer, Ber. Bunsen. Ges. Phys. Chem., 71, 685 (1967).

160. K. Takakura and B. Ranby, J. Phys. Chem., 72, 164 (1968).

161. M. S. Bains, J. C. Arthur, and O. Hinojosa, J. Phys. Chem., 72, 2250 (1968).

162. R. E. Florin, F. Sicillio, and L. A. Wall, J. Phys. Chem., 72, 3154 (1968).

163a. H. B. Brooks and F. Sicillio, J. Phys. Chem., 74, 4565 (1970).

163b. R. E. James and F. Sicillio, J. Phys. Chem., 74, 2294 (1970).

164. A. A. Galkin, I. L. Shamfarov, and A. V. Stefanishina, J. Exptl. Theoret. Phys. (U.S.S.R), 32, 1581 (1957).

165. D. E. G. Austen, P. H. Given, D. J. E. Ingram, and M. E. Peover, Nature, 182, 1784 (1958).

166. D. H. Geske and A. H. Maki, J. Am. Chem. Soc., 82, 2671 (1960).

167. A. H. Maki and D. H. Geske, J. Chem. Phys., 33, 825 (1960).

168. A. H. Maki and D. H. Geske, J. Am. Chem. Soc., 83, 1852 (1961).

169. D. H. Geske and J. L. Ragle, J. Am. Chem. Soc., 83, 3532 (1961); A. H. Maki, J. Chem. Phys., 35, 761 (1961).

170. P. H. Rieger, I. Bernal, W. H. Reinmuth, and G. K. Fraenkel, J. Am. Chem. Soc., 85, 683 (1963).

171. J. K. Dohrmann and K. J. Vetter, J. Electroanal. Chem., 20, 23 (1969).

172a. R. N. Adams, J. Electroanal. Chem., 8, 151 (1964).

172b. L. H. Piette, P. Ludwig, and R. N. Adams, Anal. Chem., 34, 916 (1962).

173. R. E. Dessy, M. Kleiner, and S. C. Cohen, J. Am. Chem. Soc., 91, 6800 (1969), and references therein.

174. R. Hirasawa, T. Mukaibo, H. Hasegawa, N. Odan, and T. Maruyama, J. Phys. Chem., 72, 2541 (1968).

175. R. Hirasawa, T. Mukaibo, H. Hasegawa, Y. Kanda, and T. Maruyama, Rev. Sci. Instr., 39, 935 (1968).

176. M. C. R. Symons, in Radiation Chemistry-II, Adv. in Chem. Series 82, Amer. Chem. Soc., (1968), 1-24.

177. F. W. Froben and J. E. Willard, J. Phys. Chem., 75, 35 (1971).

178a. C. K. Jen, S. N. Foner, E. L. Cochran, and V. A. Bowers, Phys. Rev., 112, 1169 (1958).

178b. S. N. Foner, E. L. Cochran, V. A. Bowers, and C. K. Jen, J. Chem. Phys., 32, 963 (1960).

179. C. J. Delbecq, B. Smaller, and P. H. Yuster, Phys. Rev., 104, 599 (1956).

180. A. Hausman, Z. Physik, 192, 313 (1966).

181. D. R. Eaton, S. R. Suart, J. Phys. Chem., 72, 400 (1968).

182. D. E. Holmes, N. B. Nazhat, and J. J. Weiss, J. Phys. Chem., 74, 1622 (1970).

183. S. K. Wong and J. K. S. Wan, J. Chem. Phys., 55, 4940 (1971).

184. T. G. Castner and W. Kanzig, J. Phys. Chem. Solids, 3, 178 (1957).

185. K. D. Bowers, R. A. Kamper, and C. D. Lustig, Proc. Phys. Soc., B70, 1176 (1957).

186. S. Aditya and J. E. Willard, J. Chem. Phys., 44, 833 (1966).

187. E. Wasserman, W. E. Falconer, and W. A. Yager, Ber. Bunsen. Ges. Phys. Chem., 72, 48 (1968).

188a. P. H. Dasai, Phys. Rev. Lett., 21, 67 (1968).

188b. R. Santus, A. Helene, C. Helene, and M. Ptak, J. Phys. Chem., 74, 550 (1970).

189. T. W. Martin and L. L. Swift, J. Am. Chem. Soc., 93, 2788 (1971).

190. P. W. Atkins, M. C. R. Symons, and P. A. Trevalion, Proc. Chem. Soc., 222 (1963).

191. W. G. Hodgson, J. S. Brinen, and E. F. Williams, J. Chem. Phys., 47, 3719 (1967).

192. A. J. Hay and R. L. Belford, J. Chem. Phys., 47, 3944 (1967).

193. G. Rudinger, Phys. Fluids, 4, 1463 (1961).

194. A. Treinin, in Radical Ions, E. T. Kaiser and L. Kevan, Eds., Interscience, New York (1968), 549.

195a. J. Q. Adams, J. Am. Chem. Soc., 90, 5353 (1968).

195b. J. Q. Adams, J. Am. Chem. Soc., 92, 4535 (1970).

196. P. J. Krusic and J. K. Kochi, J. Am. Chem. Soc., 91, 3942 (1969

197. R. Livingston and H. Zeldes, J. Am. Chem. Soc., 88, 4333 (1966)

198a. J. K. Kochi and P. J. Krusic, J. Am. Chem. Soc., 90, 7157 (1968

198b. J. K. Kochi and P. J. Krusic, J. Am. Chem. Soc., 92, 4110 (1970)

198c. P. J. Krusic and J. K. Kochi, J. Am. Chem. Soc., 91, 3938 (1969)

198d. J. K. Kochi and P. J. Krusic, J. Am. Chem. Soc., 91, 3940 (1969)

198e. J. K. Kochi and P. J. Krusic, J. Am. Chem. Soc., 91, 3944 (1969)

199. J. M. Riveros and S. Shih, J. Chem. Phys., 50, 3132 (1969).

200. N. Shimamoto, Y. Fujita, and T. Kevan, Bull. Chem. Soc. Jap., 43, 580 (1970).

201a. P. J. Krusic and J. K. Kochi, J. Am. Chem. Soc., 91, 6161 (1969).

201b. J. K. Kochi and P. J. Krusic, J. Am. Chem. Soc., 91, 1877 (1969).

201c. J. K. Kochi and P. J. Krusic, J. Am. Chem. Soc., 91, 1879 (1969).

202a. A. G. Davies and B. P. Roberts, Chem. Commun., 699 (1969).

202b. A. Hudson and H. A. Hussain, J. Chem. Soc. (B), 793 (1969).

203. G. A. Helcke and R. Fantechi, Mol. Phys., 18, 1 (1970).

204. D. R. Gee and J. K. S. Wan, Can. J. Chem., 49, 20 (1971).

205a. M. K. Carter and G. Vincow, J. Chem. Phys., 47, 292, 302 (1967).

205b. G. Vincow and P. M. Johnson, J. Chem. Phys., 39, 1143 (1963).

206. R. Hulme and M. C. R. Symons, J. Chem. Soc., 1120 (1965).

207a. P. B. Ayscough and F. P. Sargent, Proc. Chem. Soc., 94 (1963).

207b. G. A. Russell and E. J. Geele, Tetrahedron Lett., 1333 (1963).

208. A. Carrington, A. Hudson, and H. C. Longuet, Mol. Phys., 9, 377 (1965).

209. G. Vincow, in Radical Ions, E. T. Kaiser and L. Kevan, Eds., Interscience, New York (1968), 151-209.

210a. J. Maurani, J. A. R. Coope, and C. A. McDowell, Mol. Phys., 18, 165 (1970).

210b. J. Maurani, C. A. McDowell, H. Nakajima, and P. Raghunathan, Mol. Phys., 14, 349 (1968).

211. S. W. Charles, P. H. H. Fischer, and C. A. McDowell, Chem. Phys. Lett., 1, 451 (1967).

212. A. Hudson and H. A. Hussain, Mol. Phys., 16, 199 (1969).

213. L. Pauling, J. Chem. Phys., 51, 2767 (1969).

214. A. Hudson and R. A. Jackson, Chem. Commun., 1323 (1969).

215a. A. J. Bowles, A. Hudson, and R. A. Jackson, Chem. Phys. Letters, 5, 552 (1970).

215b. A. Hudson, Chem. Phys. Letters, 4, 295 (1969).

216. H. Zeldes and R. Livingston, J. Chem. Phys., 45, 1946, 1465 (1966).

217. R. Livingston and H. Zeldes, J. Chem. Phys., 47, 4173 (1967).

218. H. Zeldes and R. Livingston, J. Am. Chem. Soc., 90, 4540 (1968)

219. R. Livingston and H. Zeldes, J. Magnetic Reson., 1, 169 (1969).

220. P. B. Ayscough and R. E. D. McClung, Mol. Phys., 20, 35, 41 (1971).

221a. A. Hudson and K. D. J. Root, Tetrahedron, 25, 5311 (1969).

221b. A. Hudson and K. D. J. Root, J. Chem. Soc. (B), 656 (1970).

222. E. J. Baum and R. O. C. Norman, J. Chem. Soc. (B), 48 (1968).

223a. J. Q. Adams, J. Am. Chem. Soc., 90, 5363 (1968).

223b. J. Q. Adams. J. Am. Chem. Soc., 92, 4535 (1970).

224a. R. J. Wilson, Can. J. Chem., 44, 551 (1966).

224b. R. J. Wilson, J. Chem. Soc. (B), 84 (1968).

224c. R. J. Wilson, J. Chem. Soc. (B), 1581 (1968).

224d. R. S. Davidson and R. J. Wilson, J. Chem. Soc. (B), 71 (1970).

225a. S. W. Bennett, C. Eaborn, A. Hudson, H. A. Hussain, and R. A. Jackson, J. Organometallic Chem., 16, 36 (1969).

225b. S. W. Bennett, C. Eaborn, A. Hudson, R. A. Jackson, and K. D. J. Root, J. Chem. Soc. (A), 348 (1970).

226. M. Geoffroy and E. A. Lucken, Helv. Chim. Acta, 53, 813 (1970).

227a. E. J. Land and G. Porter, J. Chem. Soc., 3540 (1961).

227b. N. M. Atherton, E. J. Land, and G. Porter, Trans. Faraday Soc., 59, 818 (1963).

228. K. Adamic, K. U. Ingold, and J. R. Morton, J. Am. Chem. Soc., 92, 922 (1970).

229. R. W. Fessenden, J. Chem. Phys., 48, 3725 (1968).

230. N. Vanderkool and W. B. Fox, J. Chem. Phys., 47, 3634 (1967).

231. E. G. Janzen and B. J. Blackburn, J. Am. Chem. Soc., 91, 4481 (1969).

232a. C. Lagercrantz and S. Forshult, Acta Chem. Scand., 23, 708, 811 (1969).

232b. S. Forshult and C. Lagercrantz, Acta Chem. Scand., 23, 522 (1969).

233. I. H. Leaver, G. C. Ramsay, and E. Suzuki, Aust. J. Chem., 22, 1891 (1969).

234. A. Mackor, Th. A. Wajer, Th. J. deBoer, and J. D. W. van Voorst, Tetrahedron Lett., 385 (1967).

235. E. T. Strom and A. L. Bluhm, Chem. Commun., 115 (1966).

236. A. Mackor, Th. A. Wajer, and Th. J. deBoer, Tetrahedron Lett., 2115 (1966).

237a. R. L. Ward, J. Chem. Phys., 38, 2588 (1963).

237b. S. K. Wong and J. K. S. Wan, to be published.

238. C. Chachaty and A. Forchioni, Tetrahedron Lett., 1079 (1968).

239. D. J. Cowley and L. H. Sutcliffe, J. Chem. Soc. (B), 569 (1970)

240. A. A. McConnell, S. Mitchell, A. L. Porte, J. S. Roberts, and C. Thomson, J. Chem. Soc. (B), 833 (1970).

241. A. R. Forrester and S. P. Hepburn, Chem. Commun., 698 (1969).

242. J. J. Windle, A. K. Wiersema, and A. L. Tappel, J. Chem. Phys., 41, 1996 (1964).

243. T. Henriksen, J. Chem. Phys., 37, 2189 (1962).

244. G. S. Bogle, V. R. Burgen, W. F. Forbes, and W. E. Savage, Photochem. Photobiol., 1, 277 (1962).

245. P. S. H. Bolman, I. Safarik, D. A. Stiles, W. J. R. Tyerman, and O. P. Strausz, Can. J. Chem., 48, 3872 (1970).

246. W. Rundel and K. Scheffler, Angew. Chem., 77, 220 (1965).

247. S. M. Kavun and A. L. Buchachenko, Russian J. Phys. Chem., 42, 427 (1968).

248. T. Garofano and M. Santangelo, Ricerca Scient., 4, 75 (1964).

249. T. Shida, K. Maeda, and T. Hayashi, Bull. Chem. Soc. Jap., 43, 652 (1970).

250. S. M. Blinder, M. L. Peller, N. W. Lord, L. C. Aamodt, and N. S. Ivanchukov, J. Chem. Phys., 36, 540 (1961).

251. A. K. Piskunov, N. V. Berein, and G. A. Ozerova, Russian J. Phys. Chem., 42, 1137 (1968).

252. T. E. Gough and J. M. Gross, J. Chem. Soc. (B), 798 (1968).

253a. J. K. S. Wan, L. D. Hess, and J. N. Pitts, Jr., J. Am. Chem. Soc., 86, 2069 (1964).

253b. J. K. S. Wan, Ber. Bunsen, Ges. Phys. Chem., 72, 245 (1968).

254. C. S. Johnson and H. S. Gutowsky, J. Chem. Phys., 39, 58 (1963).

255. W. C. Daneu and T. T. Kensler, J. Am. Chem. Soc., 92, 5235 (1970).

256. E. L. Cochran, F. J. Adrian, and V. A. Bowers, J. Chem. Phys., 36, 1938 (1962).

257. M. Kamachi, K. Kuwata, and S. Murahashi, J. Phys. Chem., 75, 164 (1971).

258a. G. F. Kokoszka and F. E. Brinckman, Chem. Commun., 349 (1968).

258b. G. F. Kokoszka and F. E. Brinckman, J. Am. Chem. Soc., 92, 1199 (1970).

259. U. Schmidt, F. Geiger, A. Muller, and K. Markan, Chem. Ber., 99, 1497 (1966).

260. S. K. Wong and J. K. S. Wan, Spectroscopy Lett., 3, 135 (1970).

261. R. O. C. Norman and B. C. Gilbert, J. Phys. Chem., 71, 14 (1967).

262. B. C. Gilbert and R. O. C. Norman, J. Chem. Soc. (B), 722 (1966); 981 (1967); 123 (1968).

263. B. C. Gilbert and W. M. Bulick, J. Phys. Chem., 73, 2448 (1970).

264. F. J. Adrian, E. L. Cochran, and V. A. Bowers, J. Chem. Phys., 36, 1661 (1962).

265. E. L. Cochran, F. J. Adrian, and V. A. Bowers, J. Chem. Phys., 44, 4626 (1966).

266. F. J. Adrian, E. L. Cochran, and V. A. Bowers, J. Chem. Phys., 43, 462 (1965).

267. E. L. Cochran, F. J. Adrian, and V. A. Bowers, J. Chem. Phys., 40, 213 (1964).

268. P. H. Kasai and E. B. Whipple, J. Am. Chem. Soc., 89, 1033 (1967).

269. E. L. Cochran, F. J. Adrian, and V. A. Bowers, J. Phys. Chem., 74, 2083 (1970).

270. P. H. Kasai, L. Skattebol, and E. B. Whipple, J. Am. Chem. Soc., 90, 4509 (1968).

271. P. H. Kasai, E. Hedaya, and E. B. Whipple, J. Am. Chem. Soc., 91, 4364 (1969).

272. G. Porter and B. Ward, Proc. Roy. Soc., A287, 457 (1965).

273. J. A. Pople, D. L. Beveridge, and P. A. Dobosh, J. Am. Chem. Soc., 90, 4201 (1968).

274. W. T. Dixon, Mol. Phys., 9, 201 (1965).

275. G. A. Petersson and A. D. McLachlan, J. Chem. Phys., 45, 628 (1966).

276. J. R. Morton and W. E. Falconer, Nature, 197, 1103 (1963).

277. T. Yonezawa, I. Noda, and T. Kawamura, Bull. Chem. Soc. Jap., 41, 766 (1968); 42, 652 (1969).

278. A. A. Lamola, J. Chem. Phys., 47, 4810 (1967).

279a. M. Sharnoff, J. Chem. Phys., 51, 451 (1969).

279b. M. Sharnoff, J. Chem. Phys., 46, 3263 (1967).

279c. M. Sharnoff, Chem. Phys. Letters, 2, 498 (1969).

280. A. L. Kwiram, Chem. Phys. Letters, 1, 272 (1967).

281. J. Schmidt and J. H. van der Waals, Chem. Phys. Letters, 2, 640 (1968).

282. D. S. Tinti, M. A. El-Sayed, A. H. Maki, and C. B. Harris, Chem. Phys. Letters, 3, 343 (1969).

283. E. Wasserman, L. C. Snyder, and W. A. Yager, J. Chem. Phys., 41, 1763 (1964).

284a. C. H. J. Wells, A. Horsfield, and J. Paxton, Chem. Commun., 393 (1969).

284b. J. B. Gallivan and J. S. Brinen , Chem. Phys. Lett., 10, 455 (1971).

285. J. deLong and C. MacLean, Chem. Phys. Letters, 5, 424 (1970).

286. J. S. Vincent, Chem. Phys. Letters, 8, 37 (1971); J. Chem. Phys., 54, 2237 (1971).

287. J. S. Brinen and M. K. Orloff, Chem. Phys. Letters, 1, 276 (1967).

288. B. Smaller, E. C. Avery, and J. R. Remko, J. Chem. Phys., 46, 3976 (1967).

289. R. N. Dixon, Mol. Phys., 13, 77 (1967).

290a. H. Hayashi, S. Nagakura, and S. Iwata, Mol. Phys., 13, 489 (1967).

290b. H. Hayashi, S. Iwata, and S. Nagakura, J. Chem. Phys., 50, 993 (1969).

290c. H. Hayashi and S. Nagakura, Mol. Phys., 19, 45 (1970).

291. T. L. Gilchrist and C. W. Rees, Carbenes, Nitrenes and Arynes, Nelson and Sons, Ltd., London (1969).

292. G. Smolinsky, E. Wasserman, and W. A. Yager, J. Am. Chem. Soc., 84, 3213 (1962).

293. A. M. Trozzolo, R. W. Murray, and E. Wasserman, J. Am. Chem. Soc., 84, 4990 (1962).

294. R. W. Murray, A. M. Trozzolo, E. Wasserman, and R. M. R. Cramer, J. Am. Chem. Soc., 84, 3213 (1962).

295. E. Wasserman, A. M. Trozzolo, W. A. Yager, and R. W. Murray, J. Chem. Phys., 40, 2408 (1964).

296. A. M. Trozzolo, Acc. Chem. Research, 1, 329 (1968).

297. E. Wasserman, L. Barash, and W. A. Yager, J. Am. Chem. Soc., 87, 4974 (1965).

298. R. A. Bernheim, R. J. Kemp, J. V. Gramas, and P. S. Skell, J. Chem. Phys., 43, 196 (1965).

299. J. Higuchi, J. Chem. Phys., 39, 1339 (1963).

300. R. A. Bernheim, H. W. Bernard, P. S. Wang, L. S. Wood, and P. S. Skell, J. Chem. Phys., 53, 1280 (1970); 54, 3223 (1971).

301a. E. Wasserman, V. J. Kuck, R. S. Hutton, and W. A. Yager, J. Am. Chem. Soc., 92, 7431 (1970).

301b. E. Wasserman, W. A. Yager, and V. J. Kuck, Chem. Phys. Letters, 1, 409 (1970).

301c. E. Wasserman, V. J. Kuck, R. S. Hutton, E. D. Anderson, and W. A. Yager, J. Chem. Phys., 54, 4120 (1971).

302. G. Herzberg and J. W. C. Johns, J. Chem. Phys., 54, 2276 (1971).

303. J. F. Harrison and L. C. Allen, J. Am. Chem. Soc., 91, 807 (1969); J. F. Harrison, J. Chem. Phys., 54, 5413 (1971).

304a. P. Dowd, J. Am. Chem. Soc., 88, 2587 (1966).

304b. P. Dowd and K. Sachdev, J. Am. Chem. Soc., 89, 715 (1967).

304c. P. Dowd, A. Gold, and K. Sachdev. J. Am. Chem. Soc., 90, 2715 (1968).

305. H. M. McConnell, J. Chem. Phys., 35, 1520 (1961).

306. G. L. Closs, L. R. Kaplan, and V. I. Bendall, J. Am. Chem. Soc., 89, 3376 (1967).

307. G. Baum, R. Bernard, and H. Shechter, J. Am. Chem. Soc., 89, 5307 (1967).

308. G. L. Closs, L. R. Kaplan, J. Am. Chem. Soc., 91, 2168 (1969).

309. D. R. Arnold, A. B. Evnin, and P. H. Kasai, J. Am. Chem. Soc., 91, 784 (1969).

310. R. N. Dixon, Can. J. Phys., 37, 1711 (1959).

311. J. W. McIver and H. F. Hameka, J. Chem. Phys., 45, 767 (1966).

312a. G. Smolinsky, E. Wasserman, and W. A. Yager, J. Am. Chem. Soc., 84, 3220 (1962).

312b. E. Wasserman, G. Smolinsky, and W. A. Yager, J. Am. Chem. Soc., 86, 3166 (1964).

313. J. A. R. Coope, J. B. Farmer, C. L. Gardner, and C. A. McDowell, J. Chem. Phys., 42, 54 (1965).

314. G. Smolinsky, L. C. Snyder, and E. Wasserman, Rev. Mod. Phys., 35, 576 (1963).

315. S. Niizuma, H. Kokubun, and M. Koizumi, Chem. Phys. Letters, 7, 279 (1970).

316. D. A. Wiersma and W. C. Niewpoort, Chem. Phys. Letters, 2, 637 (1968).

317. J. Jortner, S. A. Rice and R. M. Hochstrasser, Adv. Photochem., 7, 149 (1969).

318a. M. A. El-Sayed, Acc. Chem. Research, 1, 8 (1968).

318b. W. S. Veeman and J. H. van der Waals, Mol. Phys., 18, 63 (1970).

319. A. A. Lamola and G. S. Hammond, J. Chem. Phys., 43, 2129 (1965).

320. C. A. Parker and T. A. Joyce, Chem. Commun., 234 (1966).

321. M. Guéron, J. Eisinger, and R. G. Shulman, Mol. Phys., 14, 111 (1968).

322. P. Kottis and R. Lefebvre, J. Chem. Phys., 41, 3660 (1964).

323. J. M. Lhotse, A. Haug, and M. Ptak, J. Chem. Phys., 44, 648, 654 (1966).

324. M. A. El-Sayed and S. Siegel, J. Chem. Phys., 44, 1416 (1966).

325. G. P. Rabold and L. H. Piette, Spectroscopy Lett., 1, 211, 225 (1968).

326. R. H. Clarke, Chem. Phys. Letters, 6, 413 (1970).

327a. H. Sixl and M. Schwoerer, Chem. Phys. Letters, 6, 21 (1970).

327b. D. A. Antheunis, J. Schmidt, and J. H. van der Waals, Chem. Phys. Letters, 6, 255 (1970).

328a. M. A. El-Sayed and J. Olmsted, Chem. Phys. Lett., 11, 568 (1971).

328b. M. A. El-Sayed, J. Chem. Phys., 54, 680 (1971).

328c. C. B. Harris, J. Chem. Phys., 54, 972 (1971).

328d. J. Schmidt, W. S. Veeman, and J. H. van der Waals, Chem. Phys. Lett., 4, 341 (1969).

328e. M. A. El-Sayed, J. Chem. Phys., 52, 6438 (1970).

328f. J. P. Wolfe, Chem. Phys. Lett., 10, 212 (1971).

329. G. W. Robinson and R. P. Frosch, J. Chem. Phys., 38, 1187 (1963).

330. W. Siebrand, J. Chem. Phys., 44, 4055 (1966).

331a. J. S. Brinen, F. Halverson, and J. R. Leto, J. Chem. Phys., 42, 4213 (1965).

331b. J. S. Brinen and W. G. Hodgson, J. Chem. Phys., 47, 2946 (1967).

332. P. A. Leermakers, G. W. Byers, A. A. Lamola, and G. S. Hammond, J. Am. Chem. Soc., 85, 2670 (1963).

333a. M. S. deGroot and J. H. van der Waals, Mol. Phys., 6, 545 (1963).

333b. M. S. deGroot, I. A. M. Hesselmann, and J. H. van der Waals, Mol. Phys., 10, 91 (1965).

334 J. B. Farmer, C. L. Gardner, and C. A. McDowell, J. Chem. Phys., 34, 1058 (1961).

335a. G. P. Rabold and L. H. Piette, Photochem. Photobiol., 5, 733 (1966).

335b. S. Siegel and L. Goldstein, J. Chem. Phys., 44, 2780 (1966).

336. J. K. Brown and W. G. Williams, Chem. Commun., 495 (1966).

337a. S. Siegel and H. Judeikis, J. Chem. Phys., 42, 3060 (1965).

337b. B. Smaller, E. C. Avery, and J. R. Remko, J. Chem. Phys., 43, 922 (1965).

338. R. W. Brandon, R. E. Gerkin, and C. A. Hutchison, J. Chem. Phys., 37, 447 (1962).

339a. N. Hirota and C. A. Hutchison, J. Chem. Phys., 42, 2869 (1965); 46, 1561 (1967).

339b. E. T. Harrigan and N. Hirota, Chem. Phys. Lett., 1, 281 (1967); J. Chem. Phys., 49, 2301 (1968).

340. T. N. Misra and S. P. McGlynn, J. Chem. Phys., 44, 3816 (1966); M. Kinoshita, T. N. Misra, and S. P. McGlynn, J. Chem. Phys., 45, 817 (1966).

341. D. R. Kearns, A. U. Khan, C. K. Duncan and A. H. Maki, J. Am. Chem. Soc., 91, 1039 (1969).

342. E. Wasserman, V. J. Kuck, W. M. Delavan, and W. A. Yager, J. Am. Chem. Soc., 91, 1040 (1969).

343. D. R. Snelling, Chem. Phys. Lett., 2, 346 (1968).

344. C. S. Foote, Acc. Chem. Research, 1, 104 (1968).

345. A. U. Khan, J. N. Pitts, and E. B. Smith, Environ. Sci., 1, 655 (1967).

346. K. Kawaoka, A. U. Khan, and D. R. Kearns, J. Chem. Phys., 46, 1847 (1967).

347. A. A. Westenberg, N. deHaas, and J. M. Roscoe, J. Phys. Chem., 74, 3431 (1970).

348a. L. H. Piette, J. H. Sharp, T. Kuwana, and J. N. Pitts, J. Chem. Phys., 36, 3094 (1962).

348b. E. J. Baum, J. K. S. Wan, and J. N. Pitts, J. Am. Chem. Soc., 88, 2652 (1966).

349. T. Fujisawa, B. M. Monroe, and G. S. Hammond, J. Am. Chem. Soc., 92, 542 (1970).

350. A. Terein, in Recent Progress in Photobiology, E. J. Bowen, Ed., Blackwell, Oxford (1964), 3.

351. B. Brocklehurst, W. A. Gibbons, F. T. Lang, G. Porter, and M. I. Savadatti, Trans. Faraday Soc., 62, 1793 (1966).

352. K. Shimokoshi, Y. Mori, and I. Tanaka, Bull. Chem. Soc. Jap., 40, 302 (1967).

353. P. Svejda and D. H. Volman, J. Phys. Chem., 74, 1872 (1970).

354. P. B. Ayscough, R. G. Collins, and T. J. Kemp, J. Phys. Chem., 70, 2220 (1966).

355. E. D. Sprague and F. Williams, J. Am. Chem. Soc., 93, 788 (1971).

356. M. H. J. Wijnen, J. Chem. Phys., 22, 1074 (1954).

357a. P. J. Sullivan and W. S. Koski, J. Am. Chem. Soc., 85, 384 (1963).

357b. S. B. Milliken and R. H. Johnson, J. Phys. Chem., 71, 2116 (1967).

358. K. A. Maas and D. H. Volman, Trans. Faraday Soc., 60, 202 (1964).

359a. H. E. Gunning and O. P. Strausz, Advan. in Photochem., 4, 143 (1966).

359b. D. A. Stiles, R. Kewley, O. P. Strausz, and H. E. Gunning, Can. J. Chem., 43, 2442 (1965).

360. H. L. Browning, H. D. Ackermann, and H. W. Patton, J. Polymer Sci., Al, 4, 1433 (1966).

361. A. J. Tench and P. Coppens, J. Phys. Chem., 67, 1378 (1963).

362. Y. Hama and K. Shinohara, Mol. Phys., 18, 279 (1970).

363. M. A. Long and J. E. Willard, J. Phys. Chem., 74, 1207 (1970).

364. D. E. Hoare and G. S. Pearson, Advan. Photochem., 3, 88 (1964)

365a. H. Köller, G. P. Rabold, K. Weiss, and T. Mukherjee, Proc. Chem. Soc., 332 (1964).

365b. G. P. Rabold, K. H. Bar-Eli, E. Reid, and K. Weiss, J. Chem. Phys., 42, 2438 (1965).

366. H. Zeldes and R. Livingston, J. Phys. Chem., 74, 3336 (1970).

367. R. A. Sheldon and J. K. Kochi, J. Am. Chem. Soc., 92, 4395 (1970).

368. F. D. Lewis and W. H. Saunders, J. Am. Chem. Soc., 90, 3829 (1969).

369. N. S. Isaacs and J. Paxton, Photochem. Photobiol., 11, 137 (1970).

370a. A. R. Norris, A. Breck, and J. K. S. Wan, Can. J. Chem., 46, 1823 (1968).

370b. P. B. Ayscough, F. P. Sargent, and R. Wilson, J. Chem. Soc., 5418 (1963).

371. J. N. Pitts, E. A. Schuck, and J. K. S. Wan, J. Am. Chem. Soc., 86, 296 (1964).

372. J. J. Zwolenik, J. Phys. Chem., 71, 2464 (1967).

373. M. L. Morrell and G. Vincow, J. Am. Chem. Soc., 91, 6389 (1969).

374a. S. A. Weiner and G. S. Hammond, J. Am. Chem. Soc., 91, 986 (1969).

374b. G. S. Hammond and S. A. Weiner, Intra-Sci. Chem. Rep., 3, 241 (1969).

375. K. Adamic, D. F. Bowman, T. Gillan, and K. U. Ingold, J. Am. Chem. Soc., 93, 902 (1971).

376. Biological Applications of ESR Spectroscopy, J. R. Bolton, D. Borg, and H. Swartz, Eds., Wiley-Interscience, New York (1971).

377. T. Brustad, Nature, 209, 31 (1966).

378. F. Leterrier and P. Douzou, Photochem. Photobiol., 8, 369 (1968).

379. R. B. Johns, F. D. Looney, and D. J. Whelan, Biochem. Biophys. Acta, 147, 369 (1967).

380. R. B. Johns, F. D. Looney, and D. J. Whelan, Photochem. Photobiol., 7, 65 (1968).

381. M. A. Collins and R. A. Grant, Photochem. Photobiol., 9, 369 (1969).

382. A. Meybeck and J. J. Windle, Photochem. Photobiol., 10, 1 (1969).

383. R. B. Johns and M. G. Seuret, Photochem. Photobiol., 12, 405 (1970).

384. M. D. Sevilla, J. Phys. Chem., 74, 3366 (1970).

385. Y. Lion and A. Vande Vorst, Int. J. Radiat. Phys. Chem., 2, 51 (1970).

386. H. C. Box, H. G. Freund, and E. E. Budzinski, J. Chem. Phys., 45, 809 (1966).

387. H. B. Steen, Photochem. Photobiol., 9, 479 (1969).

388. S. I. Tu and J. H. Wang, Biochem., 9, 4505 (1970).

389. G. T. Rikhireva, A. V. Umrikhina, L. P. Kayushkin, and A. A. Kranovskii, Doklady U.S.S.R., Biophys. Sect., 163, 115 (1965).

390. O. A. Azizova, Z. P. Gribova, L. P. Kayushkin, and M. K. Pulatova, Photochem. Photobiol., 5, 763 (1966).

391. C. Chachaty, D. Schoemaker, and R. Bensasson, Photochem. Photobiol., 12, 317 (1970).

392. R. O. Rahn, Photochem. Photobiol., 9, 527 (1969).

393. T. Shiga and L. H. Piette, Photochem. Photobiol., 3, 223 (1964).

394. T. Shiga, H. S. Mason, and C. Simo, Biochem., 5, 1877 (1966).

395. M. Guéron, R. G. Shulman, and J. Eisinger, Proc. Natl, Acad. Sci., U. S., 56, 814 (1966).

396. R. K. Clayton, Molecular Physics in Photosynthesis, Blaisdell New York (1965).

397. B. Commoner, J. J. Heise, and J. Townsend, Proc. Natl. Acad. Sci., 42, 710 (1956).

398. P. B. Sogo, N. G. Pon, and M. Calvin, Proc. Natl. Acad. Sci., 43, 387 (1957).

399. J. R. Bolton, R. K. Clayton, and D. Reed, Photochem. Photobiol., 9, 209 (1969).

400a. K. Cost, J. R. Bolton, and A. W. Frenkel, Photochem. Photobiol., 10, 251 (1969).

400b. J. R. Bolton, K. Cost, and A. W. Frenkel, Arch. Biochem. Biophys., 126, 383 (1968).

401. G. Sherman and E. Fujimori, Nature, 219, 375 (1968).

402. E. C. Weaver, Biochim. Biophys. Acta, 162, 286 (1968).

403. P. A. Loach and K. Walsh, Biochem., 8, 1908 (1969).

404. M. G. Morin, R. A. Uphaus, J. R. Norris, and J. J. Katz, J. Phys. Chem., 73, 1066 (1969).

405. D. H. Kohl, J. R. Wright, and M. Weissman, Biochim, Biophys. Acta, 180, 536 (1969).

406. D. C. Mukherjee, D. H. Cho, and G. Tollin, Photochem. Photobiol., 9, 273 (1969).

Primary Processes and Energy Transfer: Consistent Terms and Definitions

GERALD B. PORTER*
Centro di Studio sulla Fotochimica e Reattività degli Stati
Eccitati dei Composti di Coordinazione del C.N.R.
Istituto Chimico dell'Università
Ferrara, Italy
and

VINCENZO BALZANI and LUCA MOGGI
Laboratorio di Fotochimica e Radiazioni d'Alta
Energia del C.N.R.
Istituto Chimico "G. Ciamician" dell'Università
Bologna, Italy

CONTENTS

---

* On leave from the Department of Chemistry, University of British Columbia, Vancouver, Canada, 1970-71.

I. INTRODUCTION

During the last few years interest in photochemistry has grown vigorously. As a result, a number of books on photochemistry have appeared (1-8), and advanced courses on photochemistry are being given at most major universities. Also, sections devoted to photochemistry have been expanded in physical chemistry textbooks (9) and have been introduced even to the latest texts on organic chemistry (10). With this great expansion has come the need for well-defined, appropriate, and generally acceptable terms and definitions.

The problems concerning the "vocabulary" of photochemistry were discussed by Pitts, Wilkinson, and Hammond (11) in 1963. Although their suggestions have been generally adopted, a number of contradictory and ambiguous statements persist in the photochemical language. Further, because of the increased understanding of the details of photochemical mechanisms and the enormous success of the new techniques, some of the terms used seem to be no longer appropriate.

The redefinition of some terms and the creation of new definitions will facilitate the expression of ideas that are currently of greatest interest. As far as the chemical aspect of the problem is concerned, especially with regard to organic photochemistry, the nomenclature used is largely that of ordinary reactions. For that reason, the development here will be concerned mainly with what Noyes, Porter, and Jolley (12) referred to as the "primary photochemical process." In addition, because of the growing importance of energy transfer in the study of photochemical systems and the almost complete lack of a coherent nomenclature, this aspect will be considered in the following treatment.

II. THE PRESENT SITUATION

Much of the confusion in textbooks and to a certain extent in the photochemical literature follows, either directly or indirectly, from the ambiguity in the use of the two key terms: "primary photochemical process" and "quantum yield." We shall first review the various meanings which have been attributed to these quantities in the literature. When an example is needed, reference will be made to Figure 1, which shows schematically the most important routes by which a photoexcited molecule may lose its energy. This scheme is, however, relatively simple, and we shall have occasion later to refer to more complicated schemes.

In this and other examples used later, we have systematized the labeling of the detailed steps with two-digit numbers. The first number is the identifying sub-

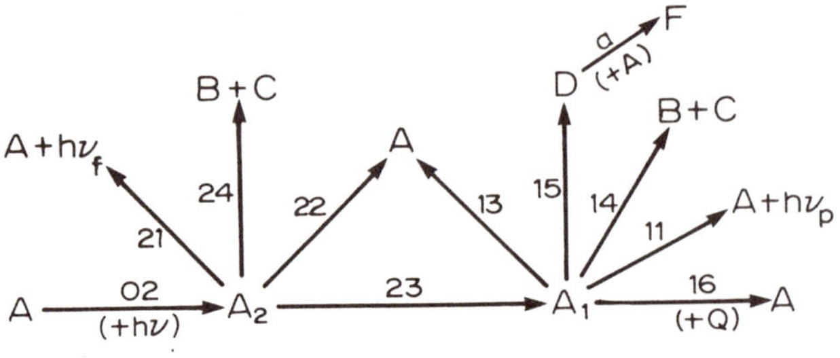

FIGURE 1. A schematic photochemical mechanism, showing some of the possible elementary transformations. For the purpose of illustration, it is assumed that the states A and A_2 have the same multiplicity, and correspond to the ground and lowest excited singlet states of most organic molecules. The state A_1 would then represent the lowest triplet state. Thus 21 and 11 are radiative transitions, fluorescence and phosphorescence, respectively, and 23 and 13 (intersystem crossing) and 22 (internal conversion) are nonradiative. All of B, C, D, and F are chemical species distinct from A. Only vibrationally equilibrated electronic states are included in this mechanism (see discussion in Section III.A.1).

script associated with the state of the molecule with which the step originates. The second number indicates the nature of the step according to the scheme:

1 luminescence
2 internal conversion
3 intersystem crossing
4 chemical reactions
5 chemical reactions
6 quenching
7 energy transfer

When the step is absorption, the first number is zero and the second indicates the state reached in the absorption step. For example, 23 identifies the intersystem crossin from state A_2 to A_1, and 02 refers to the absorption of light by A of a wavelength appropriate to reach state A_2.

A. Primary Photochemical Processes

Although it may seem obvious, it should be realized that a chemical process is one which involves the transformation of one chemical species into another. It follows that a photochemical process (which is a chemical process initiated by excitation of a molecule through absorption of light) and a primary photochemical process (which is a particular kind of photochemical process) must also imply some kind of chemical transformation.

In some physical chemistry texts, the primary photochemical process is incorrectly considered to be no more than the _absorption_ of radiation. Such a definition is not acceptable because absorption is not a chemical transformation and, more important, because it does not correspond to current usage in photochemistry. These texts then label such diverse processes as fluorescence, dissociation of an excited molecule, and chain reactions, all as different types of secondary photochemical processes. This is also unacceptable because "secondary" has come to have a specific meaning (as is discussed in Section III.A.3) which does not apply to all of these transformations, and also because some of them are not chemical. Photochemists instead use primary and secondary in the original sense of Bodenstein.

In another physical chemistry book, all of the elementary steps that involve the excited molecule are called "primary photochemical processes." Vibrational relaxation, fluorescence, energy transfer, isomerization, and dissociation are all considered to be different types of primary photochemical processes. But the first three listed do not involve any chemical change and cannot properly be described as chemical.

In their vocabulary of photochemistry, Pitts, Wilkinson, and Hammond (11) recommend the definition of Noyes, Porter, and Jolley (12): "The primary photochemical process comprises the series of events beginning with the absorption of a photon by a molecule and ending either with the disappearance of that molecule or with its conversion to a state such that its reactivity is statistically no greater than that of similar molecules in thermal equilibrium with their surroundings." This definition, which has also been adopted in two recent books on photochemistry (3,5), is somewhat too general, as it covers all of the transformations of excited molecules in a photochemical system. In Figure 1, for example, all of the events shown except "a" are part of the primary photochemical process according to this definition. Yet a number of series of events, e.g., 02-23-11 or 02-22, do not involve any chemical change and can be separately characterized on a quantitative basis by their own quantum yields (see Section III.B). Further, there are three distinct series shown, 02-24, 02-23-14, and 02-23-15, which are chemical, two of which lead to the same chemical products. Therefore, it does not seem appropriate to reserve this single term "primary photochemical process" to cover all of these series of events.

A tendency to use the term "primary photochemical process" with a narrower and more specific meaning than that of Noyes et al. is evident in most of the recent photochemistry books. Turro (1), Calvert and Pitts (2), Leermakers (4), Wayne (7), and Simons (8) more or less explicitly consider as a primary photochemical process any chemical act of electronically excited molecules.

Thus, according to these authors, there are three distinct primary photochemical processes in Figure 1, namely 24, 15, and 14. Balzani and Carassiti (6), on the other hand, define a primary photochemical process as any process starting with the absorption of a photon and ending with the disappearance of the molecule or with its deactivation to a nonreactive state. They then use "photochemical primary process" for the former and "photophysical primary process" for the latter. Thus, according to this scheme, there are three photochemical primary processes in Figure 1, the sequences 02-24, 02-23-15, and 02-23-14.

Thus, even excluding the first definition quoted at the beginning of this section, there are four different meanings attributed to the term "primary photochemical process" in these recent books.

B. Quantum Yields

The quantum yield was originally used in the photochemical sense exclusively, for the number of molecules of reactant consumed per photon absorbed. However, in this form, the quantum yield does not convey any information as to the relative contribution of different excited-state reactions or of secondary thermal reactions, which may lead to consumption of reactant. Thus, in the photochemical literature, it has become common to use two distinct types of quantum yield: (a) the experimentally measured quantum yield (often called the overall quantum yield) and (b) the primary quantum yield. The quantum yield (a) accounts for reactant disappearance (or product formation) whether it occurs directly in a primary process or in a secondary thermal reaction or in both. The second type of quantum yield is a quantitative measure of only the primary photochemical process. In physical chemistry books and in those dealing only with photochemistry, it appears that there is some confusion about the definitions of the two types of quantum yield, as well as disagreement about the symbol to be used to represent them. In addition, it has now become common to use the term "quantum yield" also for photophysical processes (including energy transfer), and this has introduced further inconsistencies and contradictions.

Just as there are a number of different definitions of "primary photochemical processes," so also does "primary quantum yield" have different meanings in the various photochemical references. It may be defined, for example, as the sum of the quantum yields of all of the events which lead to dissociation or reaction of the excited molecule. In Figure 1, these would be events 24, 15, and 14. An alternate definition is that each one of the individual quantum yields in that sum is itself one of the primary quantum yields.

The concept of quantum yield is also extensively used to characterize the various photophysical processes. A complete account of this matter may be found in the recent book by Birks (13). The "phosphorescence quantum yield" is defined there as "the ratio of the number of phosphorescence photons emitted to the number of absorbed photons." In the same book, the name "phosphorescence quantum efficiency" is given to the fraction of the molecules in the triplet state which phosphoresce. This last quantity, however, does not depend on the number of quanta absorbed, and thus the word "quantum" is inappropriate. On the other hand, the term "triplet formation efficiency" used by Parker (14) to indicate the "number of triplet molecules formed per quantum of exciting light absorbed" omits the word "quantum" from what actually is a quantum yield.

C. Energy Transfer

In the rapid development of the field of energy transfer, many of the terms commonly used to describe the various processes involved are ambiguous and are often used incorrectly. For example, Turro (1) defines the "quantum yield of sensitized phosphorescence" as the ratio of the number of sensitized phosphorescent acceptor molecules to the number of quenched phosphorescent donor molecules. But because this makes no reference to the number of absorbed photons, it is, in fact, not a quantum yield at all. Calvert and Pitts (2) define the ratio of the intensity of sensitized acceptor phosphorescence to the intensity of light absorbed by the donor to be the "sensitized yield of phosphorescence," despite the fact that it is a quantum yield. The quantum yield data for energy transfer are frequently extrapolated to or measured at high concentrations of acceptor species in order to find the value of a "limiting" quantum yield which is no longer dependent on the concentration of acceptor. Again there is no consensus on a name to be used for this quantity.

D. Lifetimes

With the advent of convenient techniques in recent years, it has become popular to measure "lifetimes" of excited states. Decay characteristics are useful adjuncts to quantum yields as they provide further knowledge about the mechanism of reaction, particularly with regard to evaluation of rate constants for excited molecule reactions. Only when the decay is exponential can a unique lifetime be defined.

Just as with quantum yields, there are basically two different kinds of lifetimes, one a result of direct experimental measurement, and the other a derived quantity. These have not always been carefully distinguished. Moreover, the same quantity has been labeled by a variety of names and symbols. For example, radiative lifetime, true radiative lifetime, natural lifetime, intrinsic lifetime, and inherent lifetime all mean the same thing: the lifetime a molecule in an excited state would have if there were no steps competing with that of spontaneous emission of radiation.

III. SUGGESTED TERMS AND DEFINITIONS

The difficulties in nomenclature described in the previous sections as well as others which arise in the photochemical "language" may be overcome by using the terms, definitions, and symbols given below. The basis of the suggested nomenclature is the clear distinction between: (a) "step" and "process", (b) "primary" and "secondary", (c) "efficiency" and "quantum yield," and (d) "experimental" and "mechanistic" quantities. Care has been used to avoid words and definitions which could be confusing or ambiguous, without generating a system too complicated to be used practically. Yet the treatment has been designed to be internally consistent and as far as possible also consistent with respect to current usage by photochemists. These recommended terms, together with others of importance in photochemistry, are defined in the Appendix.

A. The Mechanism

It must be made clear at the outset that, even for simple molecules, photoexcitation may be followed by many very complicated events. This makes it difficult, if not impossible, to obtain definitive and exhaustive information about any photochemical system under consideration. In order to interpret the available data and to suggest new lines of investigation, a mechanism must be postulated such as that of Figure 1 to consist of a number of elementary transformations of the excited and other species in the system. The whole mechanism then must be consistent with the experimental data. Any photochemical mechanism in terms of these elementary transformations is to be regarded as a "working hypothesis." Although it may be possible, with additional data, to prove a given mechanism wrong, it is difficult, if not impossible, to say that it is correct in all its details. Certainly no detailed mechanism is unique, in fact, except in the simplest of cases.

1. Primary Steps

The absorption of light by a substance causes the formation of excited-state molecules. This excitation is followed by various elementary transformations which eventually lead to the deactivation or to the disappearance of those excited molecules. The absorption of light as well as each one of the elementary transformations of the original molecule in an excited state is a primary step. Specifically, a primary step may be (a) a transformation of the excited molecule into a different chemical species, as in steps 24, 15, and 14 of Figure 1, or (b) a radiative or nonradiative transition between different energy levels of the molecule, e.g., steps 02, 21, 22, 23, 13, 11, and 16 of Figure 1. Those corresponding to (a) are photochemical primary steps, while those of (b) are photophysical primary steps.

a. Photochemical and Photophysical Primary Steps. Primary steps are classified as photochemical or photophysical depending on whether or not they result in the transformation of the original molecule into a "different" chemical species. This division of the photoinduced events may be useful in certain circumstances, and it has been adopted in most books on photochemistry and in the literature. The distinction between photochemical and photophysical steps is usually quite obvious (at least to a chemist). There can be no question, for example, that the dissociation of the original molecule in an excited state, such as step 24 or 14 in Figure 1, is a photochemical primary step, or that fluorescence emission leading to the original ground-state molecule, step 21, is a photophysical primary step.

Even so, the distinction between the two is sometimes a more subtle matter. Thus, in a photoisomerization a common excited state intermediate may undergo a transformation to either of the two isomeric cis-trans species of a planar ground-state molecule. These two transformations are virtually identical in nature, yet one leads back to the original species and is therefore a photophysical primary step, e.g., internal conversion or intersystem crossing, while the other leads to the chemically distinct isomer and should be called a photochemical primary step. As another example, the distinction between the formation of an excimer and of a photodimer lies in the instability and stability, respectively, of the dimeric species in the ground state. Excimer formation is usually considered as photophysical and photodimer formation as photochemical. These examples show that the classification of steps as photochemical and photophysical is in some cases arbitrary.

Among the more important photophysical primary steps are internal conversion, intersystem crossing, phosphorescence, fluorescence, vibrational and rotational relaxation, quenching, and energy transfer (see the Appendix). The varieties of photochemical primary steps are virtually unlimited, but we may mention dissociation, isomerization, reaction in various ways with a substrate, and molecular elimination among the many known (2).

b. Vibrational and Rotational Deactivation. Whenever a particular primary step is the unique step leading to depopulation of one of the excited states, it may be included in the preceding primary step.

In principle, in almost all photophysical primary steps, considered strictly as elementary transformations, an excited state or the ground state is formed with vibrational (and to a less important extent, rotational) energy in excess of the amount it would have at equilibrium.

In the gas phase at low pressure, vibrational and rotational relaxation are generally slow. It is now well known that a molecule in an upper vibronic state may undergo such processes as intersystem crossing, fluorescence, internal conversion, and chemical reaction before the excess vibrational energy can be removed by collisions. The different vibrational (and rotational) levels of the excited state may then have to be taken into account in the mechanism, although in detail determined by the data which are available.

In condensed media, however, because of the extraordinarily high collision rate with surrounding molecules, collisional degradation of vibrational and rotational energy is fast. It may be fast enough that no other steps can compete with such equilibration. In this case, the equilibration can be considered, without appreciable loss of generality, to be part of the preceding step. With the qualifications discussed above about the mechanism being a working hypothesis, the fact that fluorescence and phosphorescence, for example, lead first to vibrationally excited ground states can be ignored in condensed media. This is the assumption made in Figure 1. However, if deactivation is to be taken into account as a separate step, then the mechanism must be modified. Such a change is illustrated for step 22 in Figure 2.

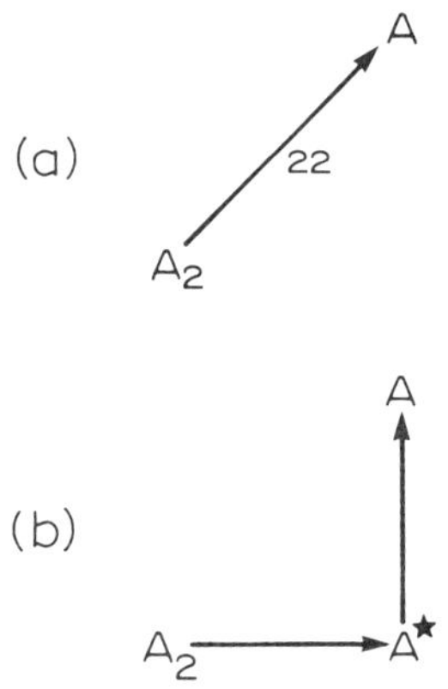

FIGURE 2. The single step of internal conversion of A_2 to the ground state: (a) as shown in Figure 1, including vibrational deactivation with the internal conversion step, and (b) in more detail, showing separately the isoenergetic internal conversion followed by vibrational deactivation. In (b) the asterisk (*) signifies a molecule with excess vibrational and rotational energy.

In Figure 2(a), step 22, the internal conversion of state A_2 to the ground state, is reproduced from Figure 1 as a single step including vibrational equilibration, as is appropriate for a photochemical system in a condensed medium. Figure 2(b) is more appropriate for a gas-phase system; the internal conversion is shown as an isoenergetic transformation of A_2 into a nonequilibrium ("hot") ground-state molecule, and is followed by the deactivation of the latter in a separate step. At low pressure, even this deactivation might have to be represented by a series of steps, showing the downward cascade through various vibrational levels. There are examples known of chemical reactions of "hot" ground-state molecules, A^*, formed by internal conversion or intersystem crossing in the gas phase, and such reactions have been suggested even in condensed media.

In the same sense, in condensed media, it is often assumed that upper excited electronic states relax rapidly by internal conversion and vibrational equilibration to the lowest excited electronic state of the same multiplicity. Thus, referring to Figure 3, which extends the mechanism of Figure 1, if two excited states, A_2 and A_3, have the same multiplicity, it is likely that the upper state, A_3, would undergo rapid internal conversion to A_2 in solution. The sequence 03-32 could then be considered as the equivalent of step 02, provided that neither of steps 34 or 35 could compete, under these conditions, with step 32. A common diagnostic test (but not proof) of this situation is independence of the quantum yield on wavelength.

c. Final Primary Steps. Among the photophysical primary steps, there are some which cannot be followed by any other primary step because they convert the excited molecule, in the words of Noyes, et al. (12), "to a state such that its reactivity is statistically no greater than that of similar molecules in thermal equilibrium with their surroundings." These may be called final photophysical primary steps. Steps 21, 22, 13, and 11 of Figure 3 are examples of final photophysical primary steps.

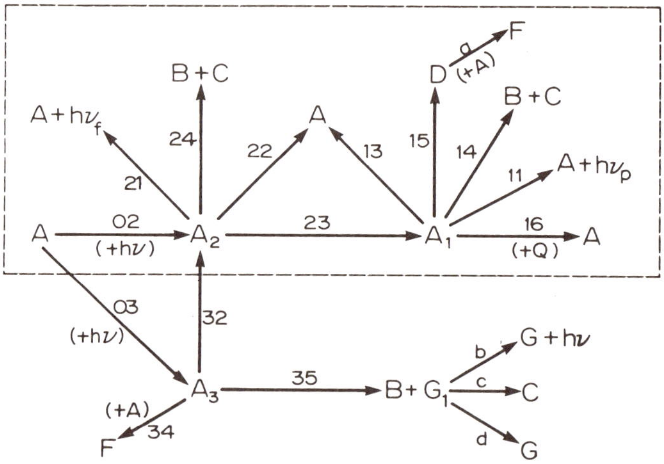

FIGURE 3. An extension of the mechanism of Figure 1 (the part enclosed in the dashed rectangle), which includes another higher excited state A_3. Both A_3 and A_2, but not A_1, can be reached by absorption of light, and both are assumed to have the same multiplicity as the ground state. G_1 is a product which is formed in an excited state and subsequently undergoes the transformations b, c, and d. As in Figure 1, vibrational deactivation of nonequilibrium levels is not taken into account explicitly.

However, especially for the nonradiative transitions to the ground state, steps 22 and 13, and especially in the gas phase, vibrational equilibration might have to be considered to represent the true final photophysical primary steps, as in Figure 2(b).

Final photophysical primary steps and photochemical primary steps cannot be followed by any other primary steps, but only by physical and/or chemical transformations of molecules other than the original excited molecules, as will be discussed in Sections III.A.3. and III.C.

d. Quenching. Certain of the photophysical primary steps, for example, step 16 of Figures 1 and 3, are bimolecular, involving a molecule originally excited by absorption of radiation and a second species, in this case Q. When these steps lead to deactivation of the excited molecule, the general term applied is "quenching." However, in some cases, the disappearance of excitation in one species coincides with the appearance of excitation in some other species in the system. Such quenching steps are more appropriately labeled "energy transfer," implying intermolecular electronic energy transfer. These, because of their great importance, will be dealt with separately in Section III.C.

2. Primary Processes

From the entire set of primary steps postulated to comprise the mechanism, a number of continuous sequences of one or more steps can be singled out which start with the absorption of light and end with any one of the primary steps. Each of these sequences represents one of the primary processes. The complete set of primary processes for the mechanism of Figure 1 (and repeated in Figure 3) is 02, 02-21, 02-24, 02-22, 02-23, 02-23-13, 02-23-15, 02-23-14, 02-23-11, and 02-23-16. These sequences include the absorption step itself as well as those processes which lead to other excited states of the molecule. Thus some of the primary processes are included in others, such as 02 and 02-23 in 02-23-11.

Each primary process may be named according to the nature of its last step. Thus a primary process which describes a sequence of primary steps leading to phosphorescence emission may be called a phosphorescence primary process, or a primary process of phosphorescence. However, different primary processes may, in the same mechanism, lead to and share their last step. For example, in the enlarged mechanism of Figure 3, there are two primary processes which have the common emission step 11: 02-23-11 and 03-32-23-11. To avoid ambiguity in such a case, the name must be supplemented by information about the sequence of steps included in the particular primary process.

As we have seen above, a photochemical primary step can itself be followed only by transformations of other than the original molecules. It follows, therefore, that a photochemical primary step can only be part of a primary process if it is the last step in the sequence. Such a primary process, ending on a photochemical primary step, is a photochemical primary process. A primary process which does not end with a photochemical primary step (and thus does not contain any such step in the sequence) is a photophysical primary process. Those which end with a final photophysical primary step are final photophysical primary processes. In Figure 1, sequences 02-24, 02-23-15, and 02-23-14 represent photochemical primary processes, while sequences 02 (absorption alone or absorption followed by vibrational relaxation as appropriate to the medium), 02-21, 02-22, 02-23, 02-23-13, 02-23-11, and 02-23-16 are each photophysical primary processes. The mechanism of Figure 3 has even more primary processes. For example, 03-32-23-15 is a photochemical primary process and 03-32-23-11 is a final photophysical primary process.

The immediate chemical products of each photochemical primary process are called primary products, e.g., B, C, and D. Products resulting from other steps are secondary products, such as F, formed by step (a). A given product may, however, be formed by several different photochemical primary processes. Thus, B is formed by, among others, the processes 02-24 and 03-32-23-14. A product may also be formed in both a primary step and a secondary step, as is C in steps 14 and (c).

3. Secondary Steps

In the primary processes, species other than the original molecule may be formed which are capable of further chemical transformations. These transformations, called secondary steps or secondary thermal reactions, are really not even properly considered photochemical steps. The most common example is the reaction of free radicals generated by photochemical dissociation.

It is also possible for an excited molecule to be formed in a photochemical primary step. It could then also undergo subsequent physical transformations. In Figure 3, the excited molecule A_3 dissociates to give two species, one a stable product, B, and the other an excited molecule, G_1, in the photochemical primary step 35. G_1 then may follow any of the paths b, c, and d, of which b and d are physical and c is chemical. These three steps are secondary steps, but not of the usual kind. Although examples of this behavior are rare, they may perhaps not remain so. It is, however, difficult to distinguish this situation from the case that G is first formed and then excited by energy transfer.

B. Quantitative Aspects of Photochemical Systems

As was pointed out in Section II.B., the concept of quantum yield was first established as the number of molecules of reactant consumed or of product formed per photon absorbed. Now, however, it is used to describe quantitatively any experimental or theoretical process, either chemical or physical, related to the absorption of light.

One major clarification of the nomenclature problem requires that the words quantum, efficiency, and yield be used only in particular ways and with quite specific meanings. Thus we propose here that "quantum" should always be used with "yield," whereas "efficiency" should be used alone, not combined with "quantum." In this way there will be two clearly distinguishable terms: efficiency and quantum yield. The first is to be reserved to describe the relative importance of one step among all the steps which depopulate a given excited state, and the second to be used only when direct reference is made to the number of absorbed photons.

The second point of clarification requires the clear distinction between experimental quantum yield and mechanistic quantum yield. The first is a quantity that can be determined directly by experiment without any assumptions about the mechanism. It should be designated by a capital phi (Φ) with an appropriate subscript. The second is a theoretical quantity, which may be related to the experimental quantum yield on the basis of a mechanism for the primary and secondary steps proposed for a given system. It should always be designated by a lower-case phi (ϕ), again with a suitable subscript.

1. Experimental Quantum Yields

An experimental quantum yield is the number of molecules which are observed to undergo a given phenomenon per photon absorbed. In order to obtain data on experimental quantum yields, there must be measured: (a) some chemical or physical change in the system, e.g., the number of reactant molecules consumed or of product formed the number of fluorescence photons emitted, etc., and

(b) the number of photons absorbed by the system.* Quantities (a) are the numerators of the expressions for the experimental quantum yields of reactant disappearance, product formation, fluorescence emission, etc., respectively, while (b) is in all cases the denominator of such expressions. The two quantities must naturally be measured in commensurate units.†

It is very important to note that the numerator and denominator of the expression for the experimental quantum yield (and thus the experimental quantum yield itself) may depend on a number of physical and chemical variables. Whenever an experimental quantum yield is reported, the precise experimental conditions under which it has been obtained should be clearly stated.

There are two different kinds of quantum yields obtainable from experimental data: the overall quantum yield and the instantaneous quantum yield.

---

*The absorbed intensity I_a may be related to the incident intensity I_o by means of the well-known Beer-Lambert law. In applying this law, one can use either decadic or natural logarithms, as well as various concentration units and measurements of path length. However, the common usage, and the one we recommend here, is as follows:

$$I_a = I_o \; (1-10^{-\varepsilon cl})$$

where c is the concentration of the absorbing species (in moles/liter), l the path length (in cm), and ε the <u>molar extinction coefficient</u> (whose units are 1000 cm^2/mole). The quantity εcl is called absorbance (or optical density). The above equation applies strictly for monochromatic light and then only if the nature of the absorbing species is independent of concentration. With polychromatic light in a photochemical experiment, the absorbed intensity is best determined *in situ*.

†Although it is not common to give units to quantum yields, they can be given as moles per einstein absorbed or molecules per photon absorbed, where by moles and molecules, reference is made to the number which react, are formed, emit, etc.

A photochemical experiment may be carried out for a given irradiation time, followed by an analysis for the total amount, n, of, for example, some product formed. The ratio of n to the total amount of light absorbed, $I_a t$ ($\int I_a dt$, when I_a is dependent on time), during the experiment is the overall quantum yield. It is usually found, provided the fractional conversion is not large, that n is proportional to $I_a t$ in experiments at different intensities and times, as is shown by the points a of Figure 4. Then the overall quantum yield is constant, as

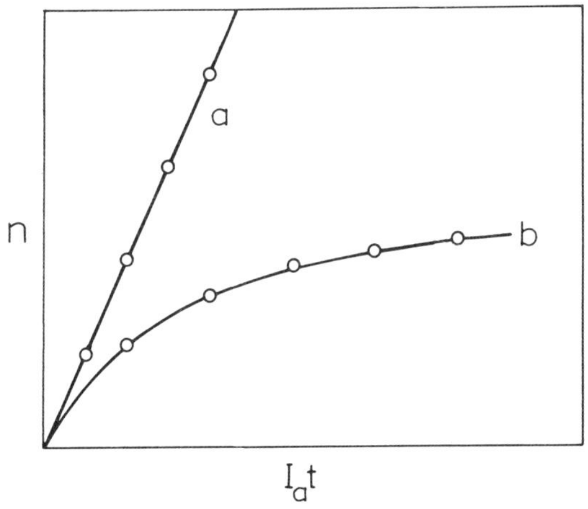

FIGURE 4. Two systems a and b, showing the amounts of a photoinduced phenomenon found by continuous in situ measurements (———) and by final analysis (o) as functions of the total amount of light absorbed.

indicated by the points a of Figure 5. For various reasons however, such as a change in the distribution of the absorbed light among the different species in the system, it may also happen that the plot of n vs. $I_a t$ is not

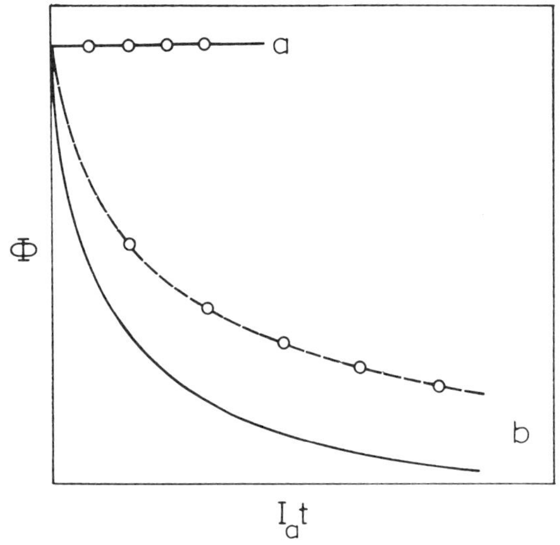

FIGURE 5. Experimental quantum yields derived from Figure 4: instantaneous quantum yields (———) and overall quantum yields (o).

linear (points b in Figure 4), which means that the overall quantum yield changes with light intensity and/or irradiation time (points b in Figure 5).

A photoinduced phenomenon may also be followed by continuous in situ measurements, as is often the case for luminescence or for concentration changes measured by absorbance, pH, etc. Such data can be treated to give the rate of the phenomenon as a function of the time. The ratio of this rate to the corresponding rate of absorption of light is the instantaneous quantum yield. It represents the slope of curves n vs. $I_a t$ (line a in Figure 4). When the instantaneous quantum yield is constant, it has then the same value as the overall quantum yield (line a in Figure 5). However, when n does not depend linearly on $I_a t$ (line b in Figure 4), both the overall and the instantaneous quantum yields will change with light intensity and/or irradiation time. Then the two quantum yields will not have concordant values, as can be seen from part b of Figure 5. They should, however, extrapolate to the same value at zero time.

Since the photoinduced phenomenon always shows some delay with respect to the absorption of light*, the instantaneous quantum yield must be zero for an initial period of time, when the absorption of light is not yet followed by the phenomenon under consideration, and is undefined for a final period of time, when the phenomenon is still occurring but light absorption has already ended. A clear example of this situation is a flash experiment, in which a given chemical product may be formed mostly after the flash. In such experiments, of course, it is only the overall quantum yield which has significance.

Although the instantaneous quantum yield may generally change with time of irradiation, in conventional experiments at constant light intensity there is often only a short initial period during which the instantaneous quantum yield increases, and then it reaches a constant value. The initial part is called the "induction period." After the instantaneous quantum yield becomes constant, the system is said to be in a stationary or steady state. Under these conditions, it is the steady-state quantum yield which is obtained. Provided that the induction period represents only a small fraction of the total irradiation time and that the quantum yield does then become constant, the overall quantum yield will also be independent of irradiation time and is practically equal to the steady-state quantum yield.

---

* The time scale of this delay may be very different depending on the nature of the photoinduced phenomenon. For example, it may be nanoseconds for fluorescence emission, hours for a secondary thermal reaction.

The overall quantum yield may depend on the time elapsed from the end of the irradiation to the instant in which the measurement is performed, since secondary thermal reactions may change the amount of the photoinduced phenomenon. The system is then said to show a "post-photochemical effect."* A meaningful overall quantum yield can then be obtained only by an extrapolation technique. An extrapolation of n to infinite post-irradiation time gives a value, n_∞ which includes the secondary reaction completely. The overall quantum yield found from this value of n_∞ is suitable for mechanistic interpretation. An extrapolation of n to zero post-irradiation time gives a value, n_0, which is unaffected by the thermal reactions occurring <u>after</u> irradiation, but which still includes the effect of the thermal reactions occurring <u>during</u> the period of irradiation. Thus a plot of n_0 vs. $I_a t$ will not generally be linear, and, as a consequence, the apparent quantum yield obtained from these n_0 values will change with the irradiation time. The extrapolation of these apparent quantum yields to zero irradiation time will finally give an overall quantum yield which does not include any effects of such thermal secondary reactions. Instantaneous quantum yields of systems which show a post-photochemical effect will also change with the time of irradiation because of the occurrence of thermal reactions <u>during</u> irradiation.

An experimental quantum yield is related to <u>all</u> of the photons absorbed by the <u>system</u>. Should there be more

---

*Strictly speaking, some post-photochemical effect is present even in systems in which there are no secondary processes, since, as previously mentioned, there is always some delay between the absorption of a photon and the appearance of its chemical or physical consequences. However, the term "post-photochemical effect" is usually reserved for the effect of secondary thermal reactions.

than one absorbing species either already present or formed in the system, appropriate corrections may be made to relate the quantum yields to absorption by one species only. But such corrections always imply mechanistic considerations. For example, to attribute the observed phenomena to light absorption by one species alone entails the assumption that energy transfer processes from other absorbing species are unimportant. Thus the quantum yields corrected in this way are not experimental but mechanistic quantum yields.

In some investigations of photochemical systems, the intensity of absorbed light is not measured, for one reason or another. The data are then reported as relative quantum yields. Such data should never be labeled as Φ (or ϕ), unless accompanied by information as to the sense in which they are "relative."

2. Mechanistic Quantities

A detailed interpretation of what happens in a photochemical system is based on a mechanism in terms of postulated steps (see Section III.A). In order to discuss the validity and the implications of any mechanism in a quantitative sense, two mechanistic quantities are involved: the efficiency of a step and the quantum yield of a primary process. Before defining and discussing these quantities, we must emphasize once again that the consistency of a postulated mechanism with the available data does not prove the mechanism correct. This also implies that any quantity which refers to a single step or to a process in the mechanism need not necessarily have a true significance. Further investigations on the same system may lead to modification or even complete revision of the mechanism. Even the values of the quantities used to characterize some of the steps and processes may then have to be changed significantly. Any discussion concerning the efficiency of a step or the quantum yield of a primary process is to be regarded in the light of these considerations.

3. The Efficiency of a Step

In a mechanism such as that shown in Figure 1, an excited state of a molecule is generally depopulated by several parallel steps. The efficiency of a step, η, describes the relative importance of that step in the depopulation of one excited state.* Thus the efficiency of a particular step, i, which leads to depopulation of a given excited state, is expressed as the ratio of the rate of step i, R_i, to the sum of the rates of all of the steps which depopulate the state involved:

$$\eta_i = \frac{R_i}{\Sigma_j R_j} \qquad (1)$$

For example, in Figure 1, the efficiency of step 21 is given by

$$\eta_{21} = \frac{R_{21}}{R_{21} + R_{24} + R_{22} + R_{23}} \qquad (2)$$

The efficiency η_{21} therefore represents the fraction of all of the molecules in the state A_2 which undergo the emission step 21. Alternatively, it expresses the probability that a molecule in state A_2 will be transformed according to step 21.

When all of the steps which depopulate a given excited state are unimolecular, the efficiency is

*The term "efficiency" can refer only to a step, either primary or secondary, never to a process.

independent of the concentrations of the various species in the system, hence

$$\eta_i = \frac{k_i}{\Sigma_j k_j} \quad (3)$$

in which the k_j are the unimolecular rate constants. Thus the efficiency of step 21 is also given by

$$\eta_{21} = \frac{k_{21}}{k_{21} + k_{24} + k_{22} + k_{23}} \quad (4)$$

Under these conditions, the efficiency depends only on those physical factors, such as temperature and nature of solvent, which may affect a kinetic rate constant.

When the depopulation of a state involves two or more steps having different kinetic order, the efficiency of each step originating with the state in question will depend on concentrations in the system. Thus in Figure 1, the efficiency of step 11 is given by

$$\eta_{11} = \frac{k_{11}}{k_{13} + k_{15} + k_{14} + k_{11} + k_{16}\,[Q]} \quad (5)$$

and depends on the concentration of the species, Q, responsible for quenching. Except in the case of bimolecular steps involving two excited molecules, the efficiency is independent of the concentration of the excited molecules. It is thus a useful concept for dealin in a quantitative sense with the various steps that one excited state may undergo. For example, the efficiency of fluorescence of an excited-state molecule is a quantity which is unaffected by the particular way in which the state is reached, other conditions being equal.

The extreme values for the efficiency of any step

are zero and one. Further, the sum of the efficiencies for all of the steps that depopulate a given excited state will be unity (when efficiencies depend on concentrations, obviously this sum will be unity only if the efficiencies refer to the same concentrations).

The concept of efficiency could also be used for the absorption step; however, because an absorption step is also a process, it is more convenient to use quantum yield (Section III.B.4).

4. The Quantum Yield of a Primary Process

As we have seen in Section III.A.2., a primary process is a continuous sequence of one or more primary steps which starts with the absorption of light by the ground-state molecule* and ends with any one of the primary steps. The quantitative characterization of a primary process must therefore be based on the number of photons absorbed by the ground-state molecule, that is, it must be a quantum yield. Just as the term "efficiency" refers only to a step, "quantum yield" refers only to a process. The quantum yield of a primary process, designated by ϕ with a suitable subscript, is given by the number of molecules which undergo the primary process per photon absorbed by the ground-state molecules.

As described in Section III.A.2., the definition of a primary process includes the special case of a single step of absorption of light. The quantum yield for the absorption primary process, ϕ', is given by the number of molecules promoted to a particular excited state per photon absorbed by the ground-state molecules of the substance studied. In the case that all levels reached by absorption

---

*For the sake of simplicity, we shall not consider the case of absorption of light by excited molecules themselves.

are classified in the mechanism as the same state, ϕ' is exactly unity. On the other hand, when several states are reached in absorption and these states are distinguished in the mechanism, each quantum yield ϕ' will be less than one, although their sum must be unity. For multiphotonic absorption, the quantum yield of absorption will also be less than one, e.g., $\phi' = 1/2$ for a biphotonic absorption.

Using the concept of the efficiency of a primary step, we can express the quantum yield of a primary process by the following equation:

$$\phi = \phi' \prod_i \eta_i \qquad (6)$$

The product is extended over all of the steps which follow that of absorption and which are part of the continuous sequence constituting the primary process being considered. For example, in Figure 1, we have, for the quantum yield of the primary process by which the primary product D is formed:

$$\phi_{02-23-15} = \phi'_{02}\eta_{23}\eta_{15} \qquad (7)$$

When <u>all</u> of the steps which depopulate <u>all</u> of the excited states involved in a particular primary process follow first-order kinetics, the quantum yield of the primary process is independent of chemical concentrations in the system. However, the quantum yield of, for example, phosphorescence in Figure 1, depends on the concentration of Q:

$$\phi_{02-23-11} = \phi'_{02}\eta_{23}\eta_{11} \qquad (8)$$

with η_{11} given by eq. (5).

When only monophotonic absorption occurs, the following relations hold:

(i) each primary quantum yield is equal to or less than one,

(ii) the sum of the quantum yields of all of the primary processes is equal to or greater than one,

(iii) the sum of the quantum yields of all of the photochemical primary processes is equal to or less than one, and

(iv) the sum of the quantum yields for all of the photochemical and final photophysical primary processes must be exactly one.

This last is a reformulation of the Second Law of Photochemistry, also known as the Stark-Einstein Law.

For a system in which only multiphotonic absorption occurs involving n photons, the same relations, (i) to (iv), still apply, but with "one" replaced by 1/n.

5. Decay of Excited States

In Section II.D. it was mentioned that useful photochemical information can be derived from the decay characteristics of excited states formed by pulse excitation. Even kinetically complicated decay curves can now be resolved with modern analytical techniques.

In the special case that the decay is a pure exponential, a unique experimental lifetime can be defined and evaluated. It represents the time required for the quantity measured to decrease to 1/e of its initial value. Although a "lifetime" can be found for almost any decay curve, it is only in the exponential case that the lifetime is independent of extent of decay. The experimental lifetime, like the experimental quantum yield, may depend on various of the chemical and physical parameters in the system. The conditions under which it is measured should be carefully specified. The experimental lifetime is denoted by an upper-case tau (T). This is a departure from current usage, but corresponds to the use of Φ for an experimental quantum yield.

A nonexponential decay may represent a sum of exponential decays, whereupon there will be more than one characteristic experimental lifetime. If the decay is other than first-order (or sum of first-order terms), the experimental lifetime is not an especially useful quantity as it is not independent of concentration of the decaying species. However, important kinetic data may still be obtained from analysis of the decay curve.

When, in a photochemical mechanism, a particular excited state is postulated to decay by steps all of which are kinetically first order, a mechanistic lifetime, τ, can be defined as the reciprocal of the sum of all of the rate constants which depopulate that state:

$$\tau = \frac{1}{\Sigma_i k_i} \qquad (9)$$

According to the definition of efficiency in eq. (3), the rate constant of a first-order step is related to this mechanistic lifetime by

$$k_i = \frac{\eta_i}{\tau} \qquad (10)$$

For example, the mechanistic lifetime of the state A_2 in Figure 1 is

$$\tau_{A_2} = \frac{1}{k_{21} + k_{24} + k_{22} + k_{23}} \qquad (11)$$

and thus the rate constant of the emission step 21, which is the reciprocal of the <u>radiative</u> lifetime, $\tau^{\circ}_{A_2}$, of A_2, is given by*

$$k_{21} = \frac{1}{\tau^{\circ}_{A_2}} = \frac{\eta_{21}}{\tau_{A_2}} \qquad (12)$$

6. The Interrelation of Mechanistic and Experimental Quantities

The evaluation of mechanistic quantities, such as the efficiency of a step, the quantum yield of a primary process, and the mechanistic lifetime from experimental quantum yield and lifetime data, may be simple or difficult depending on the complexity of the system.

In principle, the quantum yields of the primary processes may be related to experimental quantum yields through the mechanism. However, in practice such relations are simple only under particular conditions. When the reactant is the only absorbing species and is not involved in secondary reactions, the experimental quantum yield of reactant disappearance, Φ_r, is given by

$$\Phi_r = \sum_i (1 + a_i)\phi_i \qquad (13)$$

where ϕ_i is the quantum yield of the ith photochemical primary process and a_i is the stoichiometric coefficient of unexcited reactant molecules involved in the photochemical step of the primary process. Thus in Figure 1, the experimental quantum yield for disappearance of A is predicted to be

$$\Phi_A = \phi_{02-24} + \phi_{02-23-15} + \phi_{02-23-14} \qquad (14)$$

---

*It should be noted that the efficiency is used here, not the quantum yield.

Similarly, when the light is absorbed by one species only, the experimental quantum yield for the formation of a primary product P, which is neither formed nor consumed by secondary reactions, will be

$$\Phi_P = \sum_i p_i \phi_i \qquad (15)$$

where p_i is the stoichiometric coefficient for the formation of product P in the photochemical step of the ith primary process. Thus for the formation of B according to the mechanism of Figure 1, the experimental quantum yield would be expressed as

$$\Phi_B = \phi_{02-24} + \phi_{02-23-14} \qquad (16)$$

These two terms are separable in this example by variation of the concentration of Q, which is assumed to act on state A_1, but not on state A_2.

Relations between experimental and mechanistic quantum yields may also be given for photophysical processes. However, for many of these, the pertinent experimental data may be difficult to obtain, especially when the process itself cannot be measured directly.

Experimental quantum yields may depend on geometric factors, time, or intensity of the absorbed light in any one experiment. A change of the experimental quantum yield with position in the cell may occur when the light beam does not fill the cell or when the intensity changes drastically along the illuminated path because of strong absorption. A dependence on time, or extent of reaction, may indicate a change in intensity distribution through building-up of an absorbing product or simply that the system is not at a steady state. If the experimental quantum yield depends on I_a, mixed first- and second-order competing steps among the excited states or intermediate photoproducts are indicated. In all such cases, the mechanism proposed must be able to reproduce such behavior of the experimental quantum yield.

If the system contains luminescent species, reabsorption of the emitted light may have to be taken into consideration in deriving mechanistic quantities from the experimental ones.

Changes of the overall quantum yield with post-irradiation time indicate that some thermal secondary reaction occurs. In this case, the type of extrapolation performed on the experimental quantum yield (see Section III.B.1) qualifies its mechanistic interpretation. For example, according to the mechanism of Figure 3, the quantum yield for F formation obtained through the double extrapolation to zero time of irradiation accounts for the amount of F formed by process 03-34, while that obtained through the extrapolation to infinite post-irradiation time accounts for the amount of F formed by both process 03-34 and the secondary reaction a.

With regard to lifetimes, it is often the case that the experimentally measured lifetime represents the decay of a specific excited state and can thus be equated to the mechanistic lifetime of that state. However, this need not be so. For example, the decay of E-type fluorescence follows the lifetime of the phosphorescing state (14). In principle, the radiative lifetime of an excited state can be determined indirectly from the integrated extinction coefficient of the absorption transition that connects the same two states as the emission. In practice, however, several difficulties are encountered in such a determination (6,8,14).

C. Intermolecular Electronic Energy Transfer

Quenching of one excited-state species may result in transfer of electronic energy to the quenching species. Intermolecular energy transfer has become an important technique in photochemistry because it often permits the selective population or depopulation of one specific excited-state species.

Figure 6 shows four quenching steps: 16, 57, 47, and 46. Two of these, 16 and 46, do not involve electronic

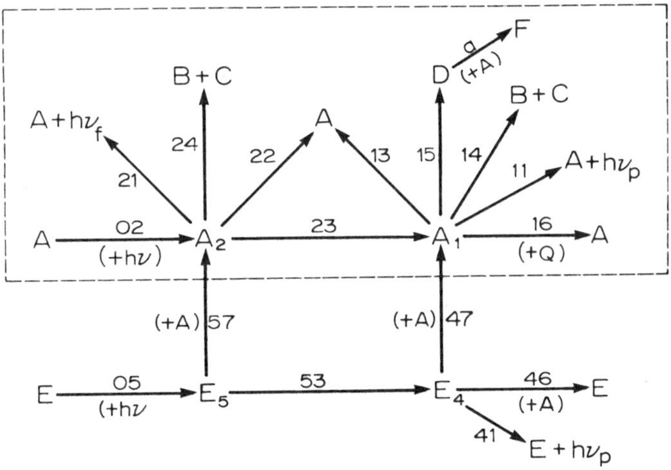

FIGURE 6. A further extension of the mechanism from Figure 1 (enclosed in the dashed rectangle), in which a second species, E, absorbs light. Steps 57 and 47 are bimolecular electronic energy transfer steps resulting in the transformation of the donor excited states E_5 and E_4 to their ground state E and excitation of the species A to the excited states A_2 and A_1, respectively. Step 46 represents quenching by A of the state E_4 without electronic excitation of A. It is assumed that the subsequent transformations of A_1 and A_2 are not affected by the presence of E and are therefore the same as in Figure 1.

energy transfer; instead, the electronic energy of A_1 and E_4, respectively, is dissipated as thermal energy in the medium. The other two do involve electronic energy transfer, with the implication that spin conservation rules apply. The species that loses excitation energy is called the donor, while the quenching species that becomes electronically excited is the acceptor.

1. Acceptor Primary Processes

Energy transfer can be simply regarded as a quenching step which depopulates one excited state of the original molecule. The photochemist's attention is thus fixed on the molecule which originally absorbed the light and on its primary processes. Energy transfer along with other quenching steps will then simply be primary steps in the mechanism for the original molecule in the presence of quencher. Much valuable information can be obtained in this way, as was discussed briefly in Section III.B.6.

However, when the photochemist is concerned mainly about the excited states of species which accept the energy, it is reasonable to refer to the subsequent transformations of the newly excited molecule as primary steps in the mechanism of that acceptor molecule. In this sense, energy transfer is used as an alternate to direct photoexcitation for the production of excited molecules. Then any continuous sequence of one or more steps which begins with the energy transfer step and ends with any one of the primary steps of the acceptor molecule is defined as an acceptor primary process. In the acceptor primary process, an excited donor molecule, such as E_5 in Figure 6, plays the same role in the formation of A_2 as do the photons in the direct excitation of A to A_2 in step 02.

2. Mechanistic Quantities

The concept of efficiency of a step can be applied equally well to the mechanisms of direct excitation and of excitation by energy transfer. Further, provided that the presence of the donor does not affect the various steps in the mechanism of A, even the values of the efficiencies may be transferred from one type of system to the other.

As far as the quantitative characterization of the acceptor primary processes is concerned, however, there is a new problem. In Section III.B.4 the concept of quantum yield of a primary process was developed to mean the yield of a primary process per quantum absorbed by the ground-state molecules of that chemical species. When dealing with processes involving acceptor excited states, the concept of quantum yield may still be used from a mechanistic point of view, but, because the quantum yield would depend on both donor and acceptor properties, it loses much of its real significance. For example, in Figure 6, in the sequence 05-53-47-14, E is the absorbing species and B and C are produced from the excited molecule A_1. The quantum yield of this process depends on the quantum yield of donor formation (ϕ_{E_4}), on the efficiency of step 14, and on the efficiency of step 47.

In order to deal quantitatively with acceptor primary processes independently of the efficiencies of the various steps of the donor primary process, we must define a quantity analogous to the quantum yield, but based on the number of excited donor molecules which have transferred energy to the acceptor molecules, rather than on the number of photons absorbed. The quantity energy transfer yield*

*It is important to realize that the term "energy transfer yield" can be applied to a primary process of the donor which ends with the energy transfer step. The yield of such a process must be referred to the number of photons absorbed, and thus is the "quantum yield of energy transfer."

is therefore defined as the number of excited acceptor molecules taking part in a given process per excited donor which transfers energy to the acceptor. The energy transfer yield, designated by a lower-case mu (μ) with an appropriate subscript, is a mechanistic quantity.

Just as a quantum yield, ϕ', can be assigned to the absorption step in direct photoexcitation, so also can an energy transfer yield, μ', be given for the energy transfer step. If only one energy transfer step originates from a given donor excited state, μ' will be one, in complete analogy to the case of absorption to only one state with $\phi' = 1$. However, unlike the case of direct photoexcitation, the efficiency of step 47 is not necessarily the same as the energy transfer yield. The efficiency η_{47} takes into account the other steps which depopulate the state, for example:

$$\eta_{47} = \frac{k_{47}[A]}{k_{41} + \{k_{47} + k_{46}\}[A]} \qquad (17)$$

In direct photoexcitation, absorption by a species other than that which leads to excited states of A is not included in ϕ'_{02}. Similarly, a step such as 46, which also does not lead to excited states of A, is not included in μ'_{47}. However, the definition of efficiency requires that all steps depopulating E_4 be part of the denominator in eq. (17). The different basis arises from the fact that while μ'_{47} refers to the acceptor primary step, η_{47} represents the efficiency of a primary step of the donor.

The energy transfer yield of an acceptor primary process is given by

$$\mu = \mu' \prod_j \eta_j \qquad (18)$$

which is the analogy of eq. (6). The product of efficiencies, η_j, in eq. (18) is extended over all of the steps in the sequence belonging to the acceptor primary process.

In the mechanism of Figure 6, for example, the sequence of steps 05-53-47-11 involves sensitized emission from the acceptor excited state A_1. The acceptor primary process includes just the steps 47 and 11, so that the energy transfer yield is

$$\mu_{47-11} = \mu'_{47}\eta_{11} \tag{19}$$

The quantum yield of this sensitized emission, however, is given by

$$\phi_{05-53-47-11} = \phi_{E_4}\eta_{47}\eta_{11} \tag{20}$$

As the efficiency of step 47 depends on the concentration of acceptor [eq. (17)], the quantum yield is also a function of [A], although the energy transfer yield is not. However, according to eq. (17), the efficiency of step 47 reaches a limiting value, independent of the concentration of A when the concentration is high enough that unimolecular steps such as 41 can be neglected. Under these conditions, the quantum yield also reaches a value independent of [A], and is referred to as the limiti quantum yield, ϕ^{∞}, in an energy transfer system.

3. Lifetimes

In energy transfer systems, lifetimes also play an important role. The mechanistic lifetime of a donor state is the reciprocal of the sum of the first-order and pseudo-first-order rate constants for all steps which depopulate that excited state, of which the latter will depend on the concentration of the acceptor. Thus in the mechanism of Figure 6, the mechanistic lifetime of the state E_4 is

$$\frac{1}{\tau_{E_4}} = k_{41} + k_{47}[A] + k_{46}[A] \tag{21}$$

The lifetime of the acceptor excited state, in the absence of energy transfer back to the donor, will, however, be the same as that found from the mechanism for direct excitation.

4. The Interrelation of Mechanistic and Experimental Quantities

In the sense that acceptor primary processes complement the primary processes of direct photoexcitation, investigations of mechanisms and their relation to the experimentally measured quantities is facilitated. This is especially so as acceptor primary processes can be examined with a variety of different donor molecules.

The experimental quantum yield for a phenomenon induced by energy transfer is defined in exactly the same way as that for direct excitation (see Section III.B.1). However, the quantum yield will usually be a function of the concentration of the acceptor species. The mechanistic quantum yield given in eq. (20) for the sequence 05-53-47-11 can be put in the form

$$\frac{1}{\phi_{05-53-47-11}} = \frac{1}{\phi_{E_4}\eta_{11}} \left\{ \frac{k_{47} + k_{46}}{k_{47}} + \frac{k_{41}}{k_{47}[A]} \right\} \tag{22}$$

If the experimental quantum yield shows the same dependence on [A] as given in eq. (22), then, in the sense that the mechanism is assumed correct, the ratio of intercept to slope of a plot of $1/\Phi$ against $1/[A]$ gives the rate constant ratio:

$$\frac{\text{intercept}}{\text{slope}} = \frac{k_{47} + k_{46}}{k_{41}} \qquad (23)$$

This is one example of the use of a Stern-Volmer plot.

If it can be shown that only one energy transfer step occurs in a system, and if the properties of the donor are established, information of great mechanistic value can be obtained. For example, if step 47 in Figure 6 could be demonstrated to be the only energy transfer or quenching step (i.e., if steps 57 and 46 are absent), the limiting quantum yield for sensitized phosphorescence is [see eq. (20)]

$$\phi^{\infty}_{05-53-47-11} = \phi_{E_4}\eta_{11} \qquad (24)$$

in which the limiting efficiency of step 47, now the only energy transfer step, is unity.

This can be compared with the quantum yield of phosphorescence upon direct excitation only into the state A_2 (so that ϕ'_{02} is unity):

$$\phi_{02-23-11} = \eta_{23}\eta_{11} \qquad (25)$$

Thus in this simple case, if ϕ_{E_4} has previously been established, the efficiency of intersystem crossing η_{23} can also be found.

The concept of the Stern-Volmer mechanism and the resolution of that mechanism can also be applied to the donor. In Figure 6, the mechanistic quantum yield for the emission step 41 is given by

$$\phi_{05-53-41} = \phi_{E_4}\eta_{41} \qquad (26)$$

The efficiency of step 41 is

$$\eta_{41} = \frac{k_{41}}{k_{41} + \{k_{47} + k_{46}\} [A]} \qquad (27)$$

Thus the ratio of the quantum yield in the absence of acceptor, $\phi°$, to that in its presence, ϕ, is

$$\frac{\phi°}{\phi} = 1 + \left\{\frac{k_{47} + k_{46}}{k_{41}}\right\} [A] \qquad (28)$$

Provided that the experimental and mechanistic quantum yields are equivalent, eq. (28) permits the evaluation of the "quenching" constant, that given in braces. The lifetime of the state E_4, eq. (21), can also be used in this sense to give directly the sum of the rate constants of energy transfer and of any other quenching.

APPENDIX

Definition of Some Common Photochemical Terms

(Terms in capitals are discussed more fully in the text.)

ABSORBANCE (A): The quantity defined as $A = \log I_o/I = \varepsilon cl$, where I_o is the intensity of the incident light, I the intensity of the light transmitted, c the concentration of the absorbing species (in moles per liter), l the path length (in cm), and ε the molar extinction coefficient (whose units are 1000 cm^2/mole).

ACCEPTOR: a chemical species which is promoted to a higher electronic state in an electronic energy transfer step.

ACCEPTOR PRIMARY PROCESS: a sequence of one or more consecutive steps starting with an energy transfer step and ending with any one of the acceptor primary steps.

ACCEPTOR PRIMARY STEP: any elementary transformation involving excited states of the acceptor formed by energy transfer.

Act: (see Step).

Actual Lifetime: (see Experimental Lifetime).

Biphotonic Absorption: the simultaneous absorption of two photons by a molecule. The sequential absorption of two photons should not be considered as biphotonic absorption.

"Dark Reaction": (see Thermal Secondary Reaction).

Deactivation: any loss of energy by an excited molecule.

DONOR: the molecule which has originally absorbed radiation and which is deactivated from an excited state in an energy transfer step.

EFFICIENCY OF A STEP (η): a quantitative measure of the relative importance of a given step involving an excited state species with respect to all of the parallel steps which depopulate that excited state.

Emission: a radiative deactivation of an excited-state species; spontaneous loss of energy as electromagnetic radiation in a transition to a lower-lying energy level (see also Luminescence, Fluorescence, and Phosphorescence).

ENERGY TRANSFER STEP: a photophysical step in which an excited state of the donor is deactivated to a lower-lying energy level while simultaneously, a different species, the acceptor, is raised to a higher energy level.

ENERGY TRANSFER YIELD (μ): the ratio of the number of molecules which undergo a given acceptor primary process to the number of excited donor molecules which have transferred energy to the acceptor molecules.

Excimer: an excited molecular complex, dissociative in the ground state, formed by interaction of an excited molecule with ground molecule(s) of the <u>same</u> species.

Exciplex: an excited molecular complex, dissociative in the ground state, formed by interaction of an excited molecule with ground state molecule(s) of a different species.

Excitation Transfer: (see Energy Transfer Step).

EXPERIMENTAL LIFETIME (τ): the time required for the measure of some observed phenomenon to decay to 1/e of its original value.

EXPERIMENTAL QUANTUM YIELD (Φ): the number of observed events of a given phenomenon per photon absorbed by the system (see also Overall Quantum Yield, Instantaneous Quantum Yield, and Steady-State Quantum Yield).

FINAL PHOTOPHYSICAL PRIMARY STEP: A photophysical primary step which converts the excited molecule to the fully equilibrated ground state.

Fluorescence: the emission of light in a radiative deactivation in which the initial and final states have the same multiplicity.

Ground State: the lowest-energy electronic state of a chemical species.

"Hot" Ground State Reaction: reaction of a molecule in an excited vibrational level of the ground state, reached indirectly as a consequence of photo-excitation.

Inherent Lifetime: (see Radiative Lifetime).

INSTANTANEOUS QUANTUM YIELD: the ratio, found by experiment, of the rate of a given phenomenon to the rate of light absorption observed at any instant during a photochemical experiment.

Internal Conversion: a radiationless transition between two states having the same multiplicity.

Intersystem Crossing: a radiationless transition between two states having different multiplicities.

Intrinsic Lifetime: (see Radiative Lifetime).

LIMITING QUANTUM YIELD: the extrapolated value of any one of the quantum yields at infinite concentration of quencher or of acceptor, as appropriate. It may be experimental (Φ^{∞}) or mechanistic (ϕ^{∞}).

Luminescence: the emission of light by an excited-state species (see also Fluorescence and Phosphorescence).

MECHANISTIC LIFETIME (τ): the reciprocal of the sum of the first-order and pseudo-first-order rate constants of all steps which lead to depopulation of a given excited state.

MOLAR EXTINCTION COEFFICIENT (ε): (see its definition under Absorbance).

Multiphotonic Absorption: the simultaneous absorption of more than one photon by a molecule (see also Biphotonic Absorption).

Natural Lifetime: (see Radiative Lifetime).

Non-Radiative Transition: (see Radiationless Transition).

OBSERVED LIFETIME: (see Experimental Lifetime).

Optical Density: (see Absorbance).

OVERALL QUANTUM YIELD: the ratio of the total amount of a given photoinduced phenomenon observed experimental to the total amount of light absorbed by the system.

Phosphorescence: the emission of light in a radiative deactivation involving initial and final states of different multiplicities.

PHOTOCHEMICAL MECHANISM: a set of steps postulated to describe the details of a photochemical system.

PHOTOCHEMICAL PRIMARY PROCESS: a primary process which ends with a photochemical primary step.

PHOTOCHEMICAL PRIMARY QUANTUM YIELD: the quantum yield of a photochemical primary process.

PHOTOCHEMICAL PRIMARY STEP: an elementary chemical transformation of the originally excited molecule.

Photochemical Reaction: a chemical transformation induced by the absorption of light.

Photodimer: a dimeric molecule, stable in its ground state, formed by interaction of an excited molecule and an unexcited molecule of the same species.

Photoexcitation: the production of an excited-state species by absorption of light.

Photolysis: a light-induced chemical reaction involving bond cleavage.

PHOTOPHYSICAL PRIMARY PROCESS: a primary process comprising a sequence of photophysical primary steps only.

PHOTOPHYSICAL PRIMARY QUANTUM YIELD: the quantum yield of a photophysical primary process.

PHOTOPHYSICAL PRIMARY STEP: an elementary transformation of the originally excited species without any chemical change.

Photosensitization: the chemical or physical transformation of a molecule brought about by absorption of light by a different molecule.

Primary Photochemical Process: a term originally used (12) to describe the entire set of photochemical and photophysical primary processes.

PRIMARY PROCESS: any continuous sequence of one or more primary steps which starts with the light-absorption step.

PRIMARY PRODUCT: a chemical species produced in a photochemical primary process.

Primary Quantum Yield: (see Quantum Yield of a Primary Process).

PRIMARY STEP: any one of the elementary transformations of an excited-state molecule of the species which absorbs light. The absorption step itself is also a primary step.

PROCESS: a continuous direct sequence of steps.

Quantum Yield: (see Experimental Quantum Yield, Quantum Yield of a Primary Process, and Quantum Yield of Photosensitization).

QUANTUM YIELD OF A PRIMARY PROCESS (ϕ): the number of molecules which undergo a given primary process per photon absorbed by the ground-state molecules of that species.

QUANTUM YIELD OF A SENSITIZED PROCESS (ϕ): the number of molecules which undergo a given photosensitized process per photon absorbed by the sensitizer.

Quenching: a general term for the deactivation of an excited-state molecule by a different chemical species in the system.

Radiationless Transition: an isoenergetic transformation between different electronic states of a molecule (see also Internal Conversion and Intersystem Crossing).

RADIATIVE LIFETIME ($\tau°$): the lifetime an excited molecule would have if the radiative step were to occur without competition. It is the reciprocal of the rate constant for the radiative step.

Radiative Transition: a change in energy level of a molecule with concomitant absorption or emission of a photon.

Rotational Relaxation: the loss of rotational energy by a molecule through collision with the surrounding medium.

SECONDARY PRODUCT: a chemical species produced in a secondary step.

SECONDARY STEP: each elementary transformation of a species that occurs without involving excited states of the originally photoexcited species.

Sensitization: (see Photosensitization).

Sensitizer: the species which absorbs light in a photosensitization.

STEADY-STATE QUANTUM YIELD: the instantaneous quantum yield in a system after it has attained a constant value in time.

STEP: any elementary transformation postulated as part of a photochemical mechanism.

STERN-VOLMER MECHANISM: a simple set of steps describing the deactivation of an excited-state molecule by first-order (and pseudo-first-order) steps and by quenching steps.

THERMAL SECONDARY REACTION: a chemical transformation of an equilibrated ground-state species which is formed in a photochemical step and which is unstable in the system.

Vibrational Relaxation: the loss of vibrational energy by a molecule through collision with the surrounding medium.

REFERENCES

1. N. J. Turro, Molecular Photochemistry, Benjamin, New York (1965).

2. J. G. Calvert and J. N. Pitts, Jr., Photochemistry, Wiley, New York (1966).

3. R. O. Kan, Organic Photochemistry, McGraw-Hill, New York (1966).

4. P. A. Leermakers, in Energy Transfer and Organic Photochemistry, P. A. Leermakers and A. Weissberger, Eds., Interscience, New York (1969), p. 1.

5. R. B. Cundall and A. Gilbert, Photochemistry, Nelson, London (197

6. V. Balzani and V. Carassiti, Photochemistry of Coordination Compounds, Academic Press, London (1970).

7. R. P. Wayne, Photochemistry, Butterworths, London (1970).

8. J. P. Simons, Photochemistry and Spectroscopy, Wiley-Interscience New York (1971).

9. See, for example, G. M. Barrow, Physical Chemistry, second edition, McGraw-Hill, New York (1966); W. C. Gardiner, Jr., Rates and Mechanisms of Chemical Reactions, Benjamin, New York (1969); W. J. Moore, Physical Chemistry, fourth edition, Prentice-Hall, Englewood Cliffs, New Jersey (1972).

10. See, for example, J. B. Hendrickson, D. J. Cram, and G. S. Hammon Organic Chemistry, third edition, McGraw-Hill, New York (1970); J. D. Roberts, R. Stewart, and M. C. Caserio, Organic Chemistry, Benjamin, New York (1971); N. L. Allinger, M. P. Cava, D. C. DeJongh, C. R. Johnson, N. A. Lebel, and C. L. Stevens, Organic Chemistry, Worth, New York (1971).

11. J. N. Pitts, Jr., F. Wilkinson, and G. S. Hammond, in Advances in Photochemistry, Vol. 1, W. A. Noyes, Jr., G. S. Hammond, and J. N. Pitts, Jr., Eds., Interscience, New York (1963), p. 1.

12. W. A. Noyes, Jr., G. B. Porter, and J. E. Jolley, Chem. Rev., 56; 49 (1956).

13. J. B. Birks, Photophysics of Aromatic Molecules, Interscience, London (1970).

14. C. A. Parker, Photoluminescence of Solutions, Elsevier, Amsterdam (1968).

Solution-Phase Photochemistry of Cyclobutanones

D. R. MORTON and N. J. TURRO
Department of Chemistry
Columbia University
New York, New York

CONTENTS

I. INTRODUCTION

The past decade has witnessed a remarkable growth in the field of organic photochemistry as evidenced by the appearance of numerous textbooks, review serials, and primary publications (1). Examination of the chemical literature reveals that this growth is due, in large measure, to research into the nature of the excited-state chemistry and physics of the carbonyl group (e.g., ketones, aldehydes, and carboxylic acids and esters). The extensive photochemical data now available for these compounds has enabled the organic photochemist to make correlations regarding structure and reactivity and has often provided him with convenient and mild methods for the construction of molecules of high energy content.

Of the various carbonyl compounds examined to date, considerable research has been performed on the photochemical transformations of cyclic aliphatic ketones (1). This review will specifically concern the photochemical transformations of cyclobutanones in the solution phase. Our purpose here is essentially threefold: (1) to present a critical review of the available literature; (2) to provide a unifying interpretation of the numerous results reported to date; and (3) to pose problems as targets for future research effort.

A. Unimolecular Reactions of Typical Cyclic Ketones

The photochemical transformations of cyclic aliphatic ketones are illustrated in Scheme I (the Norrish Type II reaction is excluded and will not be discussed in this review). We notice that these transformations may all be interpreted as involving acyl alkyl biradical intermediates (2). The ketone excited singlet state (S_1) and/or triplet state (T_1) could conceivably undergo α-cleavage to afford an acyl alkyl biradical intermediate, [1]. In general, α-cleavage occurs on the most highly substituted side of the excited carbonyl group (1h) (e.g., R = alkyl or aryl), a fact which lends support to the transient existence of [1]. Depending on the structural characteristics of the starting ketone, the biradical [1] could decarbonylate to another biradical (e.g., [2]) which, in turn, could subsequently cyclize or disproportionate. In general, photodecarbonylation of alkanones in solution phases is observed only when the biradical formed upon α-cleavage is highly stabilized as is the case for benzocyclopentenone [11] (2,3), i.e., when the loss of CO has a low activation, or when other irreversible paths for stabilization of the biradical are ineffective due to structural features of the molecule. Decarbonylation is also the favored mode of decomposition

SCHEME I

Unimolecular Photochemical Transformations of Cyclic Aliphatic Ketones

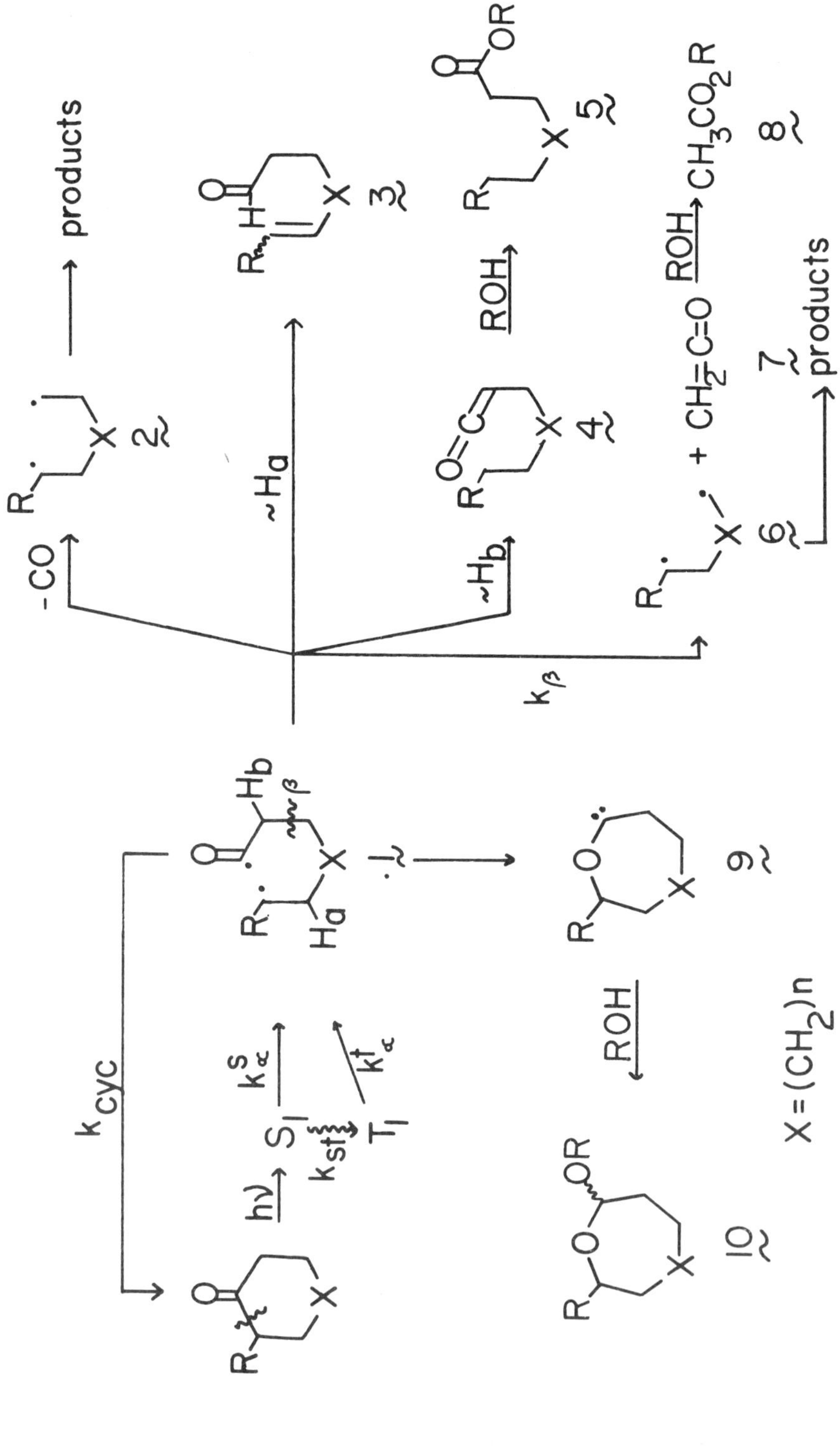

for thujone [12], where relief of ring strain, in addition to radical stabilization, is thought to be an important factor (4). Thus in this case, both cyclopropyl conjugation with the biradical centers and ring opening (which might be synchronous with CO expulsion) seem plausible explanations for the high yield of photodecarbonylated products from [12].

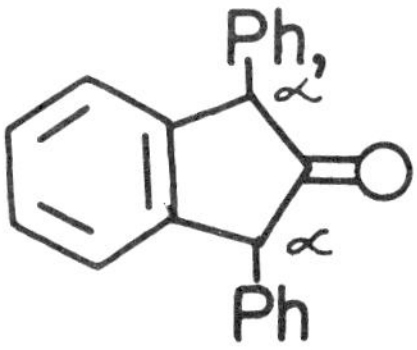

11

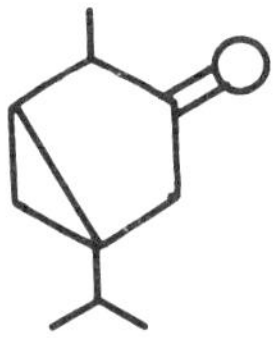

12

Intramolecular disproportionation of [1] could conceivably result in the formation of unsaturated aldehyde or ester. Both paths are well known to occur with cyclic alkanones (1a). Abstraction of H_a by the acyl radical portion of [1] would produce alkenal [3] while abstraction of H_b by the alkyl radical portion of [1] would produce ketene [4] which may be efficiently trapped in nucleophilic alcoholic solvents to yield ester [5]. The intramolecular nature of alkenal formation is supported by deuterium labeling experiments (5), while the intramolecular nature of ketene formation is supported by (a) deuterium labeling experiments (6,7), and (b) the observed decrease in ketene formation with decreasing ring size (8).

Cleavage of the β-bond in [1] results in the formation of a ketene fragment [7], and another biradical, [6]. As was the case with [4], [7] is efficiently trapped in nucleophilic alcoholic or amine solvents to yield ester [8], or a corresponding amide, while biradical [6] may cyclize or disproportionate.

A fifth unimolecular reaction involves electronic reorganization in [1] to afford a ring-expanded oxacarbene intermediate [9]. If photolysis is carried out in alcoholic solvents, [9] is efficiently trapped to afford acetal [10]. Yates has recently reviewed the photochemical ring-expansion reaction (9) and has found it to be general for tricyclo[2.2.1.0^{2,6}]heptan-3-ones (e.g., [13] → [14]).

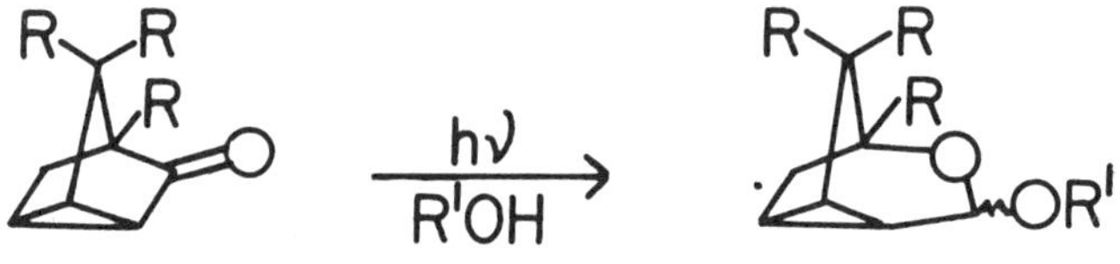

13, R = H or CH_3 14

If all of the reactions just described do not occur due to adverse conformation effects, unfavorable substitution, ring strain, or other unknown factors, biradical [1] can recyclize to the starting ketone (k_{cyc}). It should be emphasized that the existence of a discrete acyl alkyl biradical intermediate in the photolytic transformations of cyclic aliphatic ketones, while it makes for facile mechanistic rationalization, has not been directly verified by classical trapping or spectroscopic observation. Several recent reports, however, have provided strong support for the existence of [1], although only for specific cyclic ketones (10).

B. Spectroscopy of Cyclobutanones and Its Relationship to Excited-State Structure

During the past few years, considerable attention has been given to the analysis of the absorption and emission spectra of alkanones (11). The ultraviolet absorption maximum of cyclobutanone in nonpolar and polar solvents occurs near 280 nm, with an extinction coefficient of about 20 (Figure 1). Moule (11a) and Whitlock and Duncan (11b) have recently analyzed the vapor-phase absorption of cyclobutanone and conclude that the prominence of a carbonyl stretching mode (1260 cm^{-1}), and an out-of-plane wagging mode (~395 cm^{-1}) implies that the CO bond is elongated and distorted from the plane of the molecule. Chandler and Goodman (11c) have analyzed the absorption (20°C, pentane) and phosphorescence emission spectra (EPA glass, 77°K, maximum about 425 nm) of cyclobutanone. These authors also concluded that S_1 is easily distorted and that T_1 is probably also bent in the vicinity of the carbonyl function. All of these results are consistent with the spectral analyses of alkanones, and it is concluded that the $S_1(n,\pi^*)$ and $T_1(n,\pi^*)$ states of alkanones, cyclic or acyclic, <u>are nonplanar about the carbonyl group and wag back and forth between pyramidal shapes that are about 30-35° out of the molecular plane</u>. The equilibrium geometry of the first excited singlet state of cyclobutanone (11a) is pyramidal about the carbonyl carbon, and the S_1 state wags between two equivalent pyramidal shapes which are separated by a small barrier of 350 cm^{-1} (~0.1 kcal/mol) (see Figure 2).

In spite of the overall similarity of the fluorescence emission of cyclobutanone to other cyclic alkanones in the vapor phase (broad emission maximum near 400 nm), Hemminger, Rusbult, and Lee (17) report that the solution fluorescence emission of cyclobutanone is exceptional in two important ways: (a) Φ_f (excitation near 310 nm in cyclohexane) is about 1×10^{-4}, or nearly 1/20th of the vapor-phase value, and (b) the fluorescence

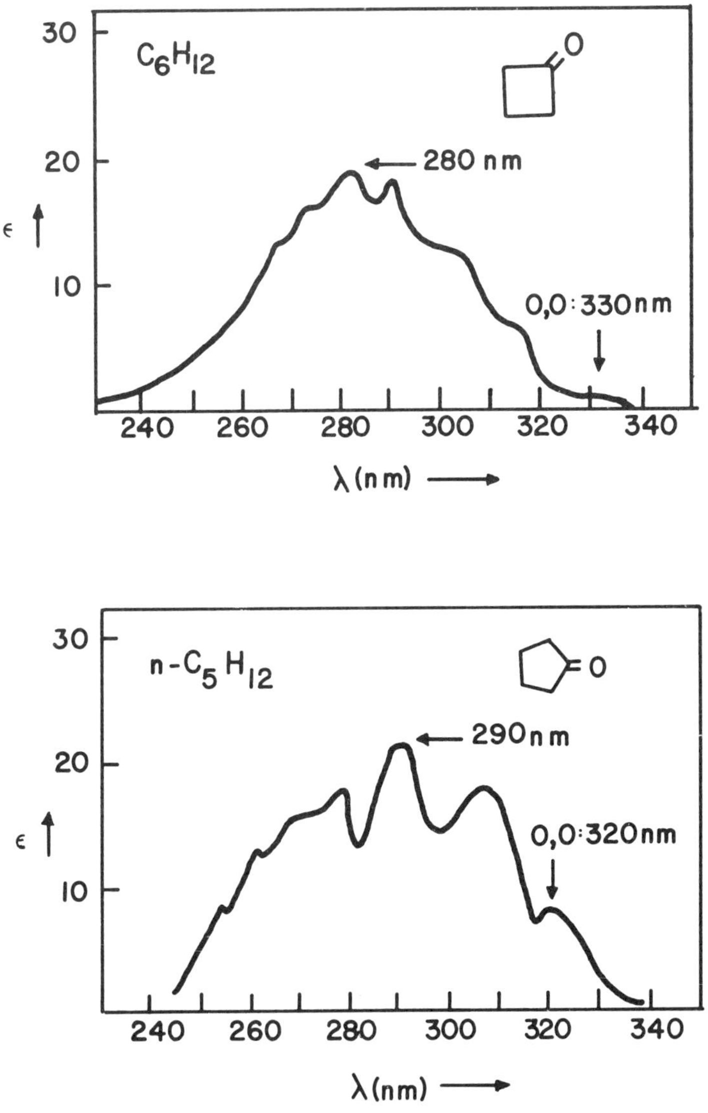

FIGURE 1. Ultraviolet absorption spectra of cyclobutanone and cyclopentanone in hydrocarbon solvents.

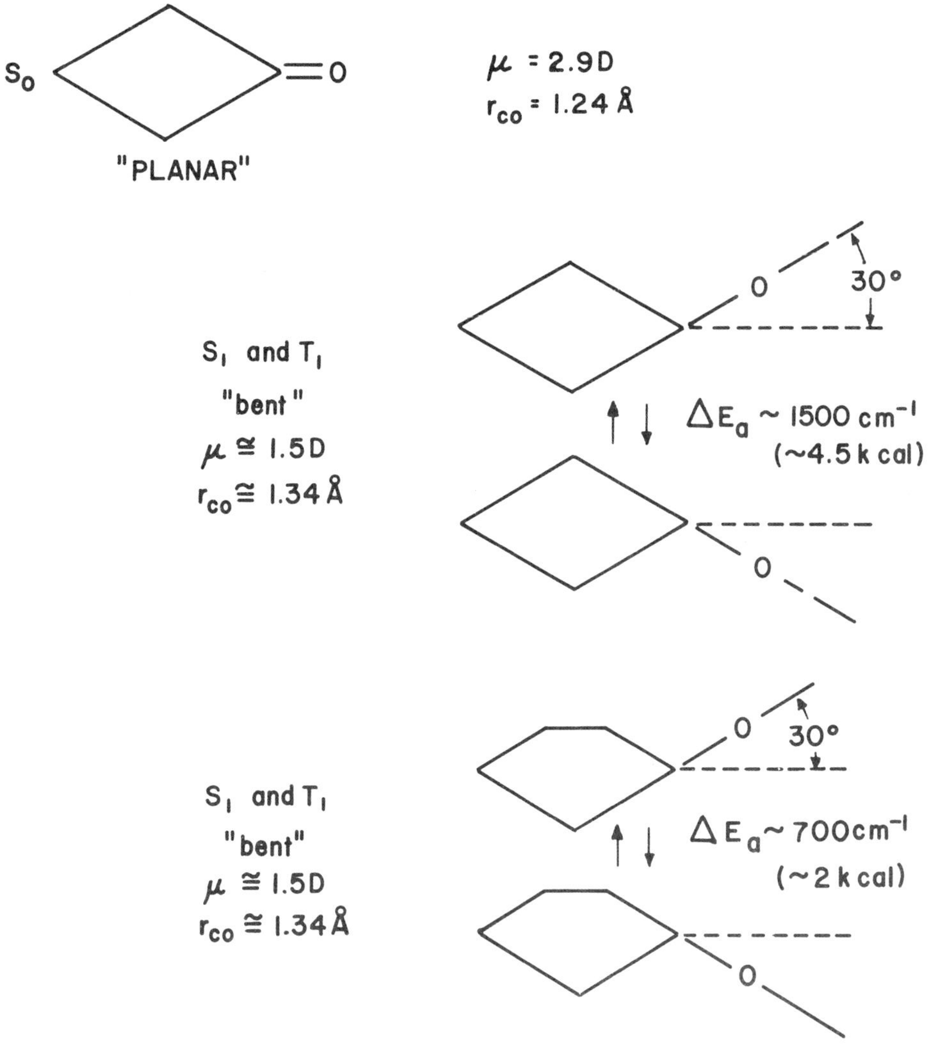

FIGURE 2. Equilibrium geometries of the excited states of cyclobutanone and cyclopentanone.

excitation spectrum essentially drops to zero below 290 nm. The fluorescence excitation spectra of acetone, cyclopentanone and cyclohexanone, on the other hand, nicely parallel their absorption spectra (see Figure 3).

The following summary can now be made with reference to the energy diagrams (Figure 4). Cyclopentanone (17,47,48),

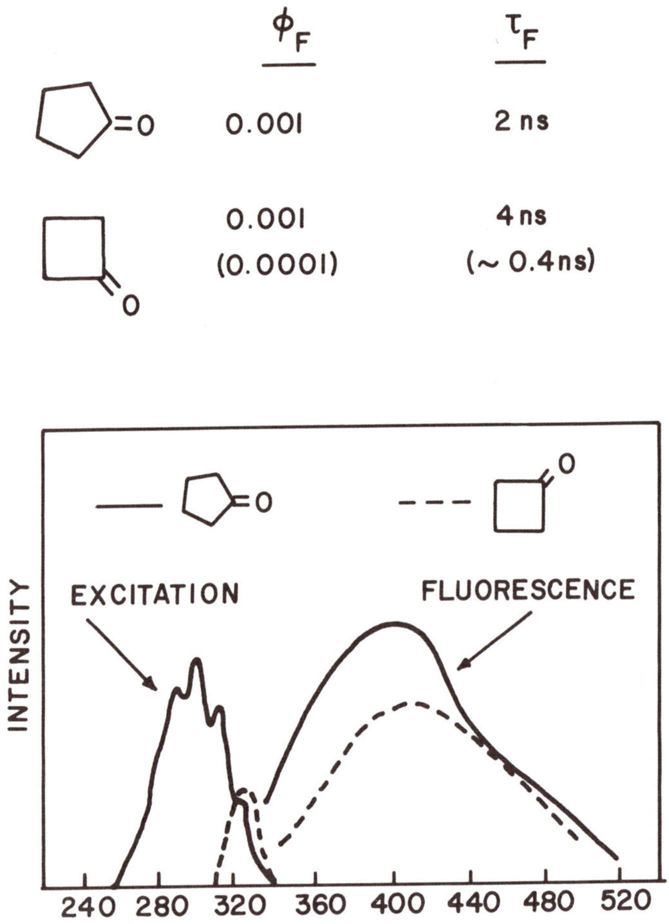

FIGURE 3. Fluorescence emission and fluorescence excitation spectra of cyclobutanone and cyclopentanone.

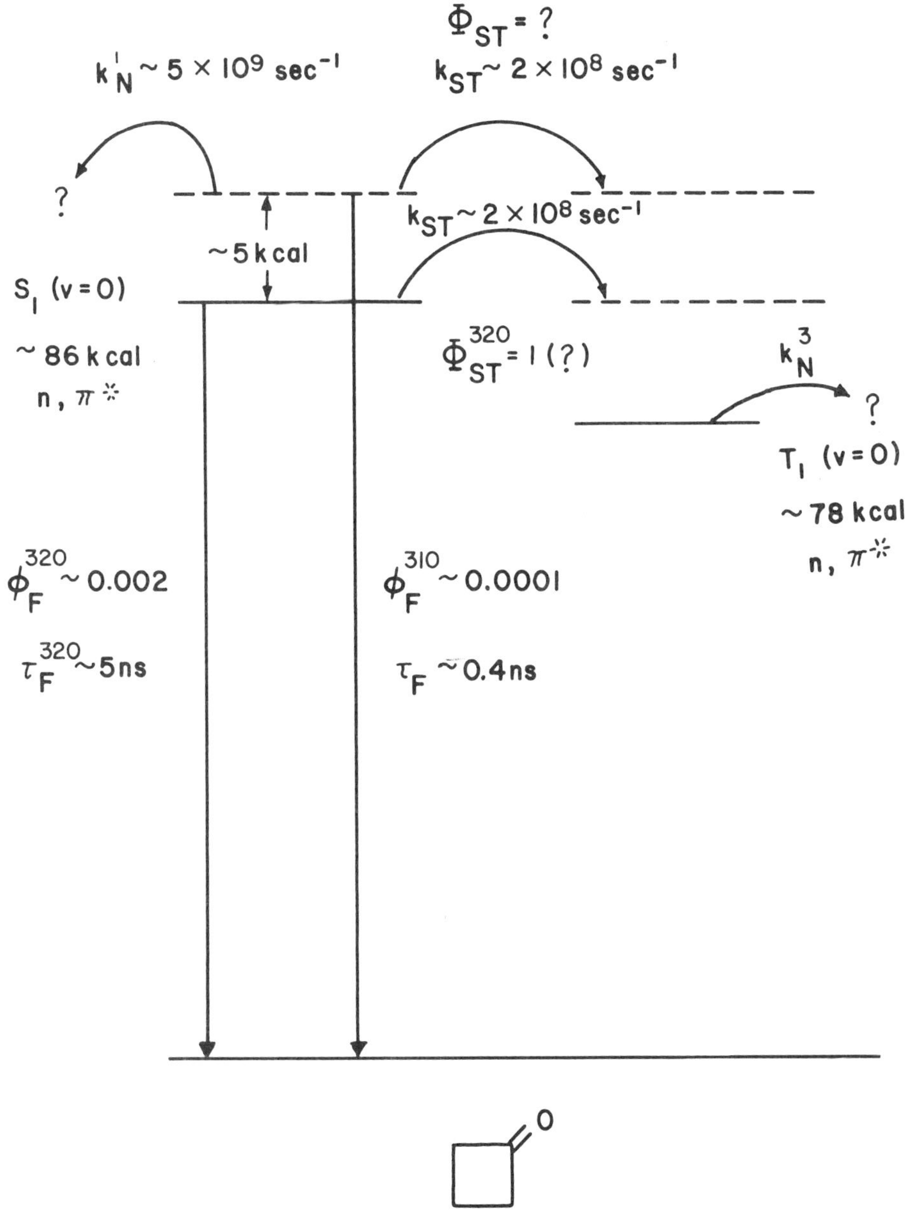

(a)

FIGURE 4. Energy diagrams for (a) cyclobutanone and (b) cyclopentanone.

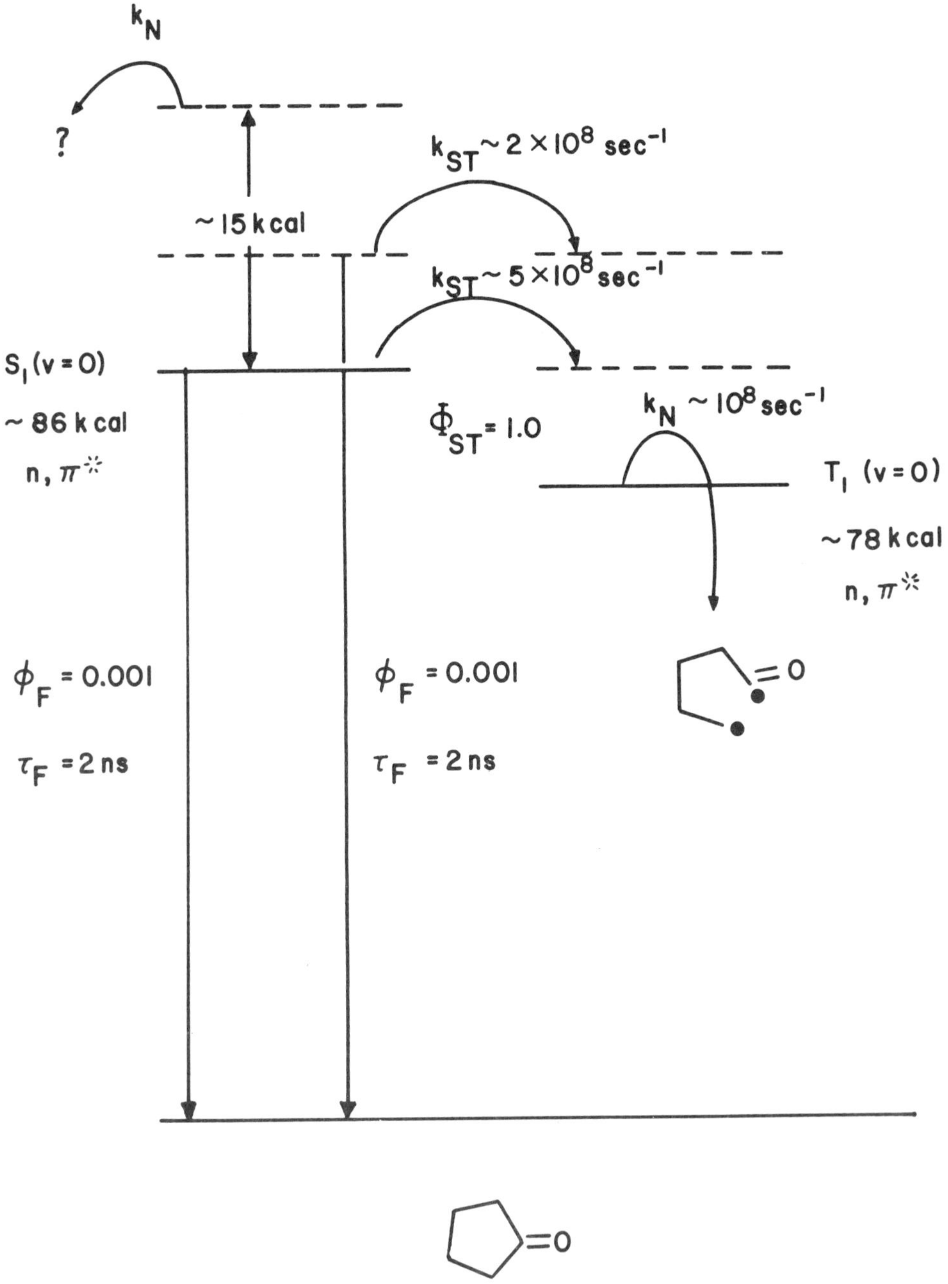
k_N
?
$k_{ST} \sim 2 \times 10^8$ sec^{-1}
~15 kcal
$k_{ST} \sim 5 \times 10^8$ sec^{-1}
S_1 (v = 0)
~86 kcal
n, π^*
Φ_{ST} = 1.0
$k_N \sim 10^8$ sec^{-1}
T_1 (v = 0)
~78 kcal
n, π^*
ϕ_F = 0.001
τ_F = 2 ns
ϕ_F = 0.001
τ_F = 2 ns

(b)

like acetone and cyclohexanone, undergoes $S_1 \rightarrow T_1$ crossing with a rate constant of about $4 \times 10^8\ sec^{-1}$. The quantum yield for intersystem crossing, Φ_{st}, is essentially unity for these alkanones. The quantum yield of fluorescence, Φ_f, for these compounds is about ten times greater than that for cyclobutanone. Since the absorption and fluorescence emission of cyclobutanone are analogous to other cyclic alkanones (Figures 1 and 3), we can conclude that k_{st} and k_f should be comparable for cyclobutanone (for excitation at 320 nm) and cyclopentanone. Figure 4a depicts a tentative energy diagram for cyclobutanone. Since Φ_f is much smaller than expected, based on the latter conclusion, some process is available to S_1 of cyclobutanone which is not available to S_1 of cyclopentanone. Two possibilities are (a) internal conversion of S_1 to a "hot" ground state and (b) chemical reaction of S_1. That (a) or (b) is operating is consistent with the observation that the absorption bands for excitation to S_1 are broadened relative to other alkanones, as is the fluorescence emission. It can be calculated that if S_1 has an excess energy of about 4-5 kcal/mol, process (a) or (b) comes into play.

We shall see that a discussion of the photochemistry of cyclobutanones in solution will help unravel the nature of the rapid new process which makes S_1 of cyclobutanones so much more reactive than other alkanones. The inherent ring strain of cyclobutanone probably triggers the predissociative behavior of the molecule in the S_1 state. This contrasts sharply with the behavior of cyclopentanone and other alkanones for which the most important primary process from S_1 is intersystem crossing.

C. Primary and Secondary Photoprocesses of Cyclobutanones

Although Scheme I illustrates six possible unimolecular photochemical transformations available to a cyclic aliphatic ketone, few, if any, ketones are found to simultaneously undergo all of these. Cyclohexanones and cyclopentanones, for example, undergo predominant rearrangement to an unsaturated aldehyde (e.g., [3] and/or a saturated ketene (e.g., [4]) (1h,8c). Seldom are decarbonylation, β-cleavage, and ring expansion observed for cyclopentanones and cyclohexanones in solution phases. Cyclobutanones, on the other hand, generally undergo decarbonylation, β-cleavage (or cycloelimination), and ring expansion to the virtual exclusion of the other transformations. For example, Hostettler observed that the bicyclo-[3.2.0]-heptenone [15] was photolytically rearranged to an equimolar mixture

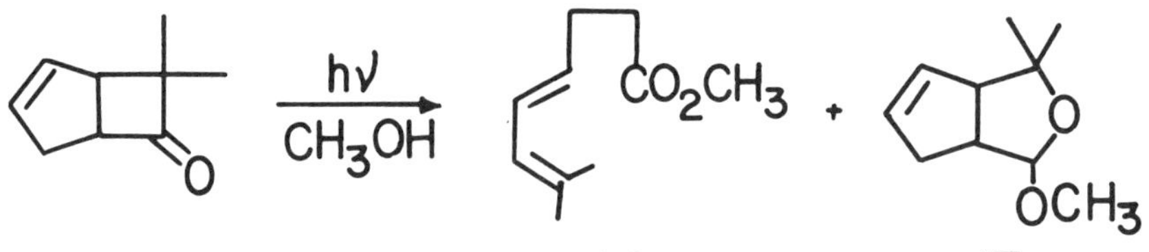

15 16 17

of the β-cleavage ester [16] and the ring-expanded acetal [17] (12). As a second illustration, the dimethylaminocyclobutanone [18] undergoes decarbonylation to [19] and ring expansion to [20] (13).

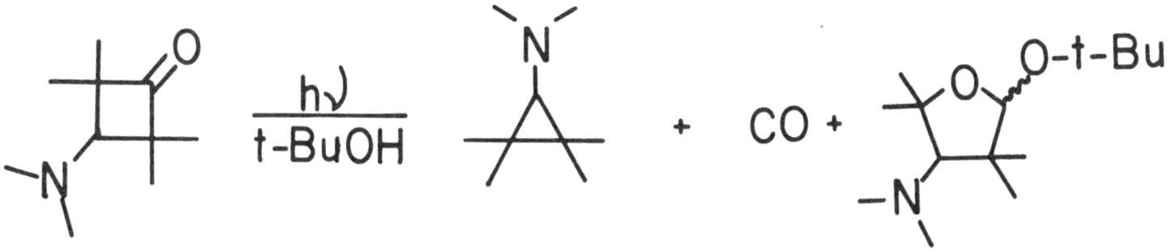

18 19, 40% 20, 43%

II.

A. Photochemical Reactions of Saturated Cyclobutanones

To date, the photochemistry of cyclobutanone has been most extensively studied in the gas phase (14), where the two processes, β-cleavage (or cycloelimination) and decarbonylation, are the only observable reactions. The solution-phase photochemistry of this parent ketone, on the other hand, has been less extensively studied. Turro and Southam (15) investigated the photochemical transformations of cyclobutanone [21] in methanol and observed products attributable to the three expected reactions. The decarbonylation product cyclopropane 23 was not analyzed quantitatively under the conditions of photolysis:

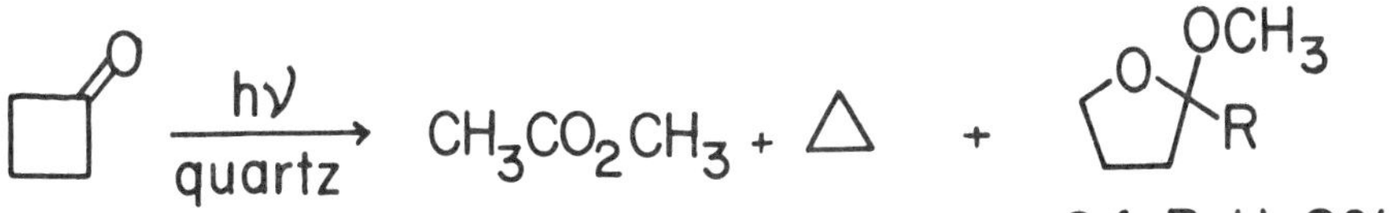

21 CH_3OH 22, 48% 23 24, R=H, 8%
25, R=D

however, its production was demonstrated by the presence of high field absorptions (9.5τ) in the nmr spectrum of the reaction mixture. When photolysis was carried out in methanol-0-d, cyclic acetal [25] was produced with one atom of deuterium located exclusively on the acetal carbon atom.

Klemm (16) and Lee and coworkers (17) have examined the effect of various solvents on the photochemistry of cyclobutanone. By monitoring the quantum yields for formation of ethylene (β-cleavage product) and cyclopropane (decarbonylation product) in different solvents, they were able to demonstrate a significant reduction in the quantum yields for product formation in methanol as compared to other hydrocarbon solvents. Whereas the quantu yield of ethylene formation was found to be essentially solvent insensitive, that for cyclopropane formation was found to be somewhat solvent sensitive. This suggested that β-cleavage and decarbonylation do not result from the same immediate precursor. Since ring-expansion derivatives have not been isolated from photolyses carried out in saturated hydrocarbon solvents, the importance of this process under these conditions remains to be determin Irradiation of cyclobutanone in the presence of 1,3-pentadiene (17,59) or 1,3-cyclohexadiene (16) did not appear to affect the quantum yields for ketone disappearance or product appearance.

In 1967 Turro and Southam reported the first systematic study of alkyl substituted cyclobutanones in methanol solutions (15). As with cyclobutanone [21], all the alkyl ketones studied ([26a-c]) exhibited products characteristic of β-cleavage, decarbonylation, and ring expansion. Interestingly, the selectivity of α-cleavage is reminiscent of similar trends observed for α-alkylated cyclopentanones and cyclohexanones in that cleavage appears to occur exclusively on the most substituted side

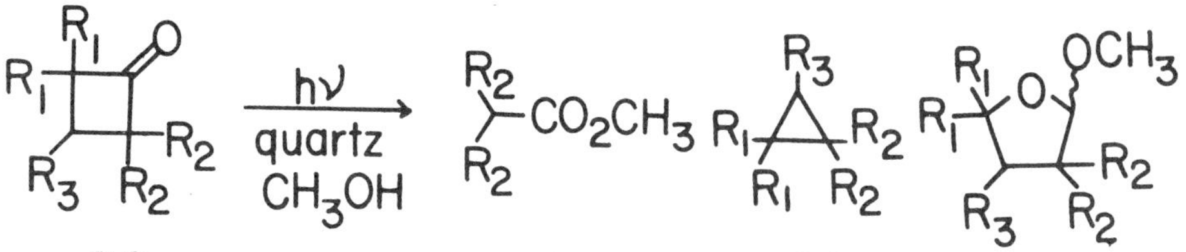

26 27 28 29

		27	28	29
a:	$R_1=CH_3$; $R_2=R_3=H$	32%	7%	41%
b:	$R_1=R_2=CH_3$; $R_3=H$	13%	11%	68%
c:	$R_1=R_2=CH_3$; $R_3=OH$	16%	17%	67%

of the carbonyl group (10). To account for these results, the mechanism illustrated in Scheme II was proposed. Using 2,2-dimethylcyclobutanone [26a], as a specific example, initial excitation to produce an excited state species followed by α-cleavage would produce the acyl alkyl biradical [30]. Subsequent decomposition of [30] would then afford ester [27a] (via ketene), cyclopropane [28a], and acetal [29a], the observed photoproducts. The intermediacy of biradical [30] was supported by (a) the nearly exclusive formation of methyl acetate (as opposed to methyl isobutyrate), (b) the exclusive formation of the 5,5-dimethyl substituted acetal [29a] (as opposed to its 3,3-dimethyl substituted isomer), (c) its role as a common intermediate for all products, and (d) analogy to the photochemistry of cyclopentanones and cyclohexanones. Recently, Wasacz and Joullie have reported that photolysis of oxacyclohexanone [32] affords a 3% yield of acetal [29a] (18). It is conceivable that the formation of [29a] occurs directly by way of biradical [30], however, prior

SCHEME II
Mechanism of Photochemical Rearrangements of Cyclobutanones

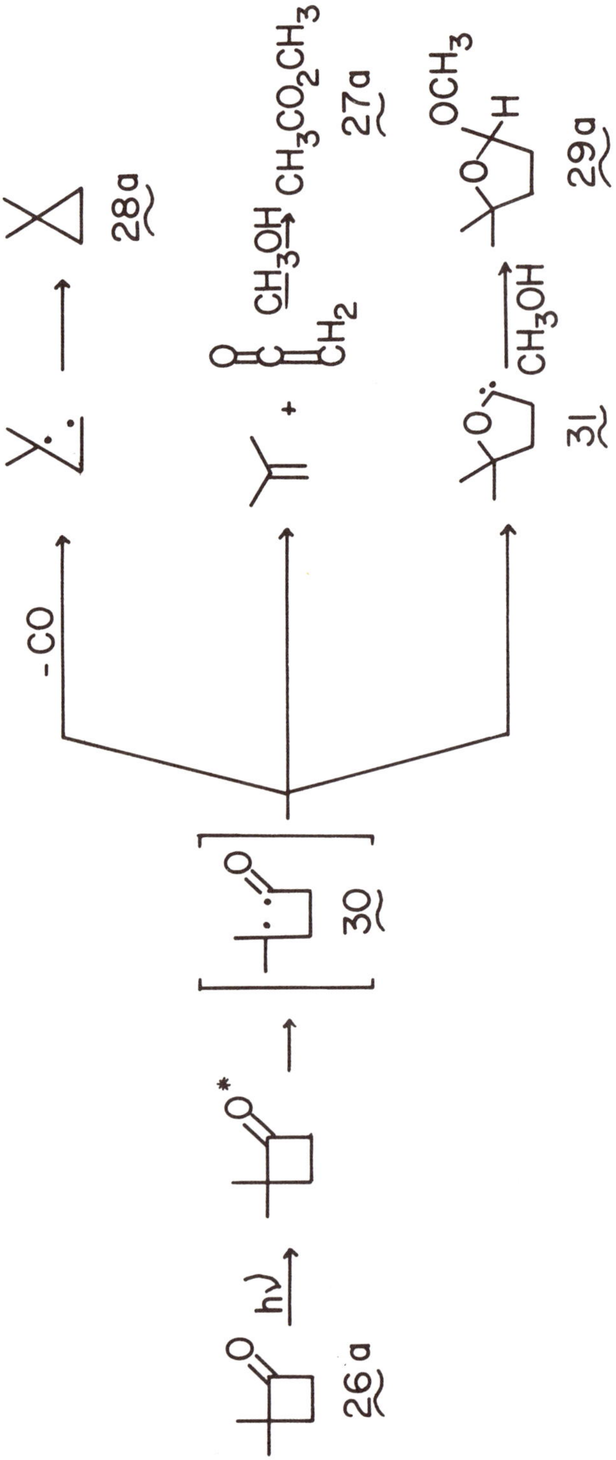

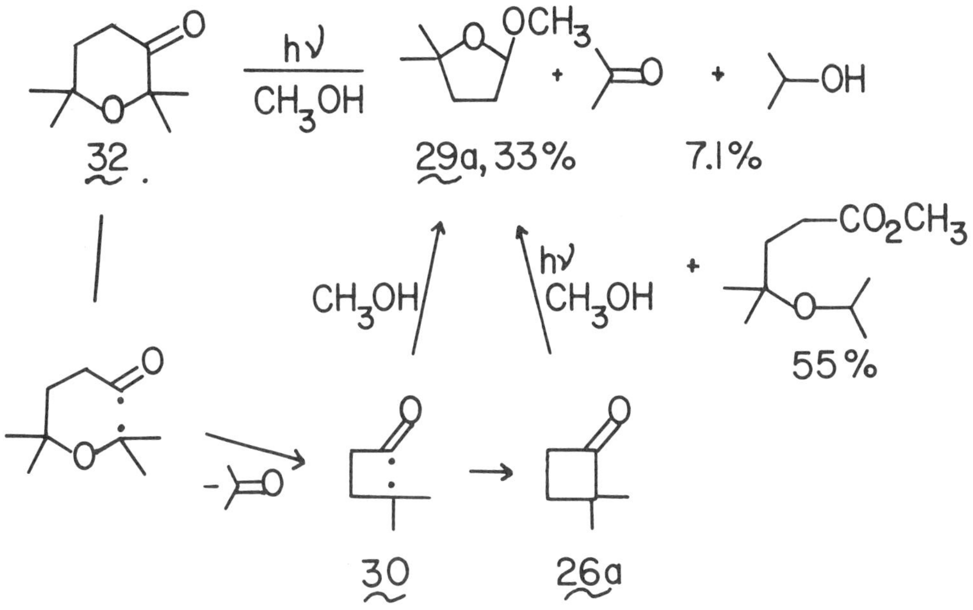

formation of 2,2-dimethylcyclobutanone [26a], followed by secondary photolysis to [29a] cannot be ruled out. In addition, the presence of β-cleavage product [27a] and decarbonylation product [28a] were not investigated.

Dowd and coworkers have also published results which seemingly support the intermediacy of an acyl alkyl biradical (19). When cyclobutanone [21] was irradiated at -78°C in neat 1,3-butadiene, an equimolar mixture of 3-vinylcyclohexanone [33] and oxetane [34] was produced. The formation of [34] is a characteristic bimolecular reaction of aliphatic ketones and 1,3-dienes and has ample precedent (20). The formation of cyclohexanone [33], on the other hand, is novel and has been suggested to arise

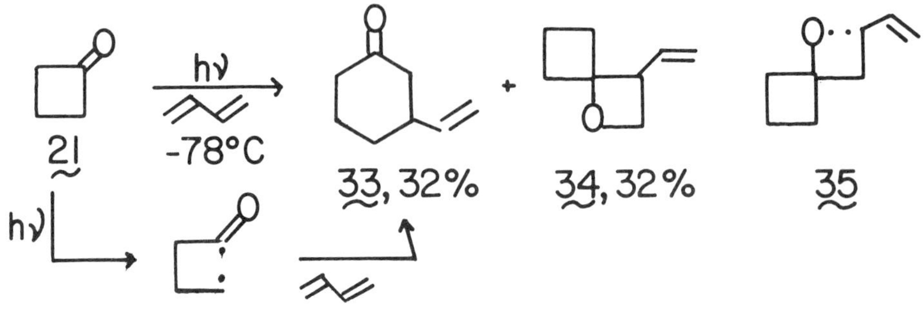

by biradical trapping by diene solvent. Alternative pathways to [33] (e.g., by way of biradical [35] or a similar intermediate) have not, however, been excluded. Whatever the mechanism of formation of [33] and [34], these products are rare examples of bimolecular adducts in which an excited state of cyclobutanone, or a derived biradical intermediate has been trapped. The low temperature may be a key here (vide infra) since the yields of addition products dropped substantially when the irradiation was run at room temperature. The only other reactions reported to date which possibly involve trapping of a cyclobutanone excited state are (a) other examples of oxetane formation (34) (viz., [69] plus the 1,3-pentadienes) and (b) cyclobutanol formation when [21] is irradiated in the presence of tri-n-butylstannane (46).

The possible intermediacy of an acyl alkyl biradical (e.g., [30]) in the photochemical reactions of cyclobutanones has intrigued several researchers. Unlike Hostettler's results with bicyclo-[3.2.0]-heptenone [15], Erman observed that α-cleavage occurred on both sides of the carbonyl group of photoexcited bicyclo-[3.2.0]-heptenone [36] (21). This result might be anticipated if an acyl alkyl biradical intermediate is produced since

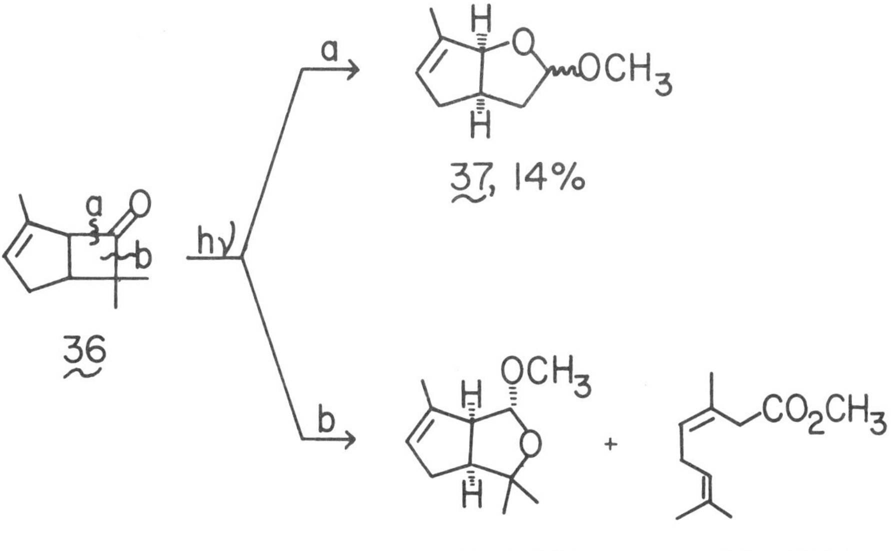

36

37, 14%

38, 25%

39, 17%

cleavage at either "a" or "b" ([36]) would produce a stabilized species. On the other hand, the work of Quinkert and coworkers (22) quickly dispelled any notion of the existence of a freely rotating acyl alkyl biradical. When cis-fused D-nor-16-ketosteroid [40] was irradiated in ethanol, cyclic acetals [41] were isolated as the major products in addition to lesser amounts of cyclohexene [42]

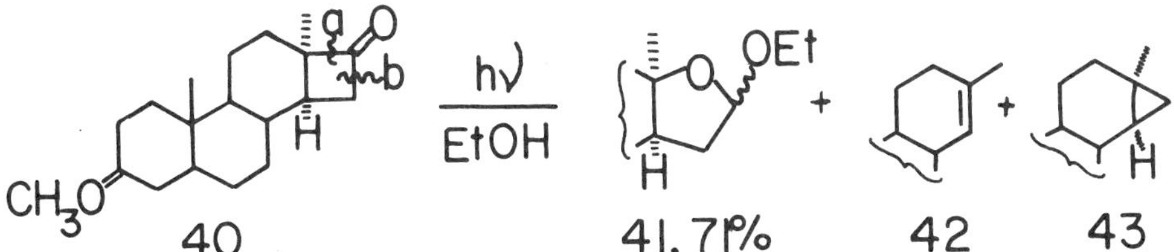

40

41, 71%

42

43

(β-cleavage product) and cyclopropane [43] (decarbonylation product). Again the pronounced regioselectivity of α-cleavage was observed (i.e., cleavage at "a" exclusively); however, the C/D cis ring junction remained intact. It can be argued that the cis-fused C/D ring junction of [41]

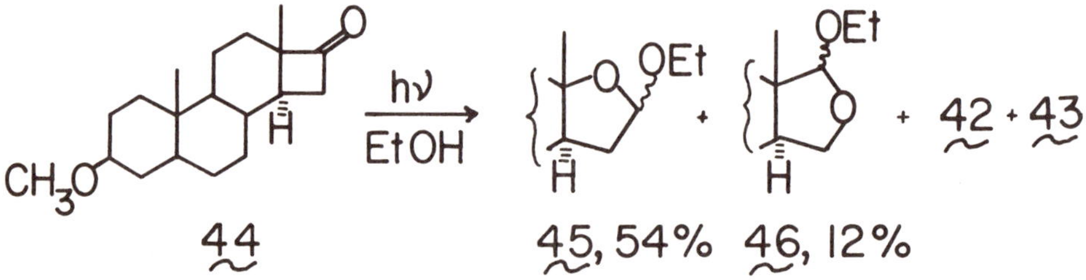

is thermodynamically favored and that a freely rotating biradical could still exist. However, an identical irradiation of the transfused isomer [44] demonstrated that the trans ring junction was retained exclusively. These results clearly preclude the intermediacy of a freely rotating acyl alkyl biradical intermediate in the ring-expansion reaction and suggest that the rearrangement may, in fact, be concerted.

Unfortunately the results of Quinkert and coworkers do not allow the stereochemical consequences of the β-cleavage and decarbonylation reactions to be studied. To this end, Turro and McDaniel investigated the solution-phase photochemistries of cis- and trans-2-ethyl-3-methoxycyclobutanones [47] and [48], respectively (23). The results are presented in Table I and clearly demonstrate that all photoprocesses are significantly stereoselective and stereospecific. Further, no interconversion of [47] to [48] or vice versa was observed after extended irradiation, and the lack of total stereoselectivity in

TABLE I

Results of Photolyses of cis- and trans-2-ethyl-3-methoxycyclobutanones in Methanol and Benzene[a,b]

Ketone	Solvent	Photoproducts[c] 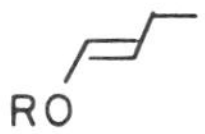				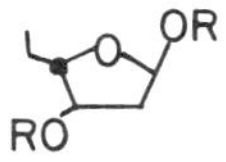	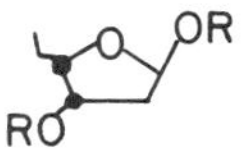
[47]	CH_3OH	0.004	0.11	0.001	0.01	<0.001	0.03
[47]	C_6H_6	0.005	0.08	0.002	0.01	--	--
[48]	CH_3OH	0.10	0.004	0.03	0.002	0.06	<0.001
[48]	C_6H_6	0.10	0.004	0.06	0.002	--	--

[a]Taken from reference (23).

[b]Irradiation at 313 nm; approximately 1M in cyclobutanone.

[c]Numbers presented are quantum yields of product formation ($R=CH_3$) based on cyclopentanone actinometry. Error limits ca. 20%.

the photoproducts could be attributed to the initial isomeric impurity in the samples of [47] and [48].

An attempt to affect the ratio of products from [47] by inclusion of ethyl iodide led to negative results. This experiment was run with the hope that intersystem crossing of some intermediate (e.g., S_1, biradical, or oxacarbene) might lead to a change in the product distribution and the reaction stereospecificity.

In a similar fashion, the photolysis of either cis- or trans-bicyclo[6.2.0]decan-9-ones, [49c] or [49t],

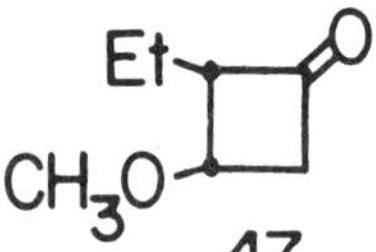

47

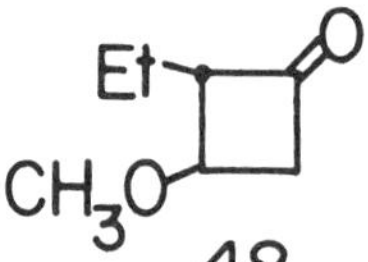

48

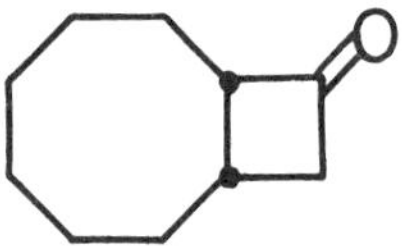
49c

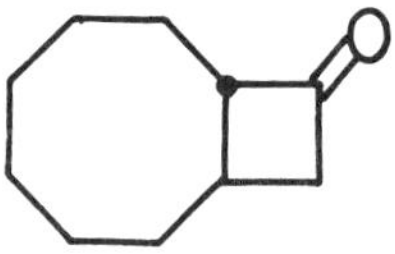
49t

in methanol or benzene solutions afforded products of retained stereochemistry about the ring junction (24).

A most interesting exception to the generalization of stereospecificity of β-cleavage and decarbonylation is given by the results (24) of excitation of cis- and trans-2,3-dimethylcyclobutanone ([50]). The key results seem to be the severe loss of stereospecificity from cis-[50]. Both ketones are regioselective in that propene is, at best, a minor product. These results are in good agreemen with those recently reported by Lee and coworkers (81) for the photolyses of [50c] and [50t] in n-heptane at 325 nm. is of great interest to see if the ring expansion of cis- and trans-[50] are stereospecific and whether the starting ketone is photoepimerized.

The general occurrence of stereospecificity in the

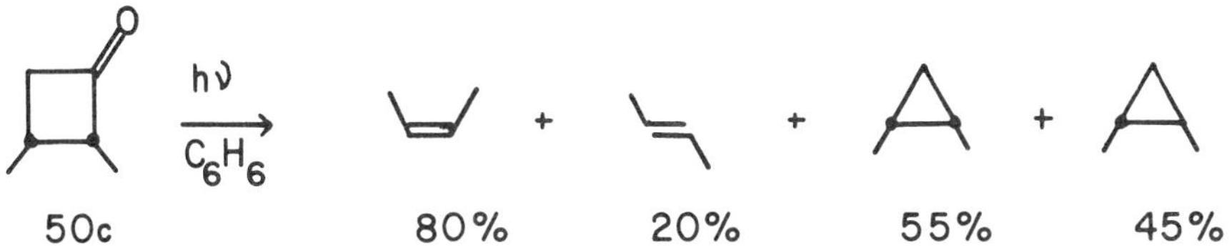

50c

50t

photochemistry of cyclobutanones requires a mechanism which incorporates one of the following ideas: (a) the product determining step is concerted, (b) a short-lived intermediate occurs which decomposes faster than it rotates about bonds, or (c) a long-lived intermediate occurs which decomposes stereospecifically because bond rotations are slow.

The guidelines for deduction of the occurrence of an intermediate are (a) energetics, (b) theoretical considerations, and (c) general chemical intuition. The calculation of the energetics indicates that formation of a 1,4-biradical would be exothermic from either S_1 or T_1 of cyclobutanone. Theoretical considerations, such as the orbital interactions which are the basis of the Woodward-Hoffman Rules, are considered to be unreliable, in general, mainly because excited-state potential energy surfaces are not known with any certainty. The general chemical intuition that we build up from scrutiny of a large range of data concerned with the nature and behavior of monoradicals and biradicals allows us to put forth the singlet biradical as a viable and attractive general intermediate in cyclobutanone photochemistry.

For example, the faster rate of reaction from S_1 of cyclobutanones relative to cyclopentanones can be qualitatively understood on the basis of greater relief of strain in S_1 for the four-membered ring versus the five-membered ring. Furthermore, since the 1,4-biradical is produced cis and in the singlet state, all of the observed reactions can occur with relatively minor nuclear reorganization and with no prohibitions.

If we use the definition that an intermediate is a distinct chemical species, detectable or inferential, along the reaction profile between reactant and isolated product, then the biradical has intellectual validity as well as predictive capabilities.

B. Spiro[2.3]hexan-4-ones and Spiro[2.3]hexan-5-ones

It is apparent from the previous discussion concerning simple alkyl substituted cyclobutanones that three types of photoreactions are generally observed: β-cleavage, decarbonylation, and ring expansion. Interestingly, the extent of ring expansion appears to be favored by increasing alkyl substitution at the α-carbon atoms of cyclobutanone. Cyclobutanone [21], for example, undergoes ring expansion in methanol to afford cyclic acetal [24] in 8% yield, whereas 2,2-dimethylcyclobutanone [26a] and 2,2,4,4-tetramethylcyclobutanone [26b] give 41% and 68% yield, respectively, of ring-expanded products under the same experimental conditions (15).

It has naturally been of interest to study the influence of other types of substituents on the nature and extent of the photochemical reactions of cyclobutanones.

In particular, substituents which would increase the ring strain already present in the four-membered ring might be expected to enhance the extent of ring expansion since this rearrangement is generally not observed for the less-strained cyclopentanones (9,25) yet is quite general for the four-membered ring ketones examined to this point. A substituent that would seem to accomplish this increase in angular strain is a spiro-fused three-membered ring (e.g., as in [51]). Since microwave studies (26-28) suggest a value of 116° for the angle between geminal substituents on a cyclopropane ring ([52]), incorporation of this

substituent at either the α- or β-carbon atom of cyclobutanone would result in a greater bond angle deformation (116° → 90° instead of 109° → 90°) and, consequently, a greater degree of ring strain.

Pacifici and Diebert have examined the solution-phase photochemistry of 1-thiaspiro[2.3]hexan-5-one[53] in methanol (29). As anticipated, ring expansion to the cyclic acetals [54] and [55] comprises the major portion

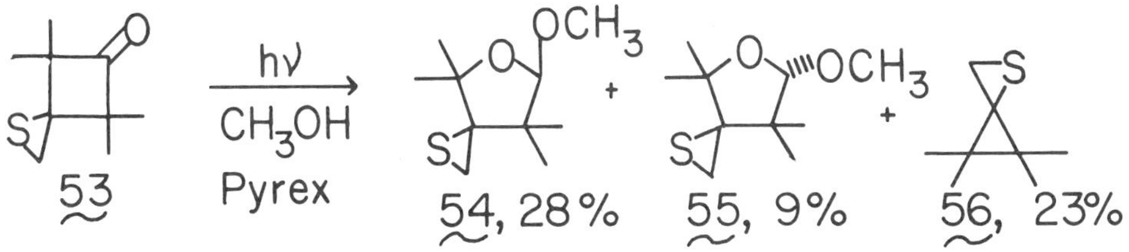

of the observed photoproducts, while no evidence for β-cleavage products was observed. At shorter wavelengths (254 nm), ketone [53] undergoes predominant n → σ* excitation followed by expulsion of sulfur to form the 3-methylenecyclobutanone [57]. The formation of lesser amounts of photoproducts [54]-[56] was attributed to the presence of 313 nm impurity in the lamp source employed.

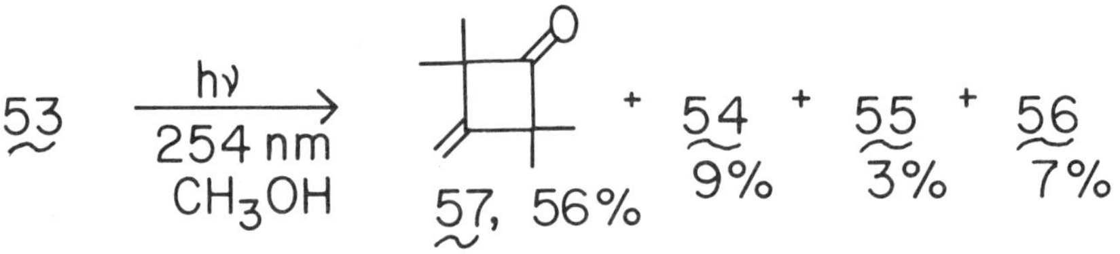

Irradiation of the oxygen analog of [53], 1-oxaspiro[2.3]hexan-5-one [58], in methanol afforded similar results (30). In this case however, ring expansion was even more pronounced. As observed for ketone [53], [58] does not exhibit products derived from the β-cleavage

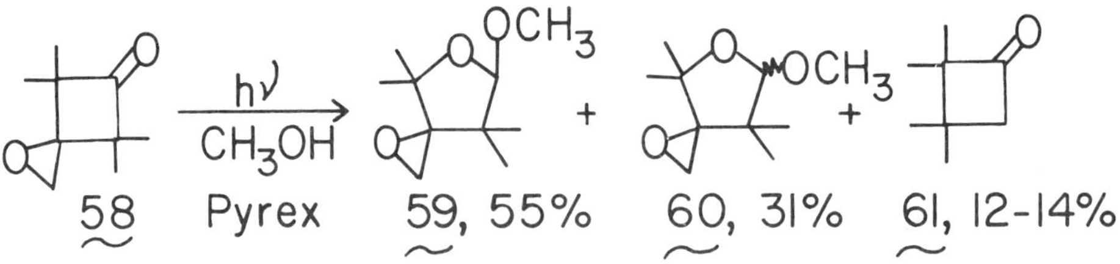

reaction, a reaction which would have afforded the allene oxide [62] and methyl isobutyrate. The formation of 2,2,3,3-tetramethylcyclobutanone [61] presumably results from photochemical decarbonylation to oxaspiropentane [63], followed by facile thermal rearrangement to [61] (31

The photochemistry of cyclobutanones possessing a spiro-fused three-membered ring at the α-position has also been reported (15b). Spiro[2.3]hexan-4-one[64a], for example, undergoes photolysis in methanol solution to cyclic acetals [65a] and [66a] in addition to β-cleavage ester [67]. The dialkylated derivative [64b], on the other hand, affords ring-expanded acetals [65b] and [66b]

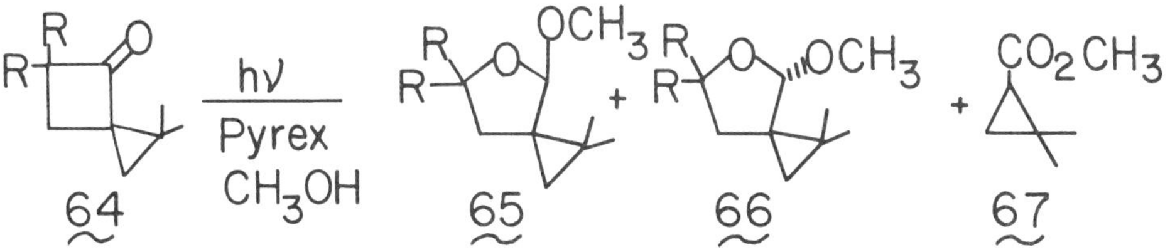

		65	66	67
[a]:	R=H	44%	44%	12%
[b]:	$R=CH_3$	50%	50%	

exclusively. Again the presence of a spiro-fused three-membered ring, which presumably increases the internal angular strain of the cyclobutanones [64a] and [64b], seems to enhance the ring expansion reaction. It is interesting to note that a greater enhancement is observed for ketones [64] where the three-membered ring is located at the α-position. This observation suggests that another factor, in addition to the internal angular strain of the cyclobutanone, favors decomposition of the ketone excited state by way of ring expansion (vide infra).

Irradiation of [68] at room temperature results in low yields of acetylene, propyne, and butyne. Dimethyltetrahedrane is a speculative intermediate in this photolysis (63).

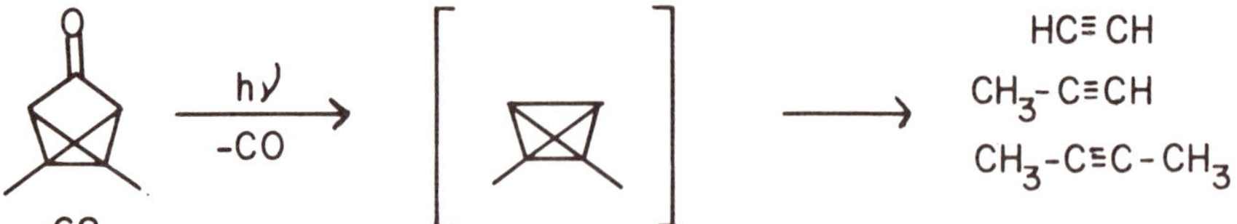

68

C. 3-Methylenecyclobutanones

In addition to serving as substrates with which to further study the effects of various substituents on the photochemical transformations of cyclobutanones, 3-methylenecyclobutanones are capable of generating the theoretically interesting trimethylenemethane biradical species (32) by way of the photodecarbonylation reaction. Indeed, Dowd and Sachdev (33) observed a triplet ESR spectrum upon photolysis of 3-methylenecyclobutanone [69]

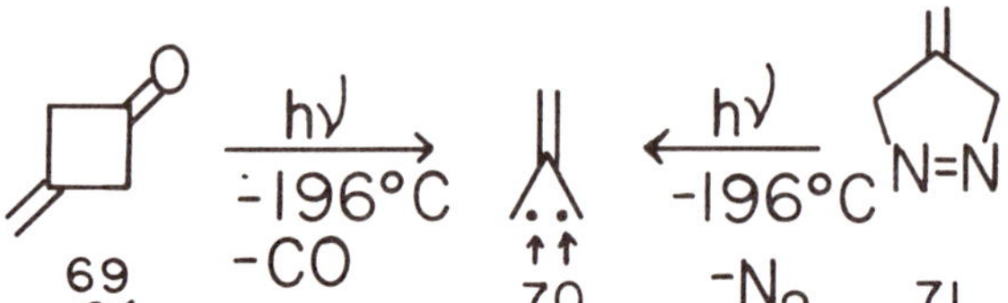

69 70 71

in a perfluorohydrocarbon medium at -196°C. This spectrum was attributed to the ground state of trimethylenemethane [70] which could alternatively be generated by photolysis of pyrazoline [71]. When ketone [69] was irradiated at low temperature in neat 1,3-butadiene, a 1:1 adduct of [70] with the diene solvent (i.e., [72]) was formed in modest yield in addition to a small amount of trimethylene

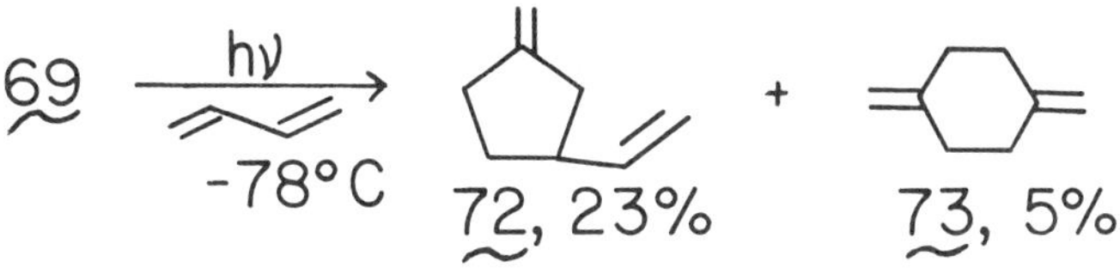

methane dimer [73], an oxetane (15-17%), and butadiene dimers (10%) (34). It is interesting to note that when

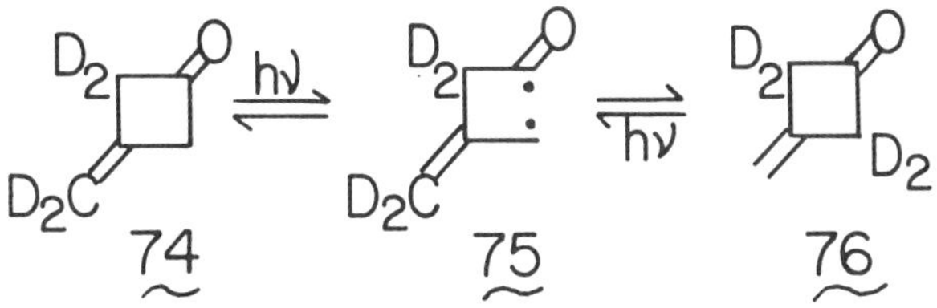

deuterated ketone [74] was used in the place of [69], the adduct [72-d_4] obtained possessed four deuterium atoms scrambled statistically among the carbon atoms of [70-d_4]. However, upon recovery of unphotolysed ketone after 75% conversion to products, it was discovered that deuterium scrambling had also occurred among the three methylene carbons. This observation naturally suggested an isomerization of [74] to [76] and vice versa, perhaps by way of biradical [75]. The [1,3] acyl shift represented by the conversion of [74] → [76] is discussed in greater detail below.

Photolysis of 3-methylenecyclobutanone in furan at 15°C afforded methylenecyclopropane in addition to trimethylenemethane dimer [73]. If the temperature was decreased to -72°C, the amount of [73] produced was increased fivefold relative to methylenecyclopropane.

This temperature effect may reflect an activation energy associated with intersystem crossing of the ground-state trimethylenemethane triplet [70] to an excited singlet species which could then undergo ring closure to methylenecyclopropane. At lower temperatures, the lifetime of [70] might be long enough to permit diffusion and subsequent dimerization with another triplet ground-state species.

While the above results for 3-methylenecyclobutanone [69] provide no information on the contributions of the ring-expansion and β-cleavage reactions, the photochemical transformations of the tetramethyl derivative [57] have been studied (35). Irradiation of [57] in methanol afforded ring-expanded acetal [77] in high yield along with two volatile hydrocarbons identified as methylenecyclopropanes [78a] and [78b] (total yield ca. 10%). In additio

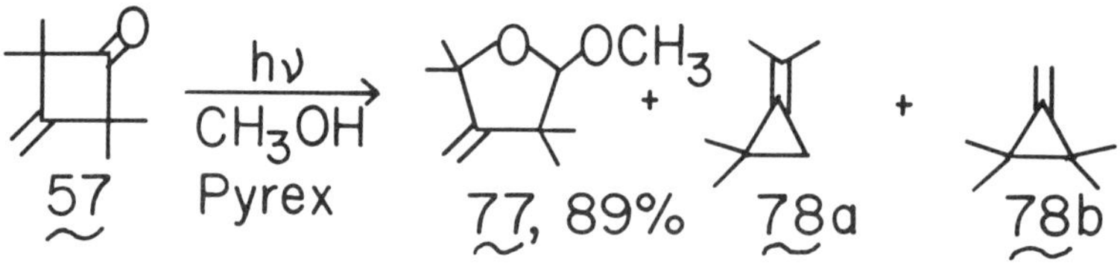

trace quantities of at least three nonvolatile compounds were also detected by gas chromatography. Although not characterized, these are most likely the isomeric trimethylenemethane dimers. Comparison of the photochemistry of [57] and 2,2,4,4-tetramethylcyclobutanone [26b] clearly demonstrates that incorporation of the exo-3-methylene group enhances the ring-expansion rearrangement. Employin similar reasoning as that used to interpret the photochemistry of the spiro[2.3]hexan-4-ones and spiro[2.3]hexa 5-ones (e.g., [53], [58], [64a], and [64b]) described above, it is conceivable that this enhancement is brought about by an increase in the internal angle strain of the four-membered ring ketone. The extent to which this interpretation is valid can be tested by examining the excited-state chemistry of cyclobutanones with exomethylene substituents at one of the α-carbon atoms (e.g., [79]). Placement of an sp^2 carbon at either the α- or β-carbon atoms should not alter the internal angle strain; however, it does generate a markedly different chromophore (i.e., an α,β-unsaturated ketone).

D. 2-Alkylidenecyclobutanones

The photochemical transformations of a series of 2-isopropylidenecyclobutanones in alcoholic media have been reported (15b,36). Irradiation of 2-isopropylidene-cyclobutanone [79a] in methanol afforded a quantitative yield of ring-expanded acetal [80a] with a quantum efficiency (at 313 nm) of 0.11. When irradiations were carried out in methanol-O-d, the acetal [80a] isolated contained >95% deuterium incorporation (nmr) at the indicated position. Completely analogous results were obtained with the 4,4-dimethyl derivative [79b] and 3,3,-

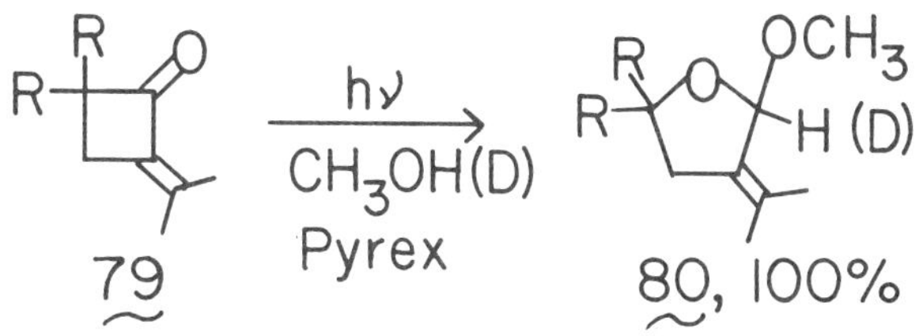

a: R=H

b: $R=CH_3$

4,4-tetramethyl-2-isopropylidenecyclobutanone (37). Apparently the transition of the exomethylene unit from the β- to an α-carbon atom of the cyclobutanone does not affect the enhancement observed for ring expansion. For the 2-isopropylidenecyclobutanones studied, in fact, ring expansion was enhanced to the point where it was the only process which gave rise to products.

Attempts to sensitize the rearrangement of [79a] to [80a] with xanthone (E_T = 74 kcal/mole) under conditions where the sensitizer absorbed essentially all of the incident radiation resulted in no observable chemical change (35). To determine if the cyclobutanone was receiving triplet energy from the sensitizer, the direct and sensitized photochemistry of syn- and anti-2-sec-butylidenecyclobutanones [81] and [82] were investigated (35). Under direct irradiation (313 nm), the isomeric acetals [83] and [84] were produced quantitatively (Φ = 0.1-0.2); yet under sensitized conditions (e.g., with xanthone, acetophenone, benzophenone, or triphenylene)

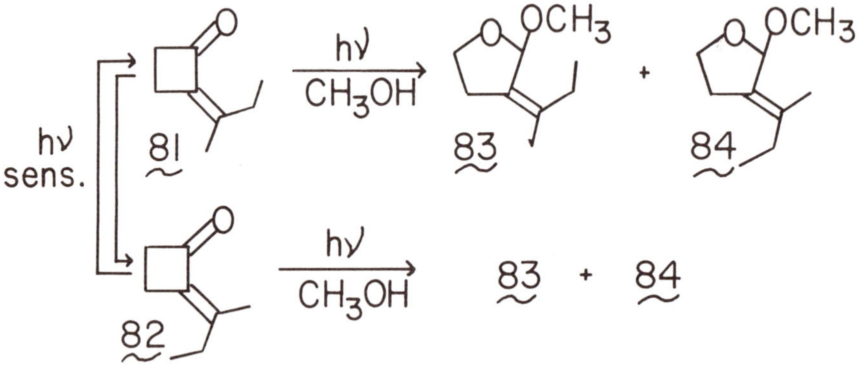

[81] and [82] were efficiently interconverted. No evidence for the production of [83] and [84] in the sensitized photolyses was obtained. Attempts to quench the isomerization process with 1,3-pentadiene met with failure. These results, therefore, suggest that the precursor of the ring-expanded acetals [80], [83], and [84] is the cyclobutanone singlet state (S_1), and that for

the 2-isopropylidenecyclobutanones [79] population of the triplet excited state (T_1) results in no net observable change due to rapid rotation about the exocyclic double bond.

The enhancement of ring expansion over decarbonylation and β-cleavage for 2-isopropylidenecyclobutanones or spiro[2.3]hexan-4-ones can be understood on the basis of stabilization of the oxacarbene intermediate or inhibition of competing reactions from the biradical intermediate. Since an sp^2-sp^3-type bond must be broken for both decarbonylation and β-cleavage, the hypothesis of inhibition of biradical reactions involving fragmentations seems attractive. A corollary to this line of thinking predicts that β-cleavage, but not decarbonylation, should be inhibited for 3-alkylidenecyclobutanones or spiro[2.3]hexan-5-ones, as is found to be the case experimentally.

The photoinertness of 2,4-diisopropylidenecyclobutanone is explained as a "free rotor" effect. In other words, in the S_1 state neither α-bond can be cleaved very facilely, so intersystem crossing occurs. In T_1, excitation is localized on the C=C bond and is deactivated

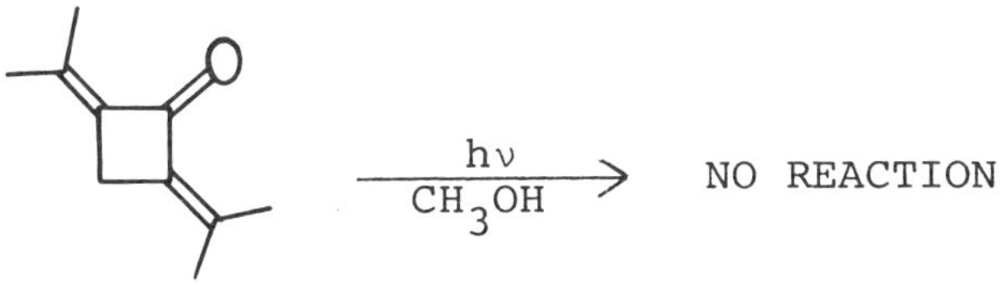

by rotational motions.

The drastically different behavior of S_1 and T_1 for [81] and [82] suggests to us that S_1 is n,π* (and therefore undergoes ring cleavage), but that T_1 is π,π* where excitation is moved out to the C=C bond, with cis-trans isomerization resulting.

E. Photochemical Ring Expansion of Cyclobutanones by Way of Oxacarbenes

The photochemical formation of ring-expanded acetals from ketones in alcoholic solvents was originally suggested by Yates and Kilmurry (9,38) to proceed by way of oxacarbenes (e.g., 31 in Scheme II) (39). The intermediacy of these species is supported by the fact that the irradiation of cyclobutanone [21] in methanol-0-d afforded acetal [25] exclusively (an analogous

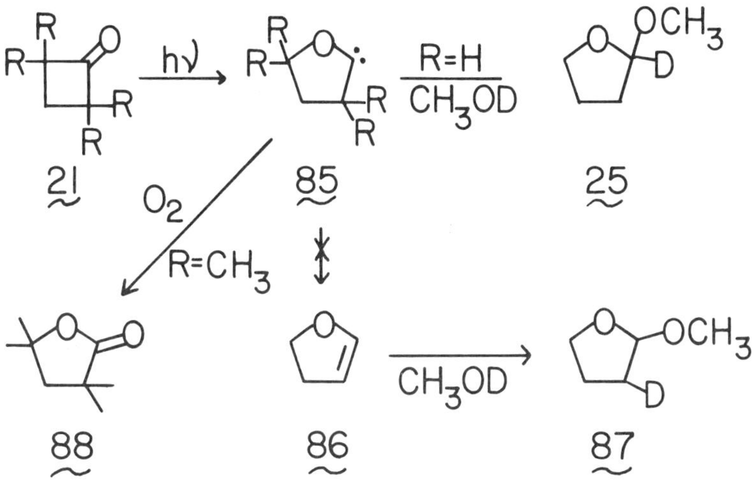

observation was made for 2-isopropylidenecyclobutanone [79a]). The cyclic acetal [87], which was independently prepared from 2,3-dihydrofuran [86], was shown to be non-identical to the isolated photoproduct [25], a result which eliminated the possibility of oxacarbene [85] undergoing prior isomerization to [86] followed by acid catalyzed addition of CH_3OD across the double bond. Irradiation of 2-isopropylidenecyclobutanone [89] in dry benzene afforded a crystalline product in moderate yield

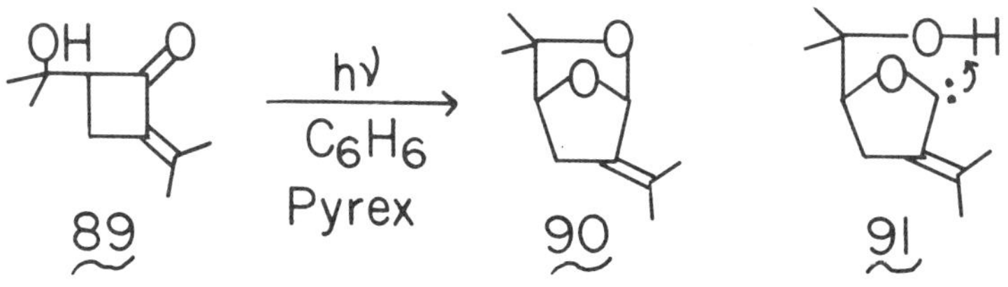

which was subsequently characterized as [90] (15b,36). The formation of [90] is most easily explained by intramolecular trapping of oxacarbene [91].

The behavior of the oxacarbenes derived photolytically from cyclobutanones is somewhat analogous to that exhibited by so-called nucleophilic carbenes (40). Dioxacarbene [92] (41), for example, is efficiently trapped by methanol to produce trimethyl orthoformate [93] as the only detectable product. In the absence of

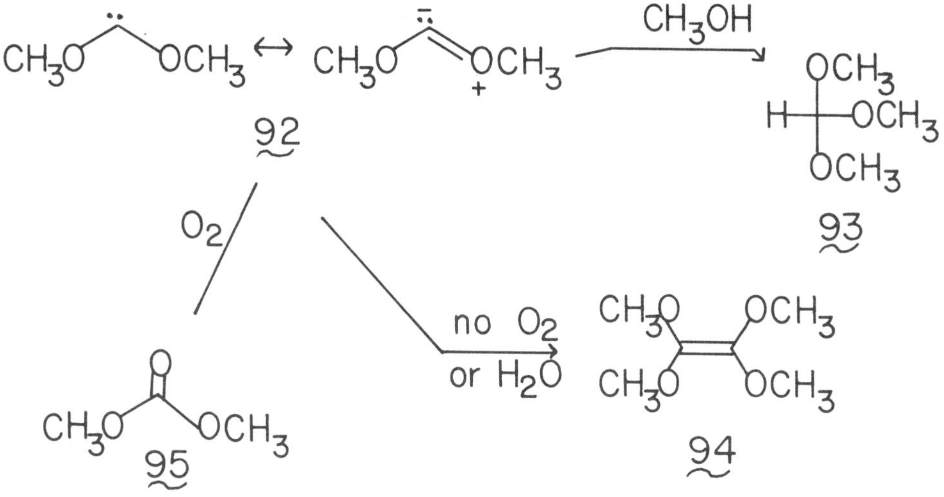

water, alcohol, and oxygen, [92] undergoes dimerization to [94], while in the presence of molecular oxygen, dimethyl carbonate [95] is produced. Although dimers of oxacarbene [85] have not been isolated to date, irradiation of 2,2,4,4-tetramethylcyclobutanone [26b] in aerated benzene has been shown to afford 2,2,4,4-tetramethyl-γ-butyrolactone [88], in addition to isobutylene (β-cleavage product) and 1,1,2,2-tetramethylcyclopropane (15). The formation of lactone [88], like that of [95], is presumably due to a reaction between the carbene species and molecular oxygen. Such reactions find precedent in the work of Bartlett and Traylor (42), where it was demonstrated that irradiation of diphenyldiazomethane in the presence of oxygen ultimatel afforded benzophenone.

Of additional interest is the observation that oxacarbenes, such as [85], are not efficiently trapped by typical olefinic substrates to produce cyclopropanes, a reaction which is usually characteristic of carbene species Yates and Kilmurry, however, have isolated the polycyclic ether [96] (in low yield) from a cyclohexene solution in which oxacarbene [97] was photolytically generated (9,38).

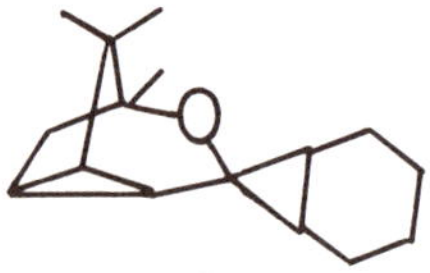

96

97

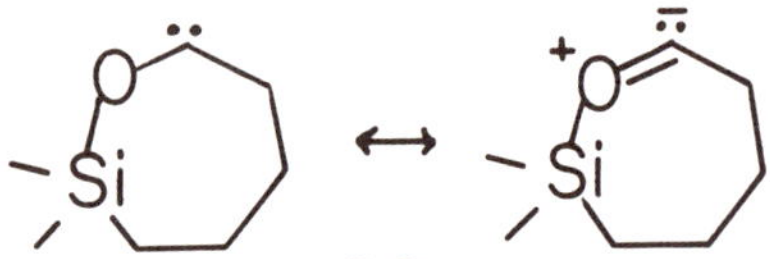

98

Further, Brook and coworkers (43) were able to trap siloxacarbene [98] by olefinic substrates, although only when these substrates were relatively electron deficient (e.g., diethyl fumarate).

Although a number of ethylenes (tetramethylethylene, dicyanoethylene, maleic anhydride) have been employed as potential carbene traps, the only successful example of such trapping is the formation of [99] when [79a] is irradiated in cyclopentadiene [see reference (13) for a

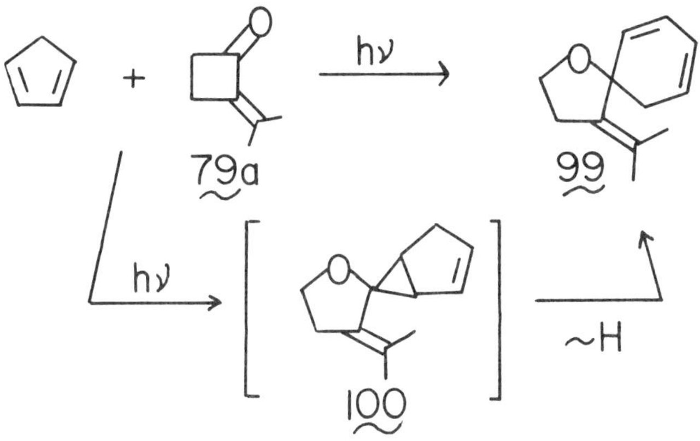

possibly related trapping].

More work on the nature of product formation from cyclobutanones, such as [79a], in inert solvents is required in order to better understand the fate of the presumed oxacarbene intermediate. At any rate, it is clear that ring expansion <u>can occur in the absence of hydroxylic solvents</u>, thereby eliminating a special role for protonic solvents.

The apparent reluctance of oxacarbenes like [85] to add to olefinic substrates is difficult to explain and must await further systematic experimentation. In any event, the intermediacy of such species in the photochemical ring-expansion reactions of cyclobutanones seems secure.

The behavior of the proposed 1,4-biradical from S_1 of cyclobutanones should be contrasted to that of the biradical produced from T_1 of cyclopentanones. The latter biradical (a) lives long enough to lose stereochemistry (10) about the C_2-C_3 bond and (b) undergoes intramolecular hydrogen abstraction and ring closure as the only significant reactions for stabilization (10). Thus the lack of ring expansion, decarbonylation, and β-cleavage is striking. Only the last reaction is easily dismissed as being "special" for cyclobutanone, since β-cleavage leads to two stable molecules in this case, but not so for cyclopentanone.

Another interesting contrast is afforded by the rarity of ring expansion of cyclopentanones. 2,2,5,5-tetramethylcyclopentanone [101], for example, shows no tendency to undergo ring expansion in methanol (15). A few examples of ring expansion for cyclopentanones are known in polycyclic systems [for an exception, see reference (25)] (9).

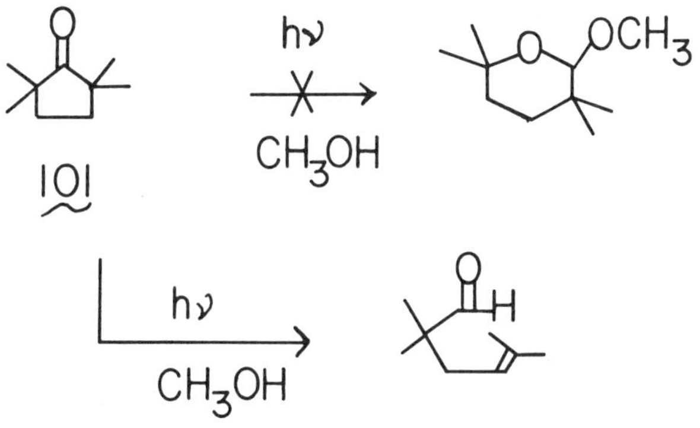

An interesting question which has not been clarified concerns whether formation of the oxacarbene intermediate is reversible, i.e., whether the reaction [31] → [26a] occurs.

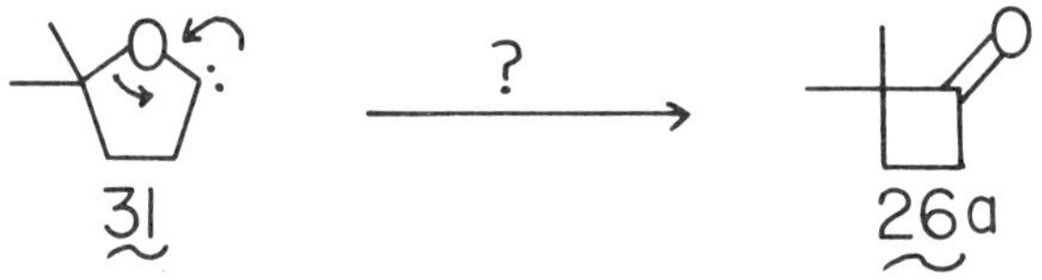

That such a reaction can occur was clearly demonstrated by the recent work of Agosta (50), who showed that thermal decomposition of the salts of tosylhydrazones, such as [102] (a process which has good precedent as a

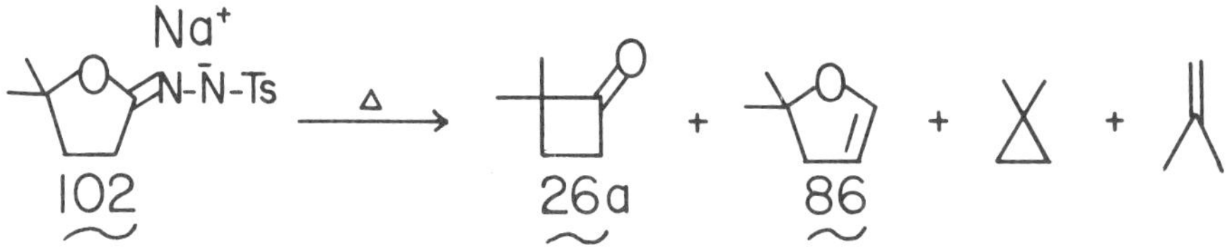

carbene-forming reaction), results in the formation of cyclobutanones. Thus it seems clear that [31] can undergo a 1,2-carbon bond shift to yield [26a]. Interestingly, the product of a 1,2-hydrogen shift [86] is also formed; however, this product is absent from photolysis of [26a].

The occurrence of cyclopropane and β-cleavage products in the thermolysis of [102] (50b) suggests that the oxacarbene might be the immediate precursor of decarbonylation and β-cleavage products when [26a] is photolysed. The possibility of decarbonylation from [31] is of particular interest, since decarbonylation of a biradical seems somewhat incompatible with published activation energies for decarbonylation of acyl radicals ($CH_3\dot{C}O \rightarrow \dot{C}H_2 + CO$, $E_a \sim 15$ kcal/mol) and with the absence of decarbonylation as an important path in cyclohexanone and cyclopentanone photochemistry (10). It might be

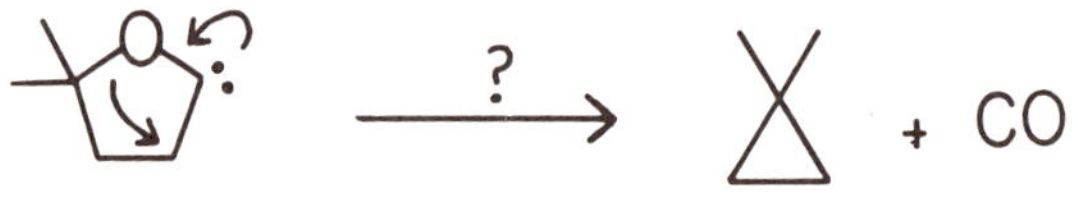

noted that the oxacarbene is isoelectronic with pyrazolines

which are well known to undergo loss of N_2 (64).

Support for reactions like [31] → [26a] is also obtained by recalling the results of Klemm (16) and Lee (17) who found that the quantum yield of ethylene formatio from cyclobutanone was less sensitive to solvent changes than that for cyclopropane formation (Table II). Furthermore, it has been shown (35) that in the case of 2-isopropylidenecyclobutanone, a case which yields only carbene-derived products in methanol, a substantial decrease in the quantum yield for disappearance of ketone is effected by going from methanol to a hydrocarbon solvent (Table III).

TABLE II
Solvent Dependence of Product Formation from Cyclobutanone Photolysis (313 nm Excitation)

Solvent	$\phi(C_2H_4)$	$\phi(C_3H_6)$	Reference
$c\text{-}C_6H_{12}$	0.18	0.10	16
$n\text{-}C_7H_{16}$	0.18 (0.12)	0.14 (0.09)	16,17
CH_3OH	0.14 (0.07)	0.08 (0.03)	16,17

These results can be interpreted as true "solvent effects" on the partition of some intermediate or, more specifically, as evidence for reversible formation of an oxacarbene, and its potential participation in decarbonylation. In any case, some revealing information might be given by experiments in which the yields for all products and quantum yields for cyclobutanone disappearance are measured in an "inert" solvent as a function of added methanol.

F. Photochemical Rearrangements and Miscellaneous Reactions of Cyclobutanones

In addition to the commonly observed photoprocesses of β-cleavage, decarbonylation, and ring expansion, rearrangements, in particular [1,3]-acyl shifts, and

TABLE III
Quantum Yields[a,b] at 313 nm for Disappearance of Ketone

Ketone	Solvent	$\Phi_{-k}(\Phi_A)^c$	Reference
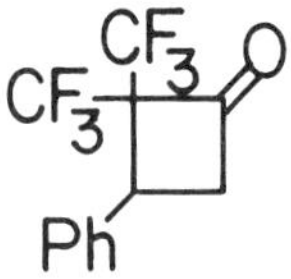	CH_3OH	0.23 (0.00)	35
	CH_3OH	0.30 (0.02)	46 (35)
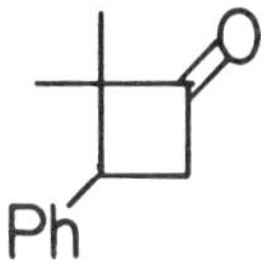	CH_3OH	0.22 (0.15)	35
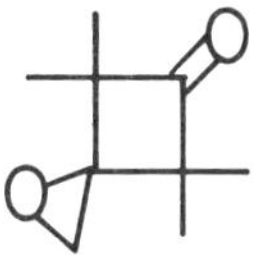	CH_3OH	0.24 (0.21)	35
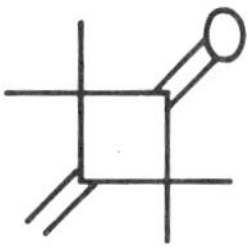	CH_3OH	0.27 (0.24)	35
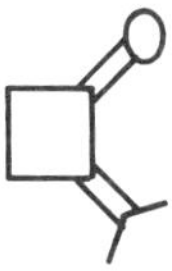	CH_3OH	0.11 (0.11)	35
	$c\text{-}C_6H_{12}$	0.07	35
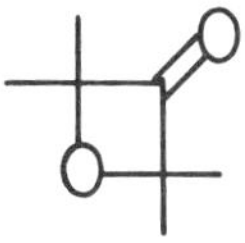	i-PrOH	1.0	45

TABLE III
(cont.)

Ketone	Solvent	$\Phi_{-k}(\Phi_A)$	Reference
	C_6H_6	0.07	46
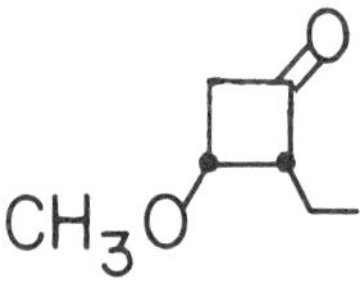	CH_3OH	0.18 (0.03)	24
	C_6H_6	0.18	
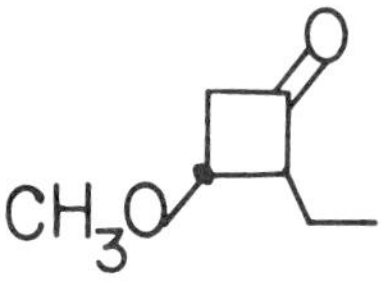	CH_3OH	0.18 (0.06)	24
	C_6H_6	0.17	
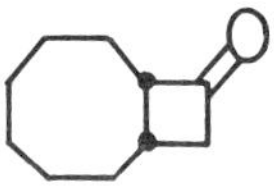	CH_3OH	1.1	24
	C_6H_6	0.64	
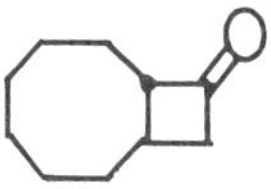	CH_3OH	0.54	24
	C_6H_6	0.48	
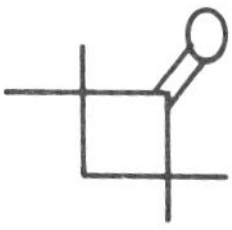	CH_3OH	(0.14)	35

[a]For nondegassed solutions of ketone in the specified solvent at 313 nm.

[b]Unless specified, the limits of error are estimated at ± 10-20%.

[c]Quantum yield for acetal from ring expansion.

unsaturated aldehyde formation have been observed for certain cyclobutanones. The rearrangement (34) of [74] to [76] was an example of a [1,3]-acyl shift. This reaction appears to be commonly encountered for β,γ-unsaturated cyclobutanones and is often competitive with the more general primary photoprocesses. For example, the cyclobutanone [103] is involved in a photoequilibrium with the enone [104]. In methanol, the decarbonylation product [105] and the ring expansion acetal [106] are also formed (65). Interestingly, no β-cleavage is observed in this case.

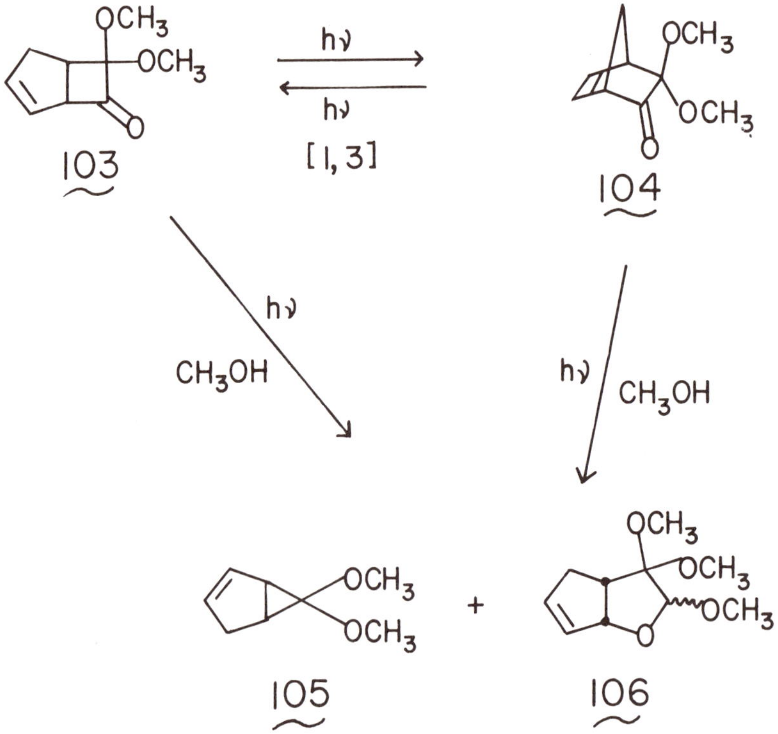

A number of examples of reversible [1,3] rearrangements of bicyclic ketones of the type [107] → [108] are known (21).

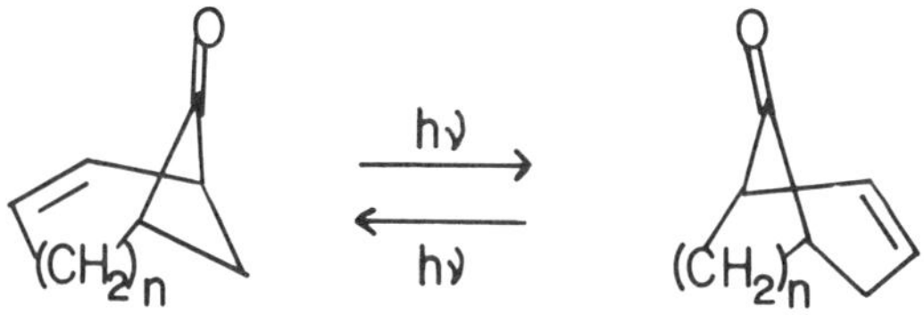

107 108

It is important that the photosensitized (triplet) (66) rearrangement of [109] leads to polycyclic ketone [111] by a [1,2] shift and not the [1,3] shift (direct irradiation) (67) product [110]. These results are analogous to the behavior of many β,γ-unsaturated ketones:

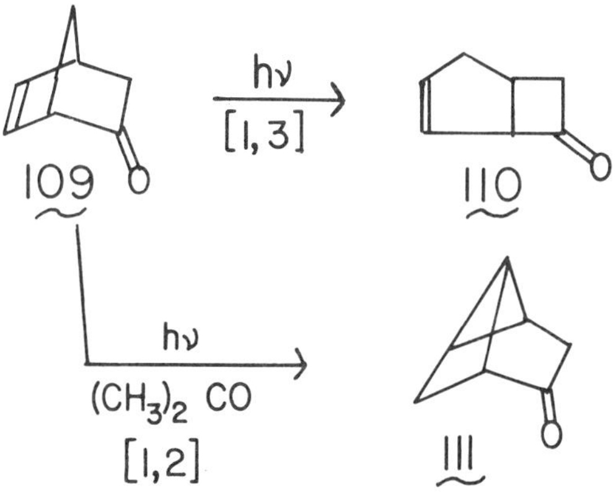

the [1,3]acyl shift occurs under direct excitation, and the [1,2] shift occurs under sensitized excitation (68). These data are interpreted as follows: the singlet state prefers a [1,3] shift, while the triplet prefers a [1,2] shift.

The [1,3] shift may be concerted (69), however, its occurrence from a wide variety of β,γ-unsaturated ketones with very different structural requirements for the rearrangement, suggests that α-cleavage occurs from S_1 to yield a biradical which can rearrange or undergo recyclization, β-cleavage, and ring expansion. The observed ratio of these paths is structure dependent.

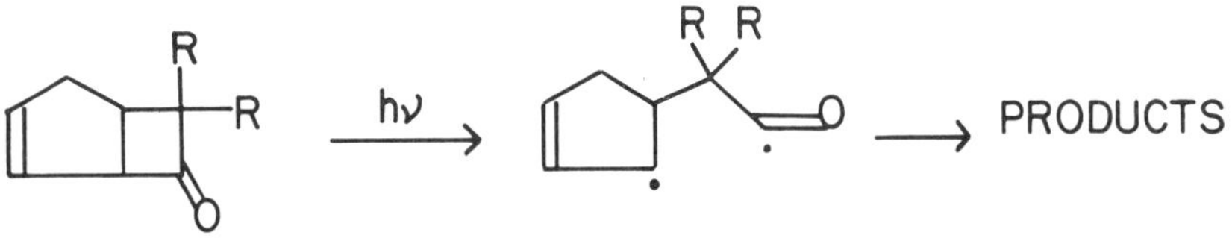

To date, no photosensitized rearrangements of β,γ-unsaturated cyclobutanones have been reported. Such studies are clearly in order.

Irradiation (44) of the hydroxycyclobutanone [112] yields the aldehyde [113]. This reaction probably proceed in analogy to the unsaturated aldehyde formation from cyclopentanones, as shown.

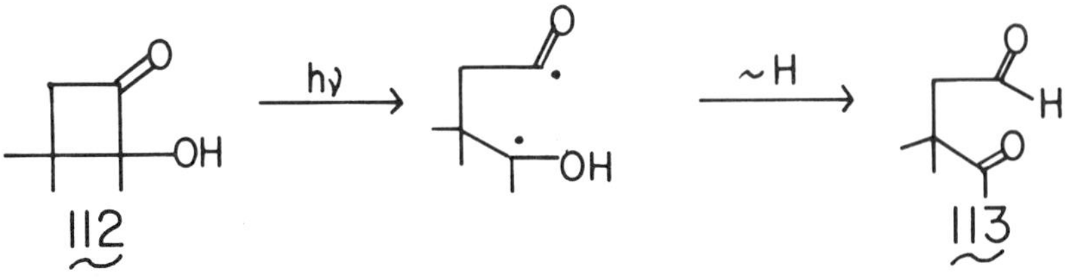

The [1,3] acyl rearrangement of cyclobutanones and the formation of ketoaldehydes both represent formal intramolecular trapping of biradical intermediates.

G. Photochemistry of 1,3-Cyclobutanediones

The photochemistry of tetrasubstituted 1,3-cyclobutanediones (70) is of considerable interest because of (a) an exceptional interaction (71) in S_1 as indicated by the absorption spectra of these compounds, which exhibit an unusual absorption band near 350 nm, and (b) the fact that decarbonylation leads to cyclopropanones (72).

Irradiation of tetramethyl-1,3-cyclobutanedione in nonhydroxylic solvents results in decarbonylation as the major course of reaction with β-cleavage being observed as a minor reaction path. A low yield of the lactone [115] has also been reported (73); however, since this material might result from dimerization of dimethylketene, it may not be a primary photoproduct (i.e., ketene dimer is isolated from irradiation of [69].

Since the cyclopropanones formed are photolabile, undergoing smooth decarbonylation to ethylenes, it is not possible to obtain high yields of cyclopropanones in inert solvents (70).

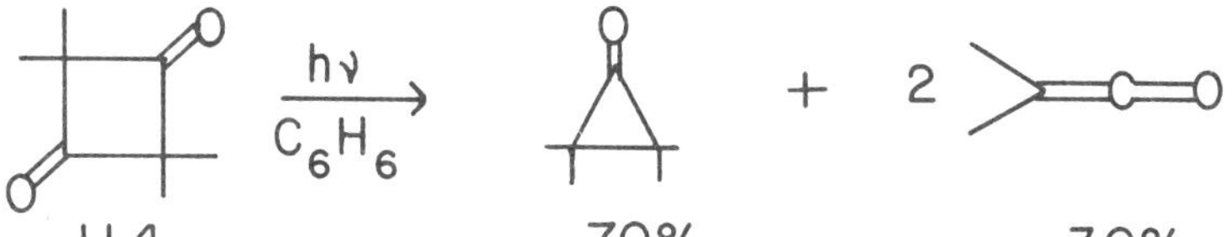

114

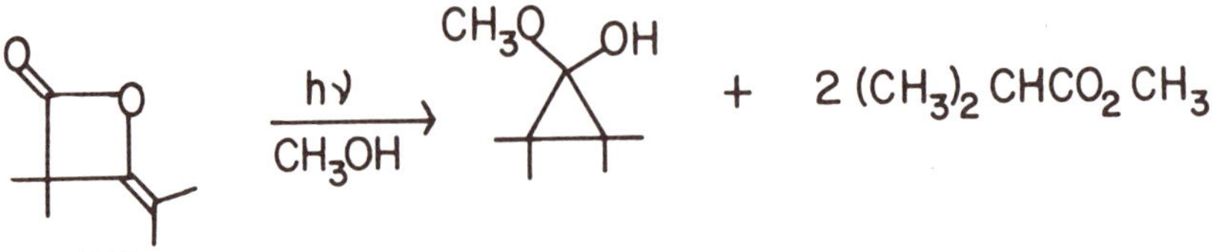

115

The lactone [115], when irradiated in methanol, yields the same product distribution as does [114] in this solvent (61). This result suggests a common intermediate in the two photoreactions, possibly the biradical [116]. Note that for the same nuclear geometry, [116a] and [116b] are simply resonance forms.

116a 116b

It is quite striking that, in contrast to 3-methyle cyclobutanones, [114] *does not undergo ring expansion in methanol*. The expected acetal product, synthesized independently from [120] was shown to be absent from the photolysis solutions by thin-layer chromatography (15b).

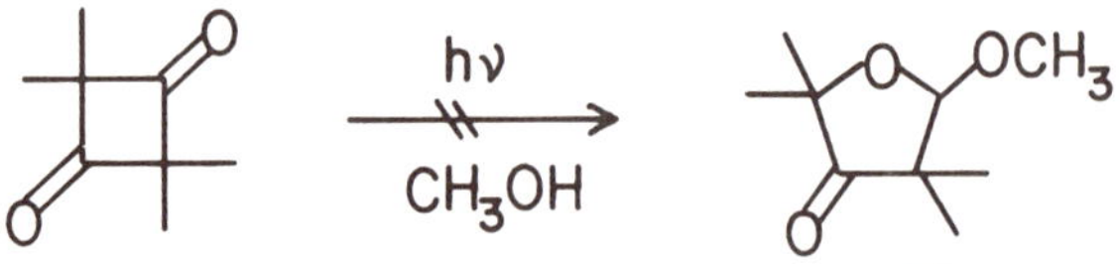

114 117

The cyclopropanone imine [119] is probably produced by photodecarbonylation of [118] in methanol since the adduct [121] is isolated (13). The ring-expansion product is also produced as the major isolated product. Irradiation of [118] in inert solvents is reported to yield no decarbonylation products (13).

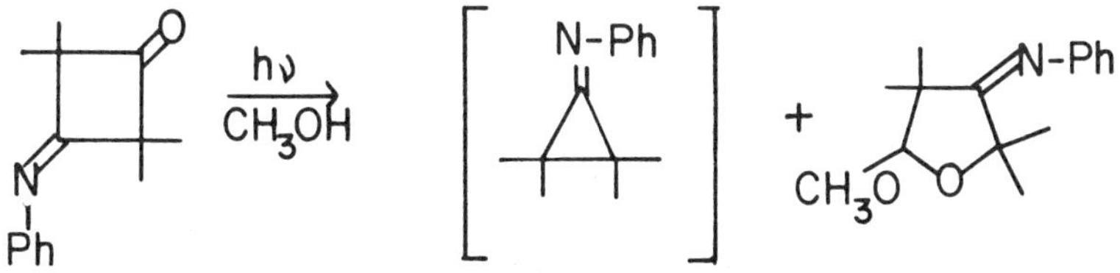

118 119 120, 30%

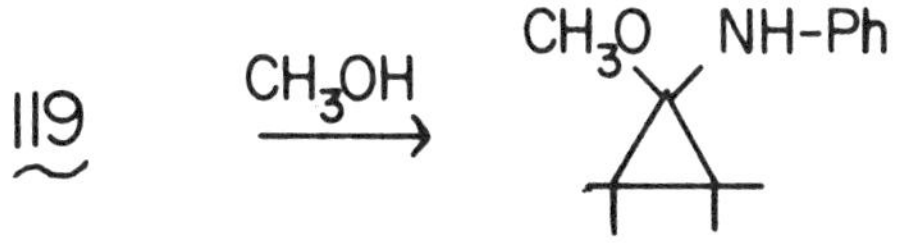

121, 20%

So far, attempts to sensitize decomposition of 1,3-diones by ketones have failed, as have attempts to quench their photochemistry (70).

H. Detection of Intermediates at -196°C by ESR and Optical Spectroscopy

There are several reports of the production of radical or biradical species when cyclobutanones are irradiated in an ESR spectrometer. We have already mentioned the detection of triplet trimethylenemethane from photolysis of [69] at -196°C. The ESR of [116b] is observed when 1,3-pentadiene is solvent.

Parenthetically, it has been reported (13) that irradiation of [114] and [26b] at -196°C leads to detection of a septet and singlet, presumably due to the acyl radical and tertiary radical. These results merit further study.

116b 26b

Irradiation (56) of tetraphenyloxetanone [122] at -196°C results in the formation of a blue intermediate identified as the ring-opened oxirane. The species is described as a ylid rather than a biradical because no ESR spectrum of the material could be detected at -196°C.

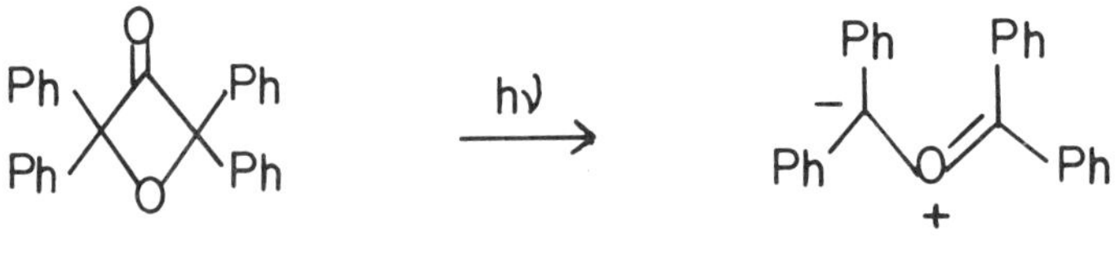

122 123

III.

A. Summary of Empirical Rules Applicable to Cyclobutanone Photochemistry

We can now consider, simply from the empirical standpoint, some "rules" which pertain to the photochemistry of cyclobutanones. (1) Three reactions are commonly observed: β-cleavage, decarbonylation, and ring expansion (15); (2) each of these three reactions is (a) stereoselective or stereospecific and (b) regiospecific, the most substituted (saturated) α-carbon cleaving preferentially (15); (3) substitution of α-alkyl agroups favors ring expansion as the major reaction course (15); (4) none of the reactions is quenched by 1,3-dienes (15) or 1,2-dicyanoethylene (35); (5) 2- or 3-isopropylidene or spiro-cyclopropyl substitution favors ring expansion (15,25); (6) β,γ-unsaturated cyclobutanones undergo a photochemical [1,3] acyl shift in addition to the three general reactions of (1) (21); (7) bimolecular photoreactions, in which the four-carbon ring is preserved, are very rare; (8) quantum efficiencies for cyclobutanone disappearance (and product formation) are less than unity (35,46); (9) fluorescence lifetimes of cyclobutanones (excitation at 313 nm) are considerably shorter than those of other cyclic alkanones (i.e., cyclopentanone) (47,48).

B. A General Mechanism Suggested by the Empirical Rules

Considering the "empirical rules" developed in the previous section, we can develop a simple general hypothesis to help us to understand all the results which conform to these rules: direct excitation of cyclobutanones results in the formation of a reactive singlet state which predominantly undergoes α-cleavage to a singlet biradical, i.e., [124] → [125] → [126].

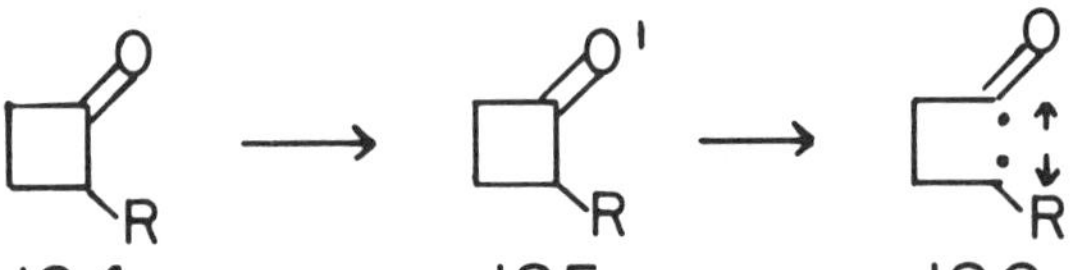

124 125 126

Consideration of α-cleavage as the key primary photochemical step explains the observed photoreactions as logical stabilization paths of [126] (Rule 1). Recyclization of [126] to the starting ketone explains the nonunit quantum efficiencies of photolysis (Rule 8). The

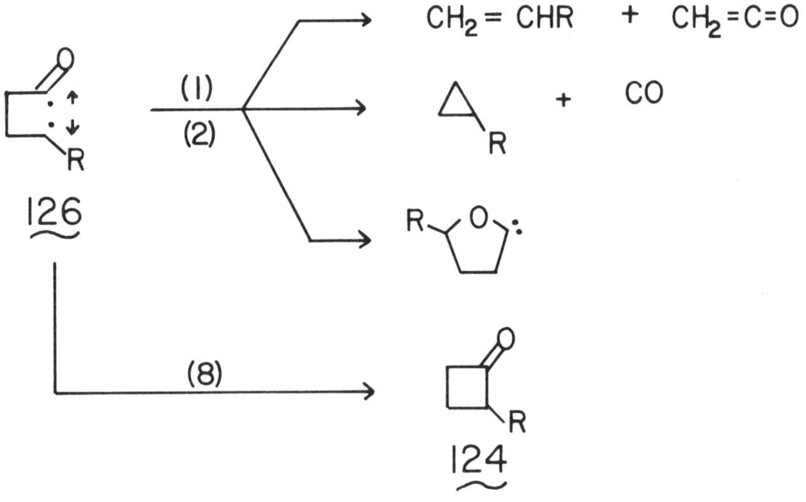

cleavage should occur to break the more substituted bond (Rule 2b), because such a cleavage produces the more stable biradical. Furthermore, since a singlet biradical is involved, its stabilization reactions are all stereospecific (Rule 2a).

Substitution probably favors ring expansion (Rule 3) because the alkyl radical makes a nucleophilic attack on the acyl radical (*vide infra*). Because ring strain is relieved in the biradical forming step, the singlet lifetime is α-cleavage limited (Rule 9), not intersystem crossing limited, and neither good singlet alkanone traps (1,2-dicyanoethylene) nor triplet traps (1,3-dienes) can effectively quench cyclobutanone photochemistry (Rule 5) [i.e., the ring is broken faster than bimolecular reactions can occur (Rule 7)]. In the case of 2- or 3-isopropylidene or spirocyclopropyl substituents, the biradical prefers ring expansion or decarbonylation because cycloelimination involves breaking an sp^2-sp^3-like bond. β,γ-unsaturated ketones possess an "intramolecular trap" for the initially formed biradical, and [1,3] shifts occur in competition with the three general reactions of cyclobutanones (Rule 6).

It thus appears that, in a rough and ready way, the simple hypothesis of α-cleavage in S_1 allows a qualitative understanding of Rules 1-9. There are, however, exceptions to the rules.

C. Real and Apparent Exceptions to the Rules of Cyclobutanone Photochemistry: Mechanistic Implications

The real test of a good general mechanism is its ability to withstand the assaults of "exceptions." If natural arguments can be employed to understand the exceptions, or if the exceptions are of an "apparent" or trivial nature, the general mechanism will be only qualitative in nature. This eventuality is demanded not only by the coarseness of the model, but also by the lack of quantitative rate data with which to make a correlation.

Among the most powerful arguments for a singlet biradical intermediate are rules 1, 2, and 3 (Section III. A), since there is much prior analogy for the correlation of radical stabilities and cleavage rates. The major exceptions to Rule 1 are (a) [1,3] acyl shifts of β,γ-unsaturated cyclobutanones (21) and (b) unsaturated aldehyde formation from 2-hydroxycyclobutanones (44). However, both of these exceptions can be turned into <u>support</u> for the singlet biradical <u>since they represent logical alternate routes for biradical stabilization which become significant due to structural modifications</u>.

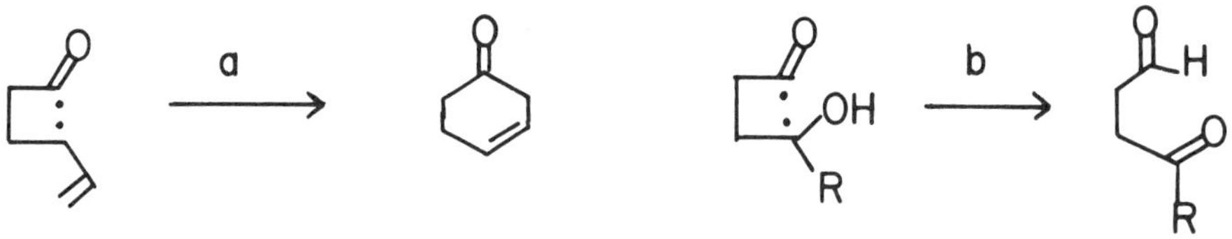

The most serious violation of Rule 2a is the non-stereospecific formation of 1,2-dimethylcyclopropane from irradiation (24) of cis-2,3-dimethylcyclobutanone ([50c]) and trans-2,4-dimethylcyclobutanone ([50t]). Interestingly the cycloelimination products, from excitation of cis- and trans-2,3-dimethylcyclobutanone, are formed in a relatively stereospecific manner (see Section II.A). We can fit this result in with our first-order mechanism if we allow for greater intersystem crossing of S_1 to T_1 of the dimethylcyclobutanones, which thereby causes the decarbonylation to pass through a triplet and (presumably) stereochemically labile intermediate. Such a possibility would be quite consistent with work on the gas-phase photolyses of dimethylcyclobutanones (74) and with the recent result of Morton, that (triplet) xylene sensitization results in a greater yield of cyclopropane products (35). In the vapor photolysis, a pronounced wavelength effect on the ratio of cyclopropane to ethylene (1.3 for λ_{ex} = 313 nm and 3.4 for λ_{ex} = 334 nm) is consistent with greater intersystem crossing from S_1 and greater yields of decarbonylation at lower excitation energies (74,75).

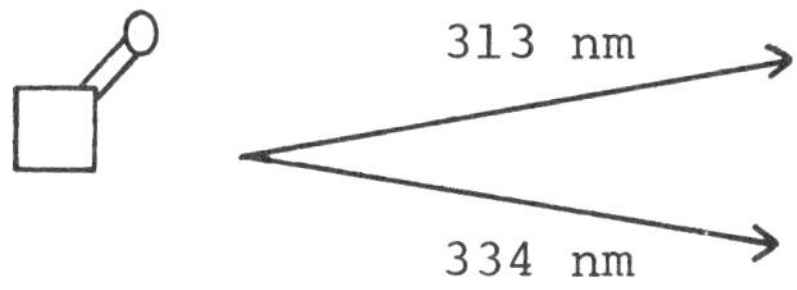

$\Delta/CH_2{=}CH_2$ = 1.3

$\Delta/CH_2{=}CH_2$ = 3.4

An interesting point concerning cyclopropane formation is whether this reaction generally occurs from a triplet state or triplet biradical. Thermolyses of cyclobutanones lead to very minor amounts of cyclopropanes, at best. However, the stereospecificity of cyclopropane formation would appear to be inconsistent with the biradical lifetime required to achieve activation for loss of CO. This conclusion derives from the report that acetyl radicals require an activation of about 15 kcal/mole for decarbonylation (76). Even assuming an A factor of about 10^{13} sec^{-1}, such an activation would produce a decarbonylation rate of about 10^3 sec^{-1} at room temperature. Although propionyl radicals will surely have a lower E_a, it seems unlikely that it will be low enough to push the rate of decarbonylation up to 10^{10} sec^{-1} (10 kcal/mole stabilization), which is apparently required to explain the stereospecificity of cyclopropane formation. More work is required to unravel this interesting point. In any event, it should be noted that the facile loss of CO from many cyclobutanones contrasts with the lack of decarbonylation in simple cyclopentanones (10), systems which most certainly proceed via triplet states and triplet biradical intermediates.

Until very recently no serious exceptions to Rule 2b, regiospecificity, were reported. The report of Miller (77) that cyclobutanones fused to bicycloheptene and bicyclo-octene skeletons do not lead to major products expected from cleavage at the most substituted α-carbon, represents a serious violation of Rule 2b.

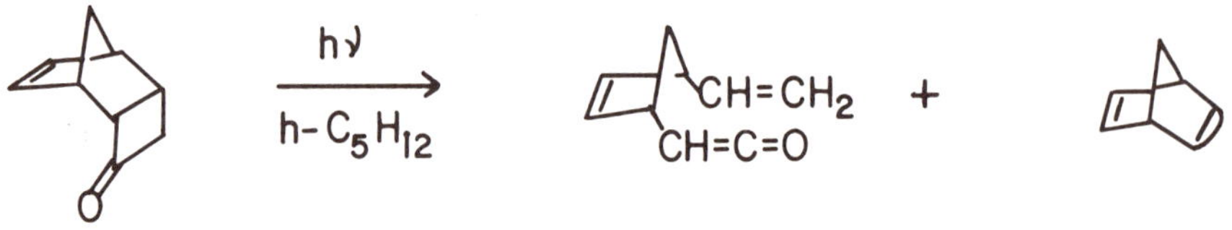

127 128, 50% 1%

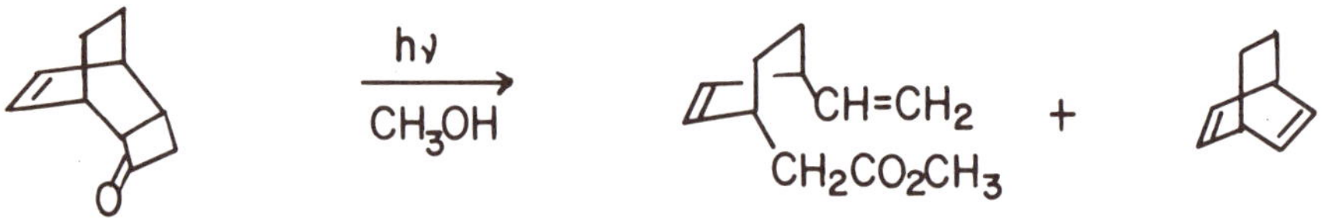

129 130, 17% 131, 7%

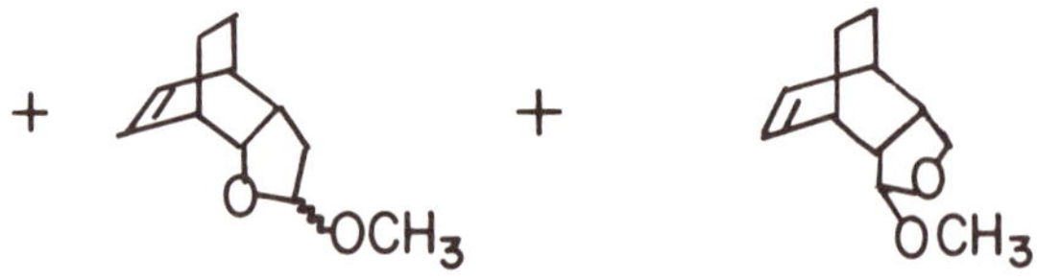

132, 53%

We can understand these violations on the basis of two possibilities: (a) cleavage does occur preferentially on the most substituted α-carbon, but this process is highly reversible relative to cleavage at the less substituted carbon, thereby allowing for a high net efficiency for product formation from the less stable biradical; (b) cleavage at the less substituted carbon is favored for some special structural feature. Both of these explanations are really connected to the idea that a torque is placed on the bond common to the cyclobutanone and bicyclic skeleton which favors concerted cleavage in the unexpected direction.

Two important exceptions to Rule 3 are found in the photochemistry of (a) 1,3-cyclobutanediones (70) and (b) 2,2-bis(trifluoromethyl)-3-phenylcyclobutanone (25). In neither case is there any evidence for ring-expansion products! However, a simply unifying argument can be presented to bring these exceptions back to the fold. The ring-expansion step evidently involves a nucleophilic attack of the carbon radical on the oxygen of the acyl moiety of the biradical intermediate. For both [116b] and [133], the nucleophilicity of the attacking carbon radical is greatly reduced by an adjacent strong electron-withdrawing function. Therefore ring expansion is inhibited relative to the other paths of biradical stabilization. This point is discussed in greater detail below.

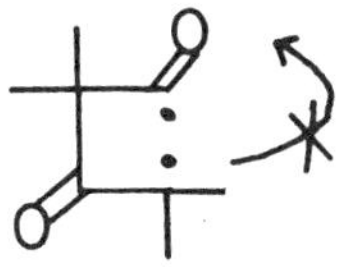

116b

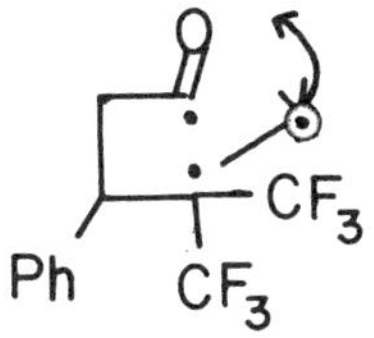

133

Finally let us consider the very unusual bimolecular addition of dienes to cyclobutanones (34), a flagrant violation of Rule 7. The reactions which were run at -78°C can be understood if we assume that the rate of cleavage of S_1 to a biradical is slowed down because of a decrease in the available thermal activation. However, a moderate yield of oxetane is formed from cyclobutanone and butadiene (19) at 10°C. From work on the quenching of fluorescence by alkanones (53), we anticipate the rate of

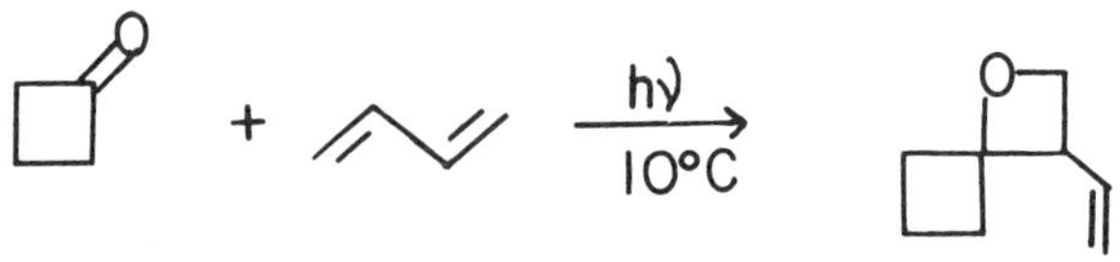

34, 13%

quenching of S_1 of cyclobutanone in neat butadiene to be about 10^8 to 10^9 sec^{-1}. Even if the larger value obtained, oxetane quantum yields are usually much less than unity, so it is quite interesting that oxetane is formed in moderate yields.

D. Causes of Quantum Inefficiency in Cyclobutanone Photochemistry: A Clue to the Mechanism

To obtain a detailed understanding of the photochemistry of cyclobutanones in solution, we must understand the cause of the general quantum inefficiency exhibited by cyclobutanones. As we have seen, there is substantial evidence to support the idea that S_1 deactivates very efficiently (for saturated cyclobutanones). We can identify four processes which might be attributed as the cause of inefficiency: (a) intersystem crossing to T_1;

(b) internal conversion to S_o* (vibrationally "hot" ground state); (c) α-cleavage to a biradical; and (d) rearrangement to an oxacarbene intermediate (Scheme III). In other words, we can imagine a direct path from S_1 back to ground state and a number of indirect paths. It is also possible to conceive of less likely paths, such as S_o* being a precursor to a biradical intermediate or an oxacarbene. Path (a) is rejected as a general pathway since the triplet of cyclobutanone apparently cleaves to yield a triplet biradical capable of scrambling starting material, stereochemistry, and product stereochemistry. Path (b) is rejected because the emission spectrum and fluorescence "lifetime" of cyclobutanone (λ_{ex} 320 nm) is very similar to that of cyclopentanone, which is known not to undergo measurable internal conversion. This leaves paths (c) and (d). Only the biradical hypothesis explains in a logical, attractive, and intuitive way, all of the general empirical rules of cyclobutanone photochemistry. Furthermore, such an intermediate is a natural extension of the photochemical α-cleavage of other cyclic alkanones (i.e., cyclopentanone and cyclohexanone). Even though a singlet biradical, B_1, should result from α-cleavage of S_1, the selectivity of α-bond cleavage corresponds precisely to that predicted by "more rapid cleavage of the bond which produces the more stable biradical." The inefficiency involved is explained because closure occurs rapidly (greater than diffusion controlled) from a system which already has the proper geometry for recyclization. Because B_1 is produced in a shape fairly close to that required for β-cleavage and ring expansion, it seems quite sensible that these processes are stereospecific. The only process which seems a little peculiar, in terms of stereospecificity,

SCHEME III
Possible Sources of Quantum Inefficiency in Cyclobutanone Photochemistry

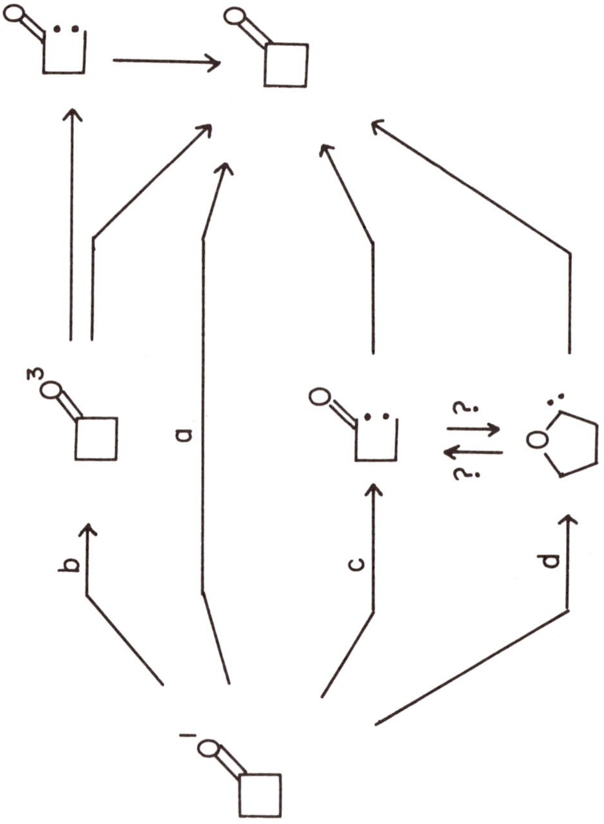

is decarbonylation (*vide supra*, Section III.C). This line of thinking allows one to imagine concerted reactions which possess an energy minimum along the reaction coordinate.

Only the *triplet* biradical from cyclobutanone would correspond to the biradical many chemists think of, i.e., one with a bond order of zero between C_1 and C_2 and of 1 between C_3 and C_4.

Of course, at this point, one can ask, "Why call the cis singlet biradical a 'biradical'?" Apparently, according to our formulation of the species, it is a "virtual biradical," since its chemical *bonding* reaction coordinate might have a minimum as one proceeds from reactants to products. This minimum might result from "loosening" up of vibrations as one proceeds along the reaction coordinate or because of some special electronic effects. In terms of bonding alone, we can imagine that the formation or breaking of one bond lags somewhat behind the other. We now see that the usual concepts of "concerted" and "biradical" can merge when the amount of bonding between two centers becomes less, but not "much less," than the bonding between two other centers.

B

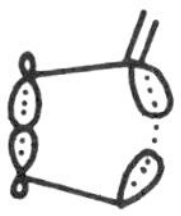

β-Cleavage

Ring Expansion

E. Biradical versus Concerted Mechanisms

The problem of deciding on the mechanism is especially challenging because it has many of the problems inherent in a discussion of a reaction pathway which may be either concerted or proceed via a biradical intermediate. To illustrate the type of problems involved, consider the reaction coordinate and bond order profile shown in Figure 5. On a free-energy curve, one can imagine that a totally

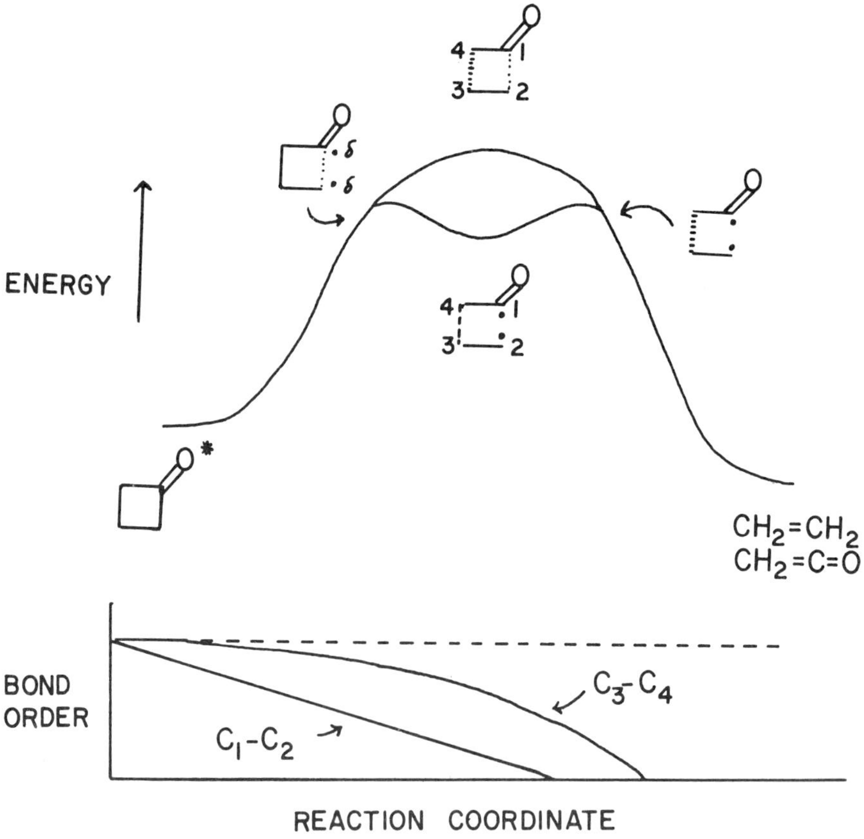

FIGURE 5. Reaction coordinate versus energy and bond order for β-cleavage.

concerted reaction is identical to that of competing reactions of S_1. While this is true, the organic photochemist can bring two important pieces of "nonground-state" evidence to bear on the problem: (a) absorption and emission spectroscopy and (b) detailed energy balance.

From a comparison of the spectroscopy of cyclobutanone to that of cyclopentanone, we see apparently similar absorption and emission predissociative behavior for cyclobutanone when $\lambda_{ex} > 313$ nm, while no such effect is found for cyclopentanone. If we suggest that three competing pathways account for the predissociative behavior, we have done little more than to offer a detailed but uninformative representation of the observations. Furthermore, we would be forced into adding more primary acts when interpreting the [1,3]-shifts and aldehyde-forming reactions discussed above.

A detailed energy balance also suggests an intermediate which can regenerate the starting ketone, since the quantum yield for all photoprocesses is generally significantly less than unity. The "competing concerted pathways" mechanism again provides no significant insight into the source of this inefficiency.

It is interesting to note that the E_a for cleavage in S_1 appears to be about 4-5 kcal/mole for cyclobutanone, but about 15 kcal/mole for cyclopentanone.

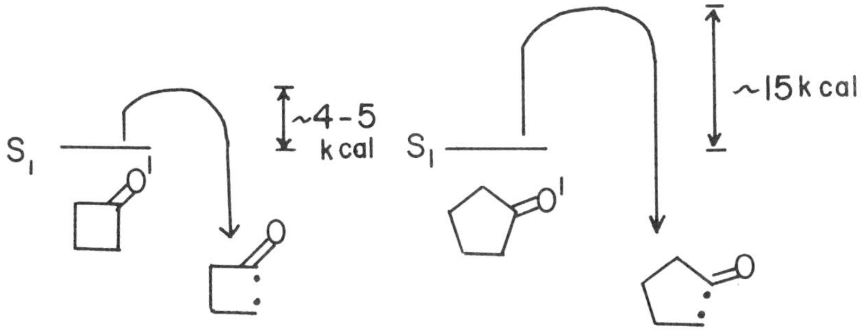

IV. MECHANISTIC CONCLUSIONS

The excited states involved in the mechanisms of the photochemical transformations of the 2-alkylidenecyclobutanones were elucidated without special difficulty (*vide supra*). These α,β-unsaturated ketones undergo only one of the reactions characteristic of cyclobutanones (i.e., ring expansion). In addition, the triplet energy of the enone chromophore is low enough that this excited state may be efficiently and selectively populated by standard carbonyl triplet sensitizers (e.g., acetophenone, xanthone, and benzophenone), thereby demonstrating that ring expansion occurs via the 2-alkylidenecyclobutanone S_1 state, while the isomerization process (i.e., [81]$\rightleftarrows$[82]) occurs via T_1.

Attempts to elucidate the photochemical mechanism(s) of the saturated cyclobutanones, on the other hand, have been less rewarding, since definitive sensitizing and quenching experiments have not been realized. It has been emphasized that a distinct mechanistic crossover occurs on going from cyclopentanone to the next lower homologue, cyclobutanone. In the former case, the only observed reactions are those due to intramolecular disproportionation of the initially formed acyl alkyl biradical intermediate to an unsaturated aldehyde and a saturated ketene (i.e., formation of [3] and [4] in Scheme I). For cyclobutanone, neither of these types of reactions is observed (44), and instead, decarbonylation, β-cleavage (cycloelimination), and ring expansion prevail. Since it is well established that cyclopentanones undergo photochemical change predominantly, if not exclusively, from their triplet-state manifold (1h,25) it is tempting to suggest that the mechanistic crossover observed on going to cyclobutanone is due to a change in the nature of the excited state leading to products. Indeed, the photochemical ring expansion of the 2-alkylidenecyclobutanones has been shown to occur by way of the ketone S_1 state, and other indirect evidence may be used to implicate S_1 reactivity in saturated cyclobutanones. In 1966, Hammond and coworkers (45) found that the rate of disappearance of oxetanone [134] at 313 nm in a variety of organic solvents could not be quenched by oxygen, naphthalene, or high concentrations of 1,3-pentadiene. In addition, all attempts to

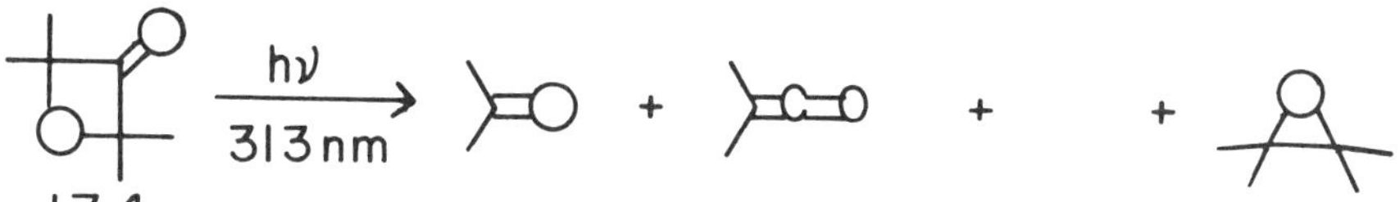

134

sensitize decomposition failed. Similarly, Klemm has found the solution-phase photochemistry of cyclobutanone [21] to be unaffected by 1,3-pentadiene (16). If 1,3-cyclohexadiene was used as a quencher, no evidence for triplet-state dimer formation was found. Finally when pure trans-1,3-pentadiene was used to quench the photochemistry of cyclobutanone, no trans-cis isomerization (a known reaction of the diene T_1 state) of the recovered diene was detected (16,17,46). Turro and McDaniel (23) have also found the photochemistry of cis- and trans-2-ethyl-3-methoxycyclobutanones, [47] and [48], to be unaffected by added 1,3-pentadiene, and Dowd and coworkers (19,34) have presented evidence which suggests that S_1 states of cyclobutanone and 3-methylenecyclobutanone are photoreactive toward addition to 1,3-dienes. In addition, cyclobutanone does not quench the Type II reaction of 2-pentanone under conditions where 1,3-

pentadiene is an effective quencher (46). All this evidence for the photoreactivity of the cyclobutanone S_1 state, however, is not totally satisfying, since reaction could still occur from a short-lived, unquenchable T_1 state. Further, sensitization results may be ambiguous, since suitable triplet sensitizers with $E_T > 78$ kcal/mole are rare.

Support for the S_1 reactivity of cyclobutanones can also be approached from a different point of view. The presence of internal angular strain in cyclobutanone has been used to explain the variation of its fluorescence quantum yield (Φ_f) with excitation wavelength (λ_{ex}) in the vapor phase (47,48). Comparison of the vapor-phase Φ_f values of cyclobutanone, cyclopentanone, cyclohexanone, and acetone at λ_{ex} = 310-320 nm suggests that all four ketones have the same rate of intersystem crossing to T_1 ($k_{ST} \sim 1\text{-}4 \times 10^8$ sec^{-1}). For cyclobutanone, however, Φ_f drops drastically for $\lambda_{ex} \leq 313$ nm while a similar but more gradual drop for cyclopentanone occurs at $\lambda_{ex} < 280$ nm. The corresponding Φ_f values for cyclohexanone appear to be independent of exciting wavelength. Lee and coworkers (47, have interpreted the sharp decrease in Φ_f for cyclobutanone to be due to a competing predissociation mechanism from vibrationally excited S_1. This facile predissociation is described as α-cleavage of the vibrationally and electronically excited ketone to form an acyl alkyl biradical which subsequently decomposes to products. Interestingly, the rate of this cleavage must be extremely rapid ($k \sim 10^{10}$ sec^{-1}), since it must compete with vibrational relaxation ($k \sim 7 \times 10^{12}$ sec^{-1}) (49). From the variation of Φ_f with λ_{ex}, it has been suggested that vapor phase predissociation will effectively compete with intersystem crossing in cyclobutanone when S_1 has $\geq$ 4 kcal/mole excess vibrational energy (47,48). In similar fashion, the S_1 states of cyclopentanone and cyclohexanone must possess ~15 kcal/mole and ~ 25 kcal/mole excess vibrational energy, respectively, for predissociation (i.e., α-cleavage from S_1) to become competitive. The lower energy requirement for predissociation of cyclobutanone relative to its higher homologues is consistent with the greater degree of internal angular strain present in the former.

Lee and coworkers (17) have also investigated the solution-phase photochemistry of cyclobutanone and have found Φ_f (λ_{ex} ~ 310 nm) to be 1/20th of that obtained in the vapor phase. This reduction has been attributed to enhanced predissociation by the interaction of vibrationally excited S_1 with neighboring solvent molecules. If this solvent interaction in addition to ring strain already present in the cyclic ketone both facilitate α-cleavage from the vibrationally excited S_1 state, relative to intersystem crossing to T_1, then it is not unreasonable to propose the cyclobutanone S_1 state as the precursor to some, if not all, of the observed photoproducts.

Any attempt to present a meaningful mechanistic scheme to describe the excited-state chemistry of cyclobutanone, therefore, must be capable of accounting for (a) the apparent photoreactivity of the S_1 state, (b) the regiospecificity of α-cleavage (i.e., that direction which would afford the most stable acyl alkyl biradical intermediate), (c) the apparent stereoselectivity of all reactions, and (d) the inherent inefficiency of all photoreactions, including that for ketone disappearance (see Table III). Consideration of all the available evidence suggests the mechanism set forth in Scheme IV. Excitation ($n \rightarrow \pi^*$) of the ground-state ketone (S_o) initially produces the lowest electronically excited

SCHEME IV

Mechanistic Scheme for Cyclobutanone Photochemistry

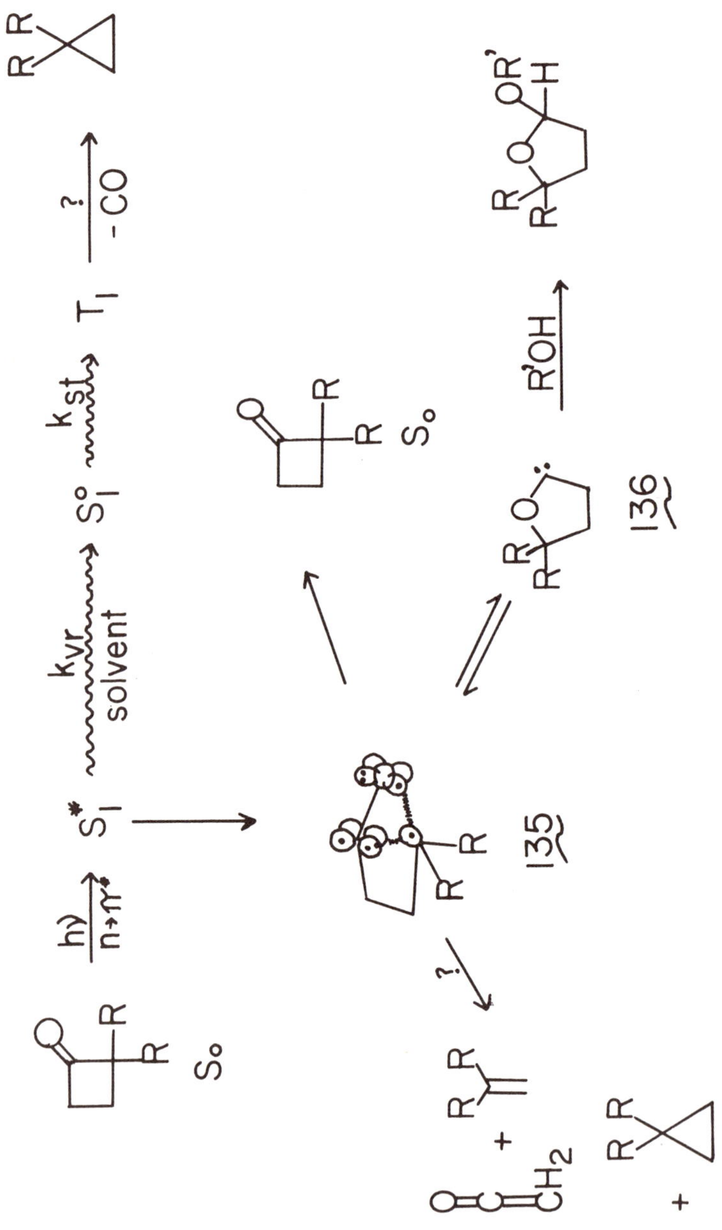

singlet state containing excess vibrational energy (S_1*). From this state, ring strain facilitated predissociation to a "biradical-like" transition state [135] or vibrational relaxation (k_{vr}) to S_1° may occur. It is also conceivable that transition state [135] could be produced directly from S_1°. Alternatively, molecules in the S_1° state could intersystem cross (k_{ST}) to the triplet manifold (T_1). For 2-alkylidenecyclobutanones, T_1 reactivity is manifested in isomerization about the exocyclic carbon-carbon double bond, while for the saturated cyclobutanone derivatives studied, definitive evidence for solution-phase T_1 reactivity is not available. If analogy is again made to the vapor-phase photochemistry of cyclobutanone [21], T_1 reactivity could conceivably result in decarbonylated products. Indeed, preliminary evidence has been obtained from sensitization experiments employing m-xylene as triplet sensitizer that decarbonylation of a saturated cyclobutanone is enhanced by selective population of its T_1 state (35).

The postulation that the "biradical-like" transition state [135] (and not a freely rotating acyl alkyl biradical intermediate) is the precursor of the oxacarbene intermediate [136] is made primarily to accomodate the fact that the ring-expansion reaction is stereoselective. Transition state [135] could also decay by β-cleavage and/or decarbonylation [both stereospecific (23), although definitive evidence concerning this point is not available in the solution phase. Finally [135] could decompose back to the starting cyclobutanone, which would explain the observed lack of efficiency in the previously described photolyses. (See Section II.E for a further discussion of this point.)

With respect to the mechanism described above, the observed effects of increased angular strain in the cyclobutanone molecule can be interpreted in at least three ways: (a) increased angular strain would facilitate dissociation to [135]; (b) increased angular strain would shift the postulated [135] $\rightleftarrows$ [136] equilibrium in favor of [136]; and (c) factors which increase ring strain may also decrease the feasibility of competing reactions (e.g., decarbonylation and/or cycloelimination).

For alkylidene substituents and spiro-fused three-membered rings attached to the α-carbon atom of cyclobutanone (e.g., [64a], [64b], [79a], and [79b]), the observed enhancement in ring expansion appears to be greater than that expected from simple strain arguments. Comparison of the photochemistry of [79] and [57], for example, reveals that an alkylidene substituent at the α-carbon position results in a higher yield of cyclic acetal (100%) than an analogous substituent at the β-carbon position (89%). A similar trend is observed if one compares spiroketones [64] and [58]. This additional enhancement might be attributed to a resonance stabilization of the oxacarbene intermediate as illustrated for [138]. Interaction of carbene centers with adjacent

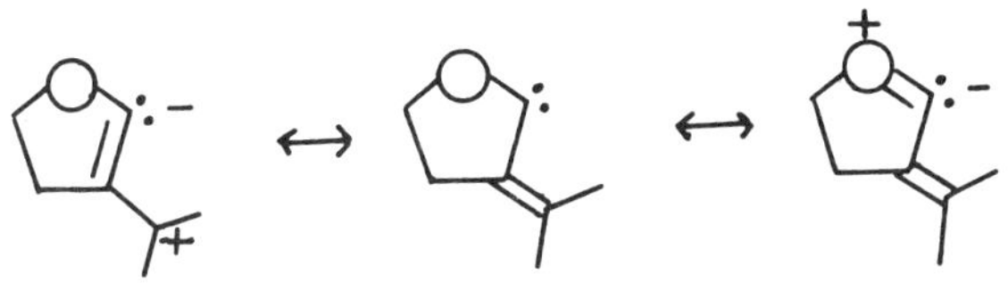

137 138 139

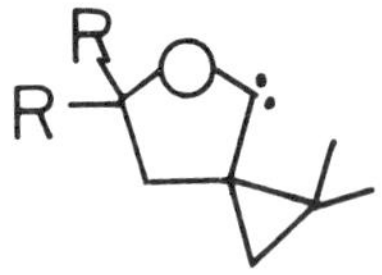

140

R=H or CH

cyclopropyl groups is also well documented in the literature (51), and such interaction could conceivably result in stabilization of oxacarbenes, such as [140].

There appears to be at least one factor, in addition to angular strain and resonance stabilization, which determines the relative importance of photoinduced decarbonylation, β-cleavage, and ring expansion of cyclobutanones. It has already been demonstrated that increasing α-alkylation of cyclo-butanone results in an increase in the extent of cyclic acetal formation (i.e., ring expansion). Comparison of [64a] and [64b] reveals a similar trend. The 3-phenyl-cyclobutanones [141] and [144], however, provide the most spectacular effect of α-substitution (25). In both cases, α-cleavage occurs exclusively at the most highly sub-

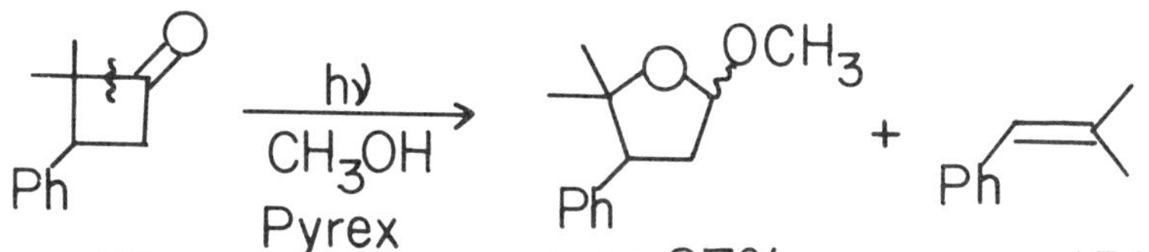

141 142, 87% 143, 15%

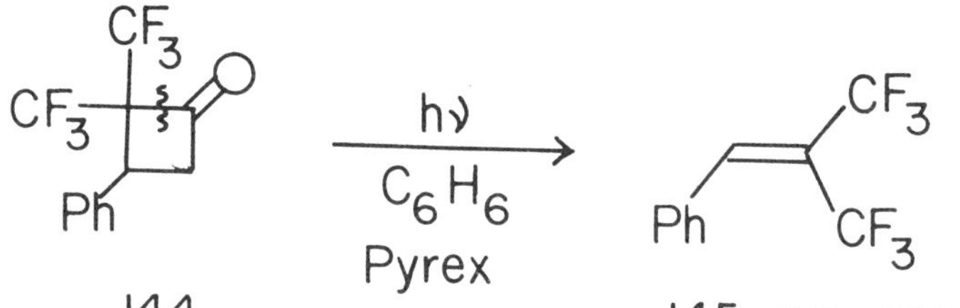

144 145, 94-6%

stituted bond. The dimethyl derivative [141], as expected, afforded cyclic acetals [141] in high yield; however, the bis(trifluoromethyl) ketone [144] underwent essentially exclusive β-cleavage to β,β-bis(trifluoromethyl)styrene [145] (94-96%), whether irradiated in methanol or benzene. These results strongly suggest that the migrating center in transition state [135] cannot be electron deficient. As additional support for this conclusion, Kaplan and coworkers (52) have investigated the photochemistry of bicyclo[3.2.0]hept-2-en-6-one, [146], and reported only the presence of esters [147] - [150] in the photolysate. Since esters [149] and [150] are the expected β-cleavage products if initial α-cleavage occurs at bond "b," the absence of cyclic acetals in the product mixture again

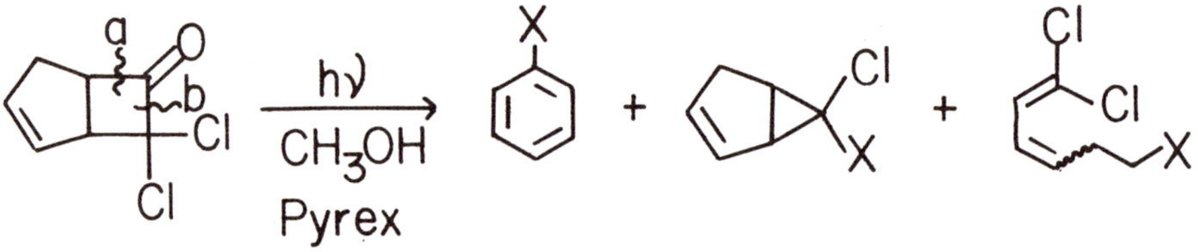

146 147 148 149, cis 150, trans

$X = CO_2CH_3$

suggests that electron-deficient α-carbon centers do not undergo migration by way of transition state [135]. It should be emphasized that dichloroketone [146] is structurally analogous to the dimethylated derivative [15] which was found to undergo exclusive cleavage at the most substituted α-bond to afford a 50:50 mixture of ring-expanded acetal [17] and the expected β-cleavage ester [16] (12).

To account for these observations, recourse is made to the simple but surprisingly effective model of the carbonyl n,π* excited state (53). When applied to the present case, n,π* excitation of the ground-state cyclobutanone molecule (S_0) produces an excited state which may be illustrated as [151] in Scheme V. The carbonyl model then predicts a nucleophilic region above

SCHEME V
Effect of Electron Density at the Migrating Terminus

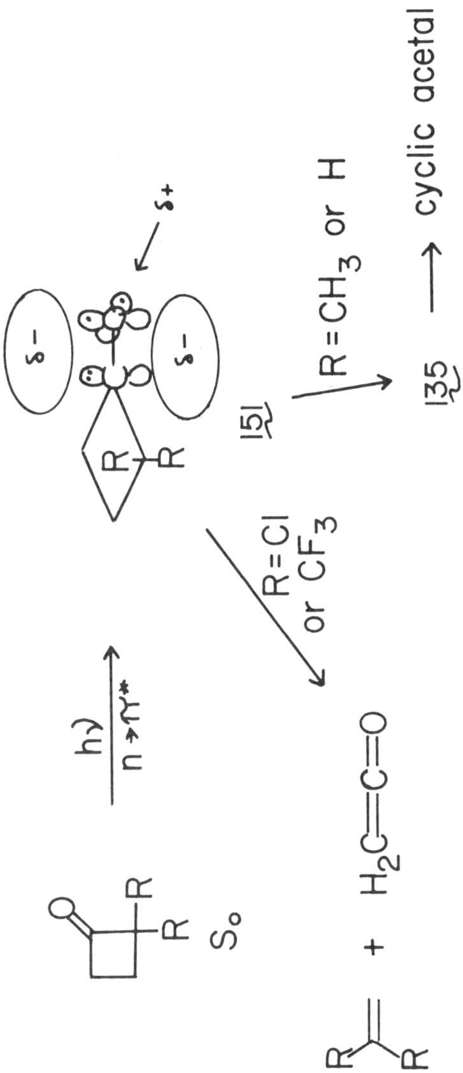

and below the molecular plane and an electrophilic and radical-like region in the plane of the carbonyl group. When R=CH_3 or H, the migrating center is capable of undergoing migration and concomitant nucleophilic attack on the half-vacant n-orbital, while for R=Cl or CF_3, a corresponding nucleophilic attack on the carbonyl oxygen atom is not feasible. Instead, β-cleavage to an olefin and ketene predominates.

V. SUGGESTIONS FOR FUTURE RESEARCH EFFORT

This review has demonstrated that, unlike their higher homologues, cyclobutanones undergo photolyses in a variety of solvents to afford products of β-cleavage, decarbonylation, and ring expansion. Even though a significant effort has been exerted to elucidate the mechanism(s) by which these products are generated, the general area of solution-phase photochemistry of cyclobutanones still contains several voids which can be filled only by future research.

We have expressed our opinion that the cyclobutanone S_1 state is generally the precursor of the ring-expanded oxacarbene species and perhaps the other photoproducts as well, and have presented a variety of data (albeit indirect) which supports this opinion. Unfortunately, our opinion is still an opinion, and must be confirmed by experimentation. What is needed is a definitive sensitization experiment using a triplet sensitizer which has a high triplet energy ($E_T \geq 78$ kcal/mole) under conditions such that competitive singlet-singlet energy transfer is excluded or minimized as much as possible. If the T_1 state of cyclobutanone can be specifically populated, an examination of the photoproducts will reveal which, if any, of the three types of reactions originates in this manifold. It may turn out, however, that both S_1 and T_1 lead to all photoproducts. Since quenching experiments have also proved negative in all reported instances, it would also be of interest to determine the lifetimes of both the S_1 and T_1 states of cyclobutanone in solution.

The work of Lee and coworkers (47) has revealed that cyclobutanone undergoes predissociation when S_1 has $\geq$ 4 kcal/mole excess vibrational energy. An interesting experiment would involve the design of a light source which would allow selective excitation of the vibrational levels of the cyclobutanone S_1 state. In analogy with gas-phase studies, one might anticipate a change in the ratios of the photoproducts as the lowest vibrational level of S_1 is approached.

As a final suggestion for future research, cyclobutanones have also provided the organic photochemist with the opportunity of investigating the existence of unusual and reactive intermediates: oxacarbenes, trimethylene biradicals, trimethylenemethane biradicals, acyl alkyl biradicals, and ketenes. Evidence for the intervention of oxacarbenes in the ring-expansion reaction is quite compelling; however, their unusual behavior relative to "typical" carbenes (e.g., failure to form cyclopropane adducts with some olefinic substrates) makes them prime subjects for further study and characterization. Unlike oxacarbenes, the existence of acyl alkyl biradicals (e.g., [30]) is tenuous at best. Ideally, an experiment should be performed using a molecule capable of generating this species independently. Thermal or photochemical elimination of nitrogen from [152] or carbon dioxide from [153] could conceivably generate [30], although in the case of a photoelimination, secondary

30

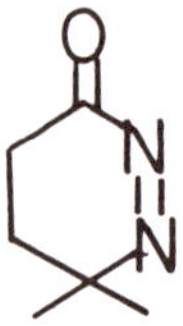

152

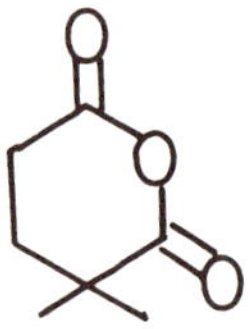

153

photolysis of 2,2-dimethylcyclobutanone must be excluded.

VI. SUMMARY OF SOLUTION-PHASE PHOTOCHEMICAL REACTIONS OF CYCLOBUTANONES

Table IV provides a summary of current data for these systems.

TABLE IV. Solution-Phase Photochemical Reactions of Cyclobutanones

Ketone	Solvent	$\lambda_{ex.}^{a}$	Photoproducts[b]				Ref.
			Decarbonylation	β -Cleavage	Ring Expansion	Other	
	Several solvents	313 nm		C_2H_4	--	--	16 17
	CH_3OH	Quartz	 (-)	$CH_3CO_2CH_3$ (48%)	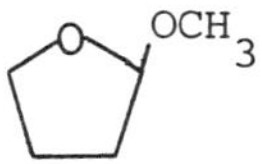 (8%)	--	15
	 (10°C)	Pyrex	--	CH_3CO_2H (19%) 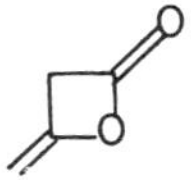 (32%) 1:1 adduct (5%)	--	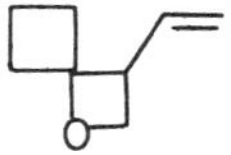 (13%)	19
	 (-78°C)	Pyrex	--	--	--	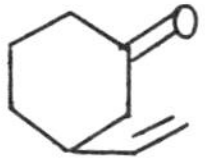 (32%) 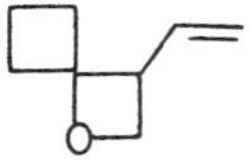 (32%)	19

	$(n\text{-Bu})_3SnH$ C_6H_6	313 nm	--	--	--	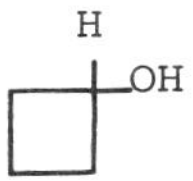	46
B. α-Alkyl Substituted Cyclobutanones							
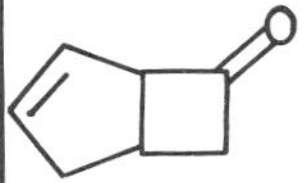	$c\text{-}C_6H_{12}$	Pyrex	--	 $CH_2{=}C{=}O$	--	--	62
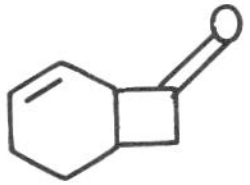	Et_2O	Pyrex	--	 $CH_2{=}C{=}O$	--	--	82 83
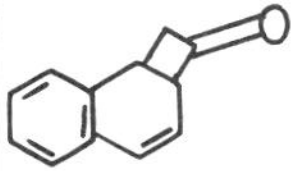	EtOH	Pyrex	--	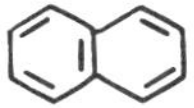 CH_3CO_2Et	--	--	82 83
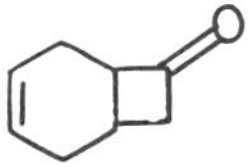	$n\text{-}C_5H_{12}$	--	--	 (60%)	--	--	77

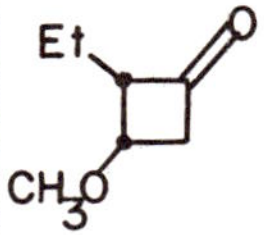	CH_3OH	313 nm	 (4.1%) (o.4%)	$CH_3CO_2CH_3$ (37%) 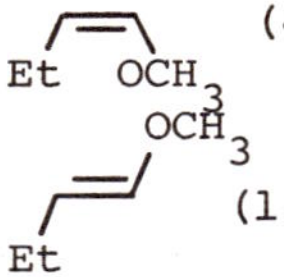 (45%) (1.6%)	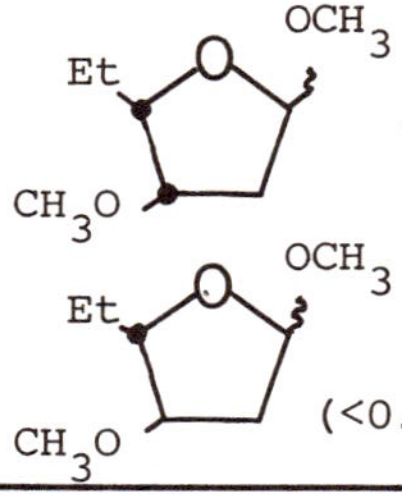 (12%) (<0.4%)	--	23
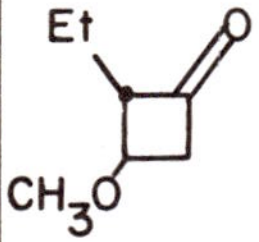	CH_3OH	313 nm	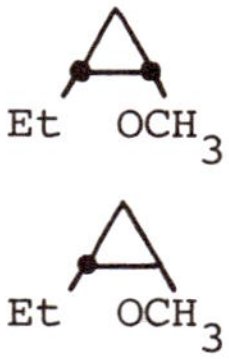 (0.7%) (10%)	$CH_3CO_2CH_3$ (31%) 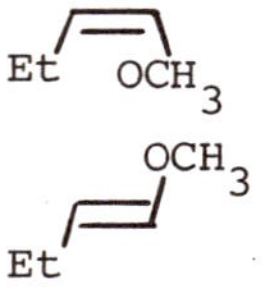 (1.4%) (35%)	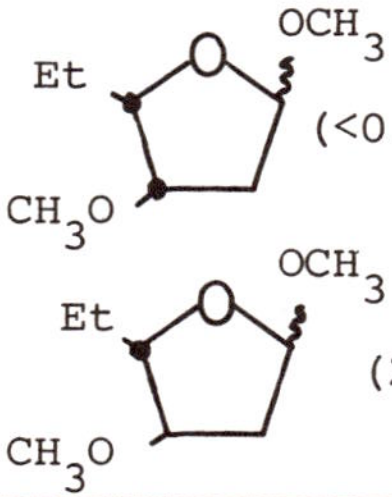 (<0.3%) (21%)	--	23
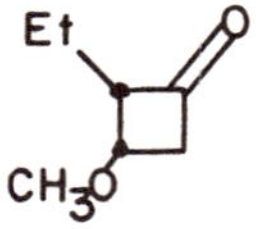	C_6H_6	313 nm	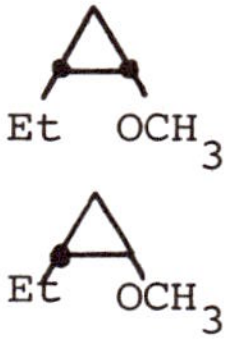 (10%) (2.1%)	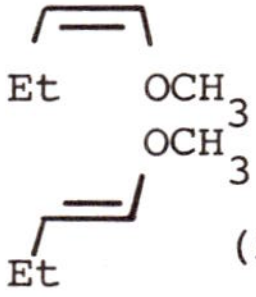 (82%) (5.2%)	--	--	23
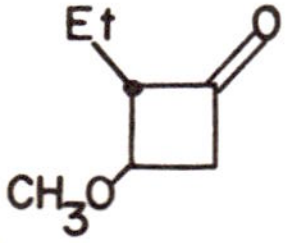	C_6H_6	313 nm			--	--	23

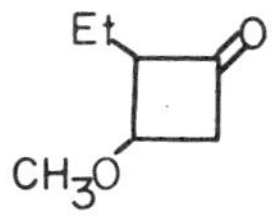	n-C_5H_{12} O_2	Pyrex	--	--	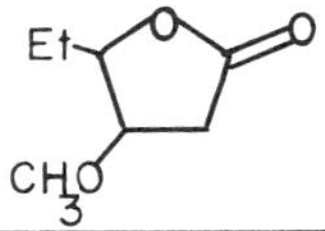	--	24
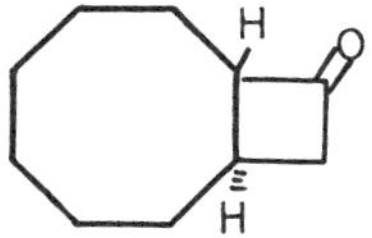	CH_3OH	313 nm	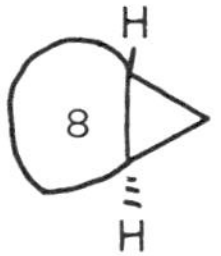 (~15%)	 (~10%)	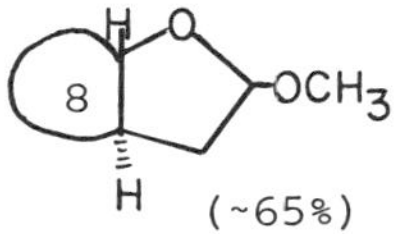 (~65%)	--	24
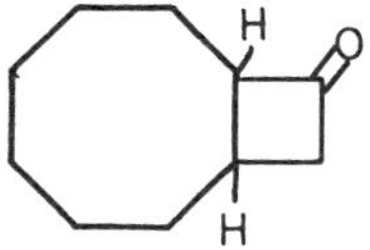	CH_3OH	313 nm	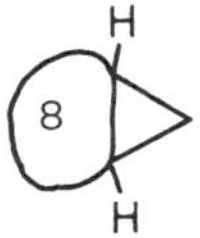 (~10%)	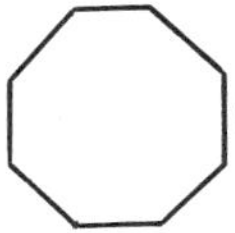 (~50%)	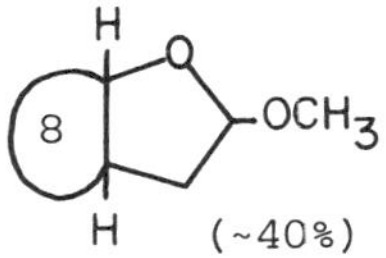 (~40%)	--	24
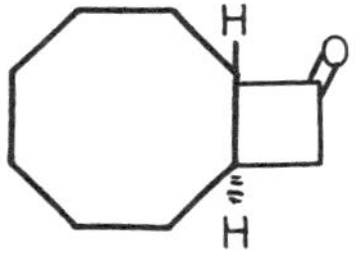	C_6H_6	313 nm	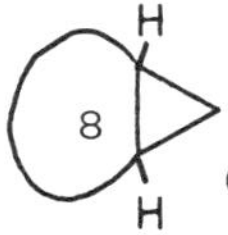 (~88%)	 (~6%)	--	--	24
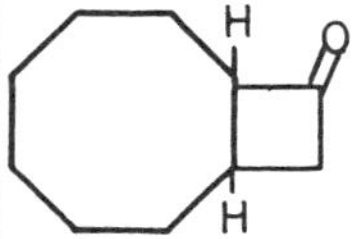	C_6H_6	313 nm	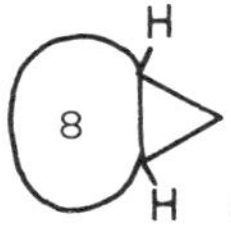 (~30%)	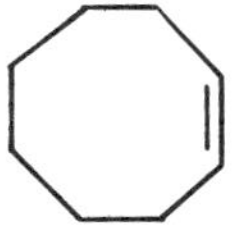 (~70%)	--	--	24

	C_6H_6	Pyrex	 (45%)[c] (55%)[c]	 (20%)[c] (80%)[c]	--	--	24
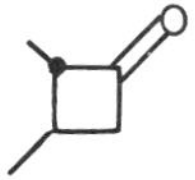	C_6H_6	Pyrex	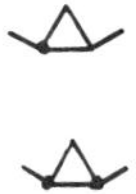 (90%)[d] (10%)[d]	 (97%)[d] (3%)[d]	--	--	24
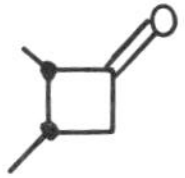	$n\text{-}C_7H_{16}$	325 nm	 (9%) (10%)	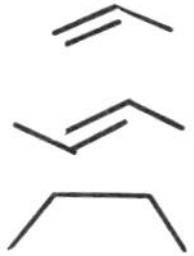 (5%) (8%) (68%)	--	--	81
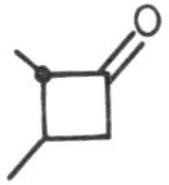	$n\text{-}C_7H_{16}$	325 nm	 (18%) (2%)	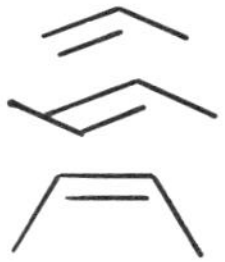 (3%) (74%) (3%)	--	--	81
	$n\text{-}C_7H_{16}$	325 nm	 (12%) (3%)	 (85%)	--	--	81
	$n\text{-}C_7H_{16}$	325 nm	 (17%) (12%)	 (71%)	--	--	81

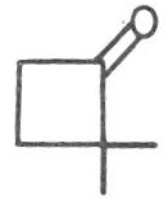	CH_3OH	Quartz	 (7%)	$CH_3CO_2CH_3$ (32%) 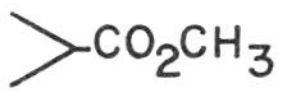 (3%)	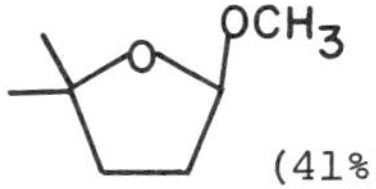 (41%)	--	15
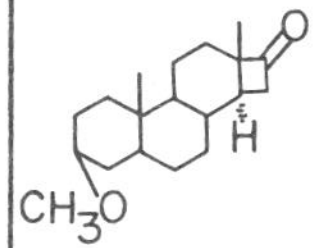	EtOH	Quartz	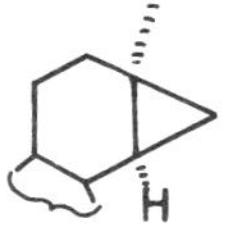 (16%)	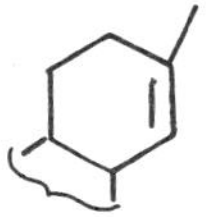 (10%)	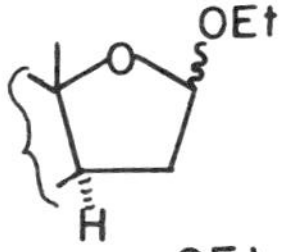 (54%) 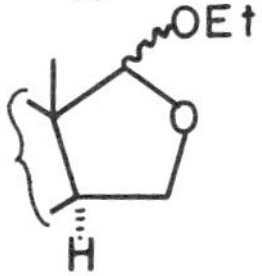 (12%)	--	22
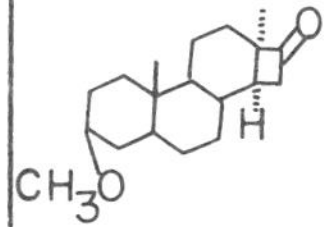	EtOH	Quartz	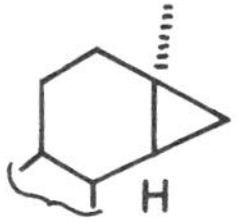 (1.5%)	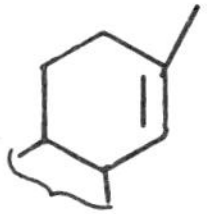 (17.5%)	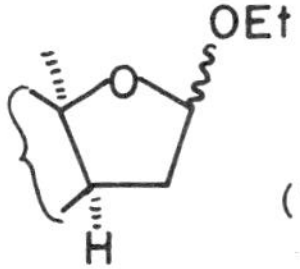 (71%)	--	22
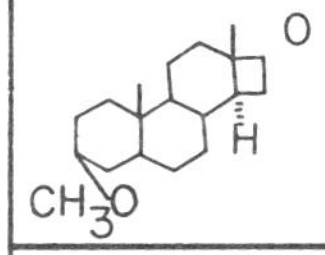	C_6H_6	Quartz	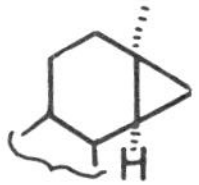 (16%)	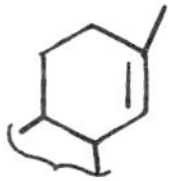 (10%)	--	--	22
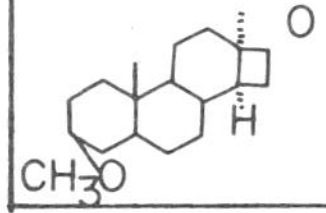	C_6H_6	Quartz	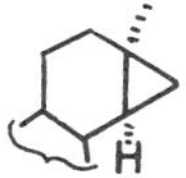 (1.5%)	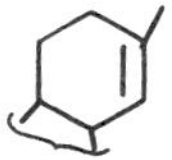 (17.5%)	--	--	22

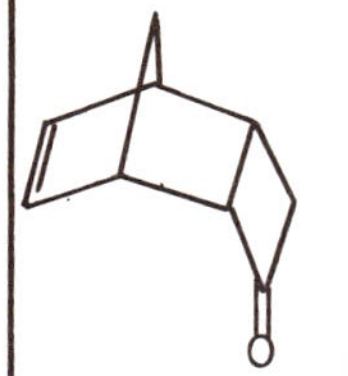	$n\text{-}C_5H_{12}$	300 nm	< 2%	 (<1%) 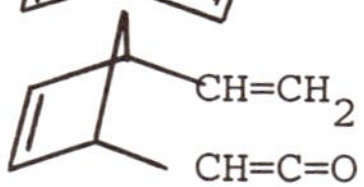 (50%)	--	--	77a
	$n\text{-}C_5H_{12}$	300 nm	< 2%	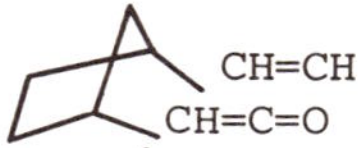 (45%) 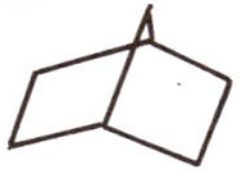 < 2%	--	--	77a
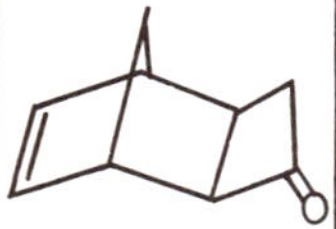	$n\text{-}C_5H_{12}$	300 nm	< 2%	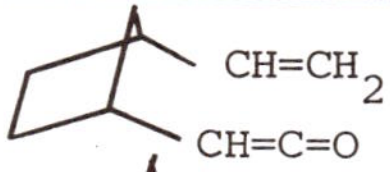 (10%)  (5%)	--	--	77a
	$n\text{-}C_5H_{12}$	300 nm	< 2%	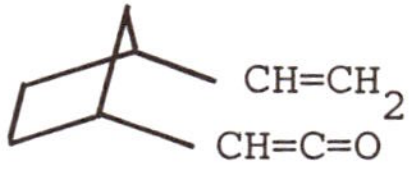 (30%)	--	--	77a
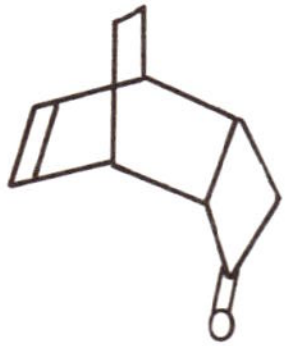	CH_3OH	300 nm	--	 (10%) 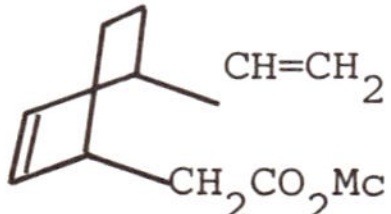 (18%)	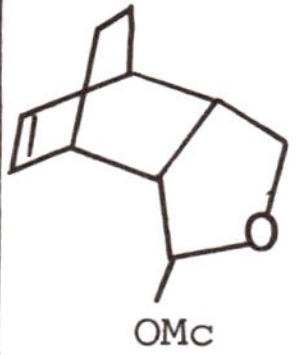 (60%)	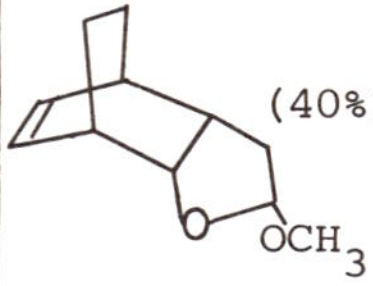 (40%)	77a 77b

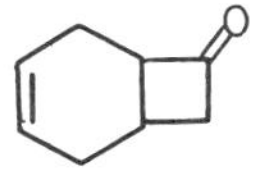	CH_3OH	300 nm	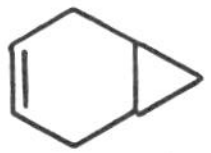 (3%)	 (50%)	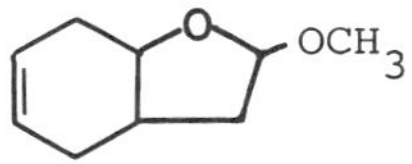 (15%)	77b
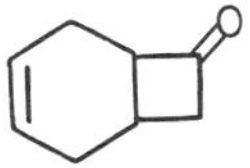	n-C_5H_{12}	300 nm	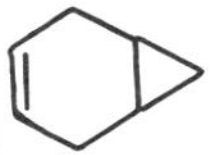 (6%)	 (58%)	--	77b
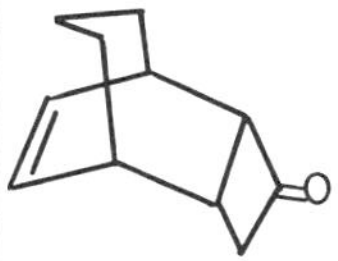	n-C_5H_{12}	300 nm	--	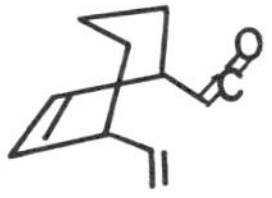 (19%)	--	77b
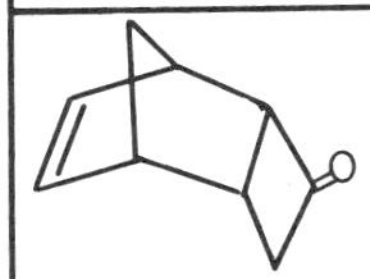	CH_3OH	300 nm	--	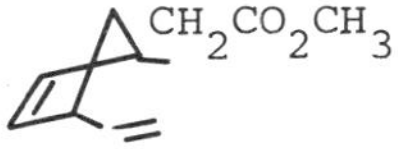 (72%)	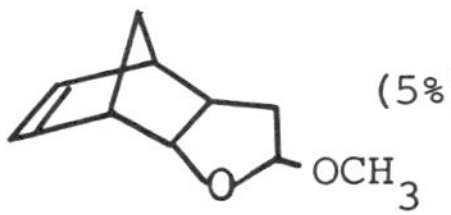 (5%) 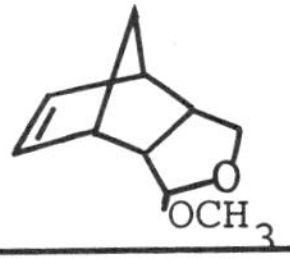 (5%)	77b
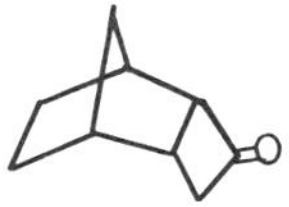	CH_3OH	300 nm	--	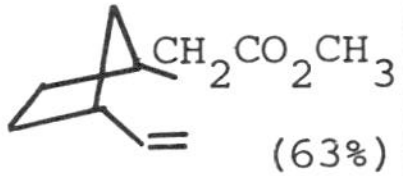 (63%)	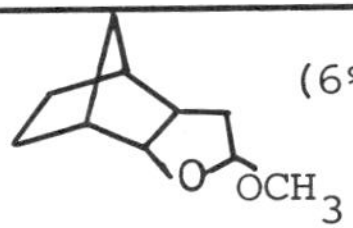 (6%) 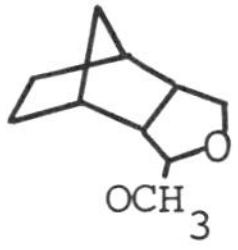 (8%)	77b
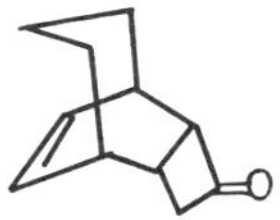	CH_3OH	300 nm	 (2%)	 (15%)	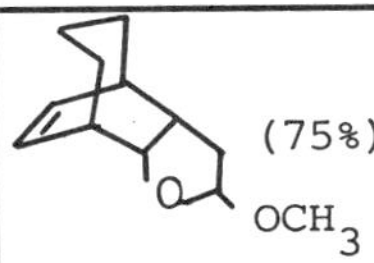 (75%) 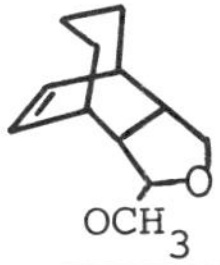 (25%)	

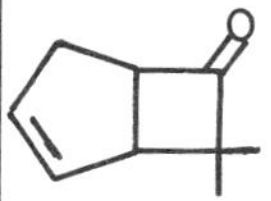	CH_3OH[e]	--	--	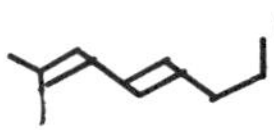 CO_2CH_3 (50%) cis + trans	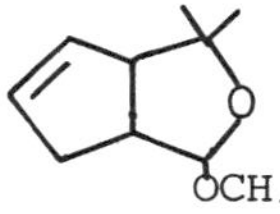 OCH_3 (50%)	--	12
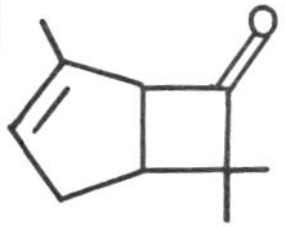	CH_3OH	Vycor	--	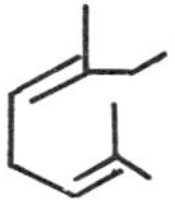 CO_2CH_3 (17%)	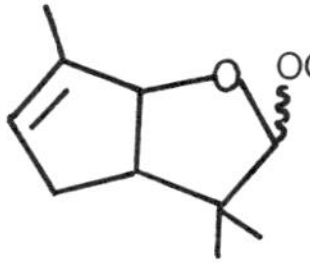 OCH_3 (14%) 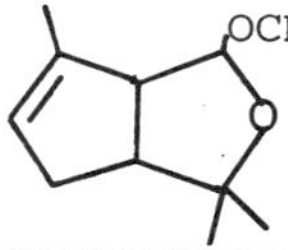 OCH_3 (25%)	--	21
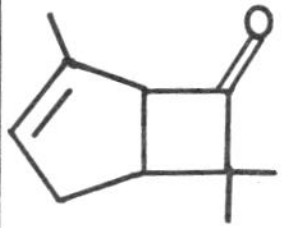	$c\text{-}C_6H_{12}$	Vycor	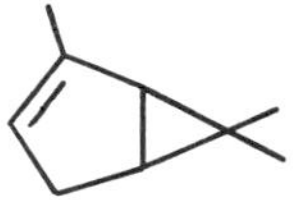 (10%)	--	--	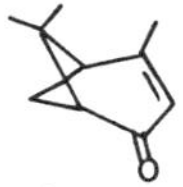 (8%)  (2.1%) 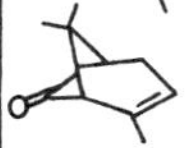 (1.8%)	21
	CH_3OH	Quartz	 (11%)	 CO_2CH_3 (13%)	 OCH_3 (68%)	--	15
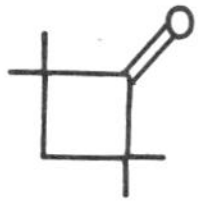	C_6H_6 O_2	Quartz	 (12%)	 (20%)	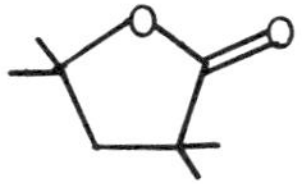 (-)	--	15

	Et_2O	Quartz			--	--	84
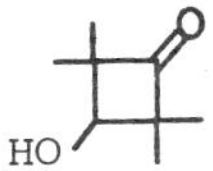	CH_3OH	Quartz	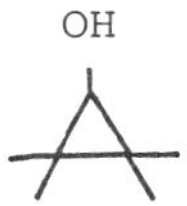 (17%) CO (24%)	 CO_2CH_3 (15%) CHO (15%)	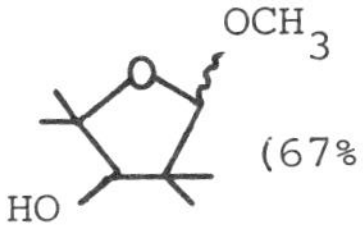 (67%)	--	13 15
	CH_3OH	Pyrex or 3ì3 nm			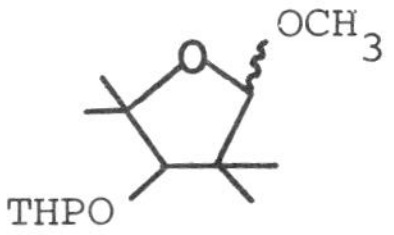	--	35
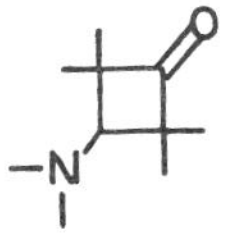	Et_2O	Quartz	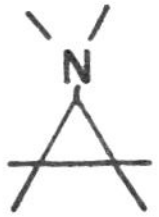 (38%) CO (46%)	--	--	--	13
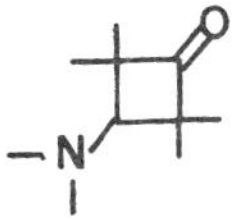	CH_3OH	Quartz	 (25%) CO (40%)	--	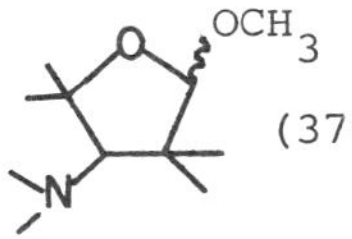 (37%)	--	13
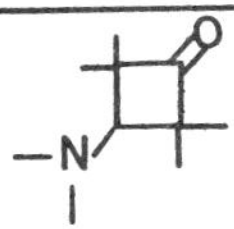	t-BuOH	Quartz	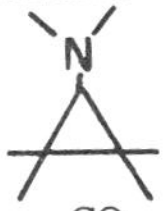 (40%) CO (40%)	--	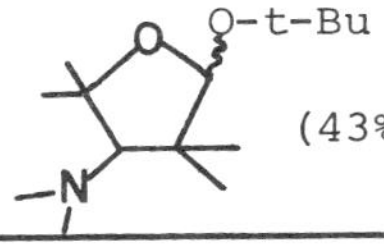 (43%)	--	13

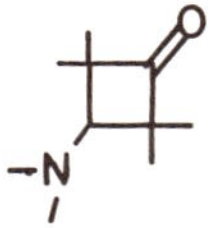	2N HCl	Quartz	 (30%) CO (49%)	--	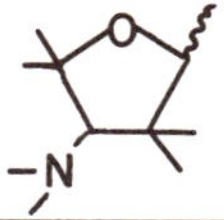 (50%)	--	13
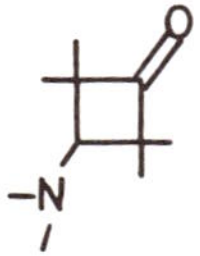	$Et_2O:H_2O$ (25:1)	Quartz	 (45%) CO (45%)	--	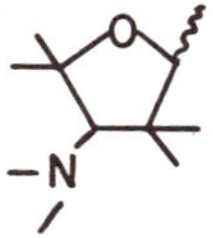 (41%)	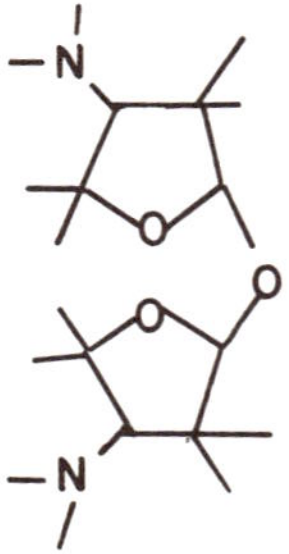 (11%)	13
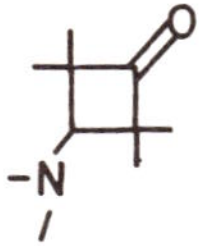	Et_2O:HOAc (25:3)	Quartz	 (36%) CO (52%)	--	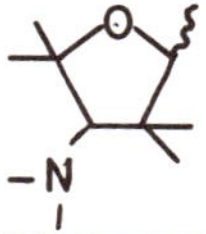 (25%)	--	13
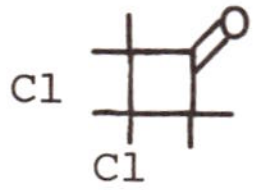	$c-C_6H_{12}$	Quartz	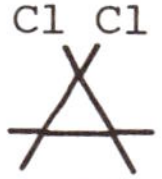 (33%) CO (65%)	--	--	--	13
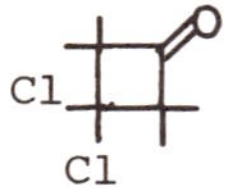	CH_3OH	Quartz	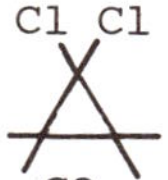 (35%) CO (50%)	--	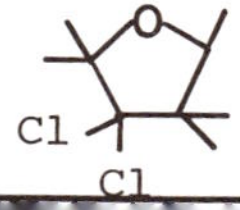 (30%)	--	13

	Vapor	Corex	--	--	--	$CH_3C{\equiv}CH$ $HC{\equiv}CH$ $CH_3C{\equiv}CCH_3$	63
D. 3-Methylenecyclobutanones and 2-Alkylidenecyclobutanones							
	CF_3 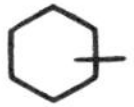 F_{11} (-196°C)	270-310 nm		--	--	--	33
	 (-78°C)	Pyrex	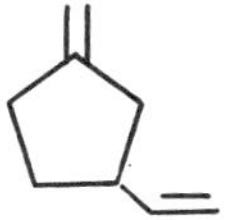 (23%) 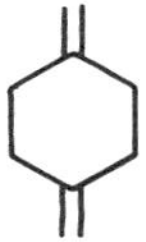 (5%)	--	--	g 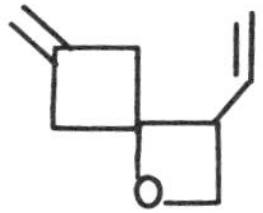 (17%)  Dimers (10%)	34
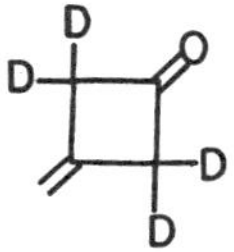	 (-78°C)	Pyrex	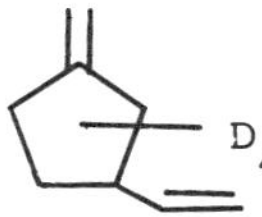 D_4 h	--	--	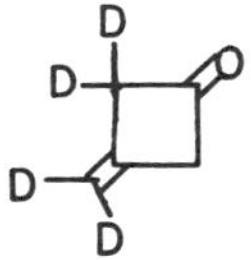 l	34
	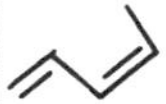 (-78°C)	Pyrex	--	--	--	 Major	34

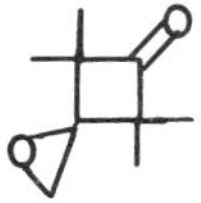	CH_3OH	Pyrex	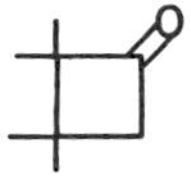 (12-14%) CO	--	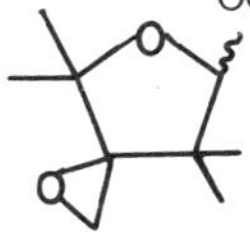 OCH_3 (86%)	--	30
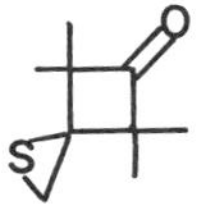	CH_3OH	280 nm	 (22.5%)	--	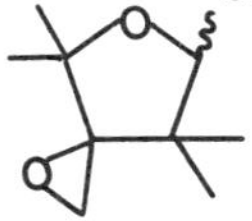 OCH_3 (36.5%)	--	29
	CH_3OH	254 nm	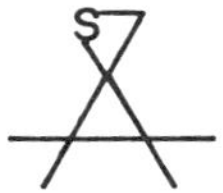 (7.3%)	--	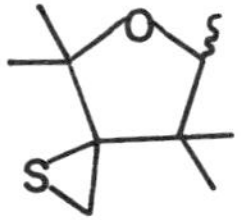 OCH_3 (11.7%)	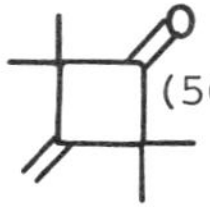 (56.1%)	29
	CH_3OH	Pyrex	--	 CO_2CH_3 (12%)	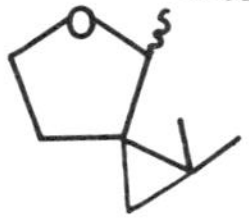 OCH_3 (88%)	--	15b
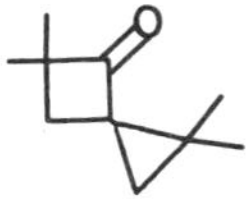	CH_3OH	Pyrex	--	--	 OCH_3 (100%)	--	15b

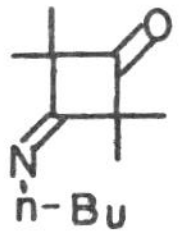	i-PrOH	Quartz	CO (34%)	--	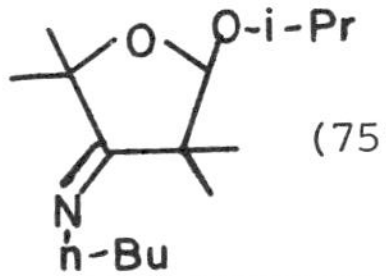 (75%)	--	13
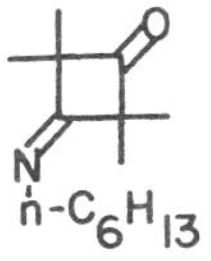	CH_3OH	Quartz	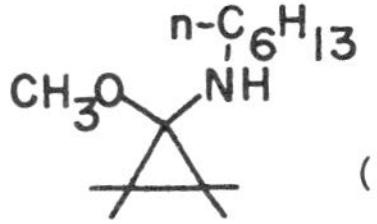 (17%) 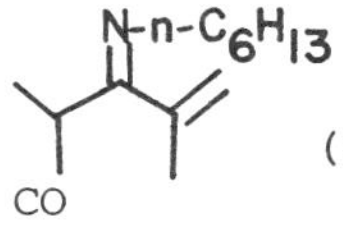 (28%)	--	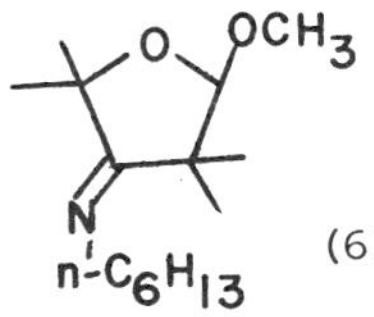 (63%)	--	13
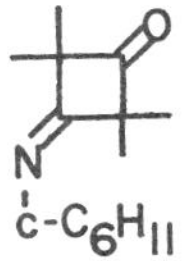	CH_3OH	Quartz	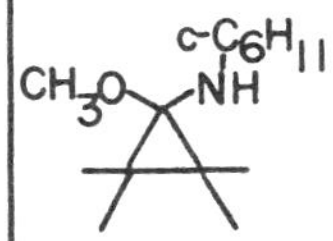 (22%) 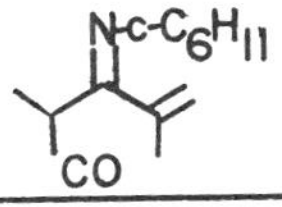 (22%)	--	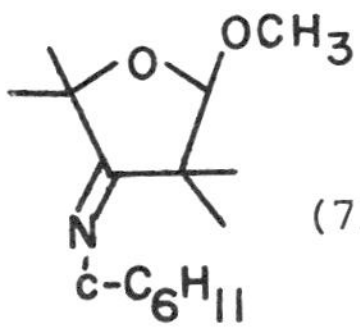 (71%)	--	13
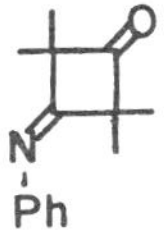	CH_3OH	Quartz	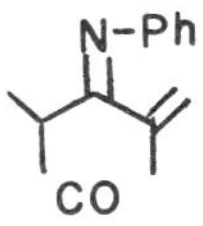 (20%) (36%)	--	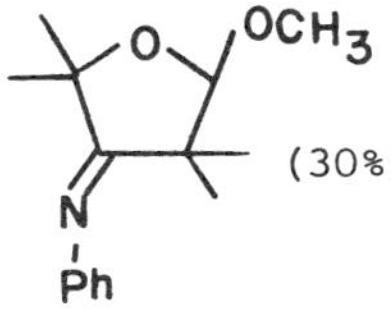 (30%)	--	13

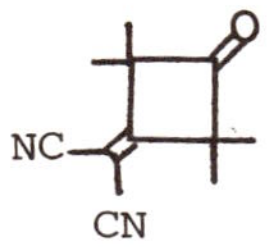	C_6H_6	Quartz	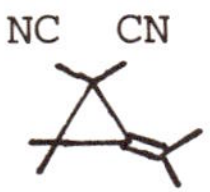 (35%) 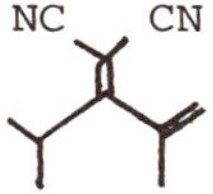 CO (35%)	--	--	--	13
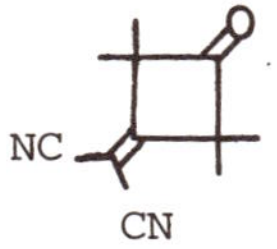	CH_3OH	Quartz	CO (40%)	--	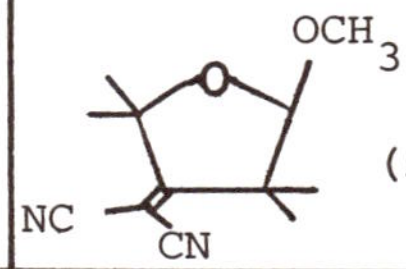 (57%)	--	13
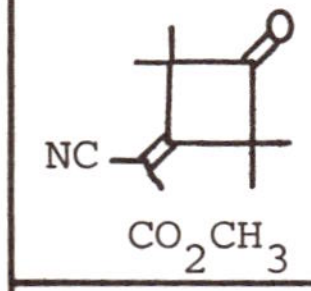	CH_3OH	Quartz	CO (40%)	--	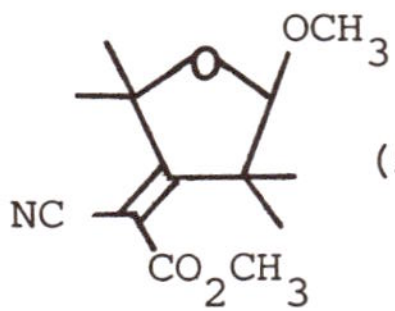 (58%)	--	13
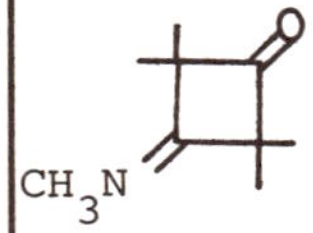	CH_3OH	Quartz	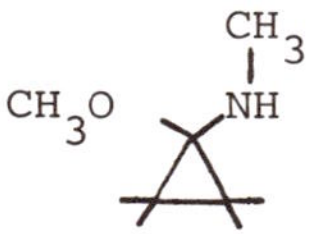 (23%) CO (26%)	--	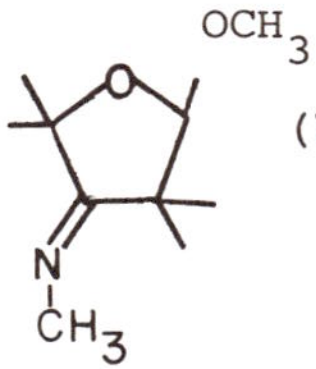 (73%)	--	13

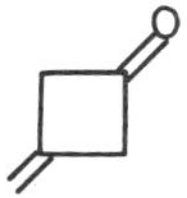	 (-78°C)	Pyrex	--	--	--	 Major	34
	 (-72°C; 15°C)	Pyrex	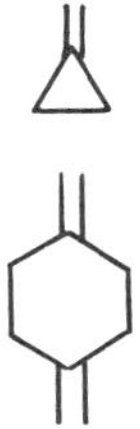 (60%)	--	--	--	34
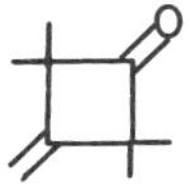	CH_3OH	Pyrex	 (11%)	--	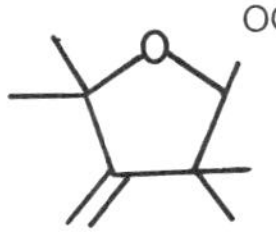 (89%)	--	35
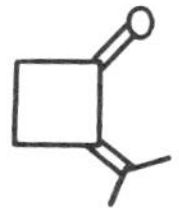	CH_3OH	Pyrex	--	--	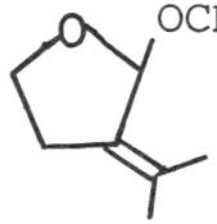 (100%)	--	15b 36
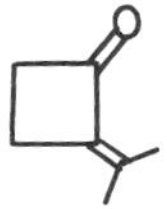	n-$C_5H_{12}O_2$	Pyrex	--	--	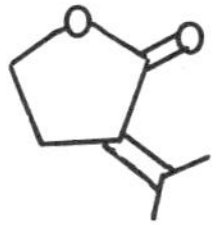 (-)	--	15b 36

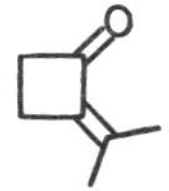	 C_6H_6	Pyrex	--	--	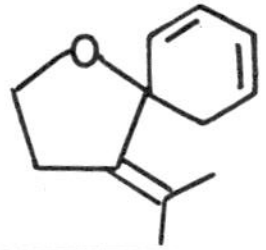	--	35
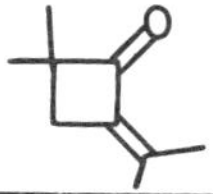	CH_3OH	Pyrex	--	--	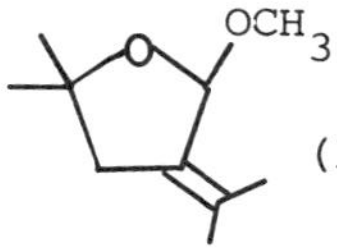 (100%)	--	15b
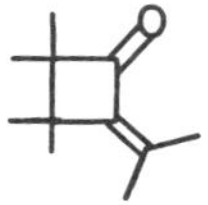	CH_3OH	--	--	--	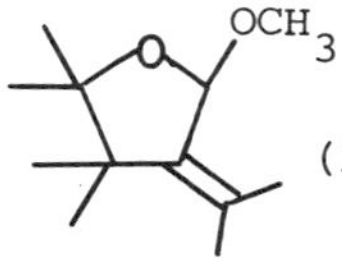 (100%)	--	37
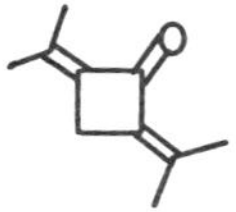	CH_3OH	Pyrex or Quartz	--	--		No reaction	15
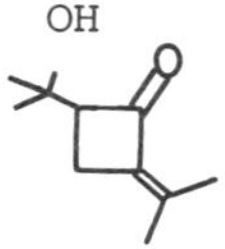	CH_3OH	Pyrex	--	--	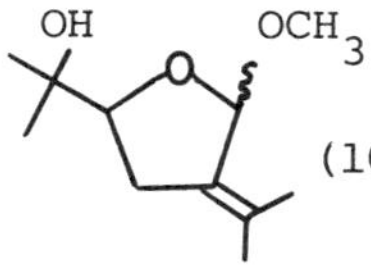 (100%)	--	15b 36
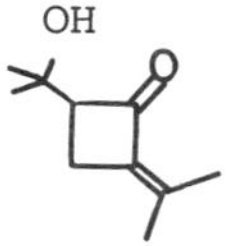	C_6H_6	Pyrex	--	--	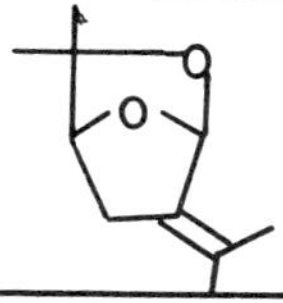 (60%)	--	15b 36

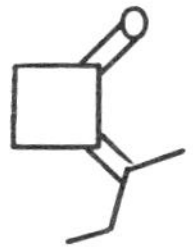	CH_3OH	313 nm	--	--	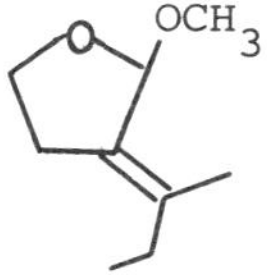 (50%) 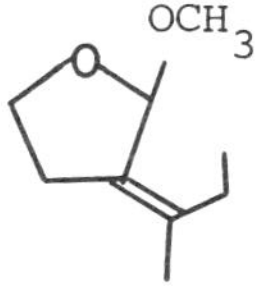 (50%) j	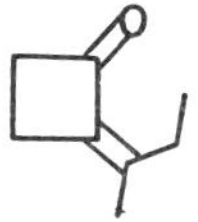 (-) k	35
E. Miscellaneous Cyclobutanones							
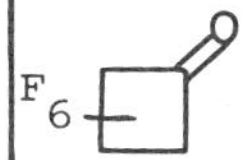	--	--	 (100%)	--	--	--	54
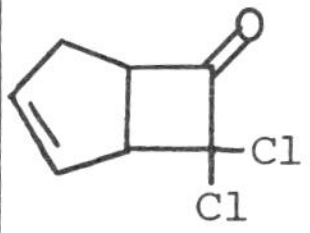	CH_3OH	Pyrex	--	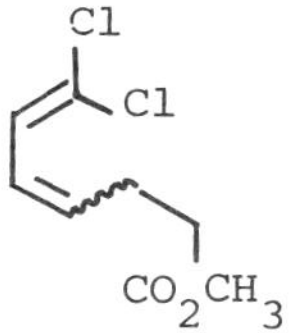	--	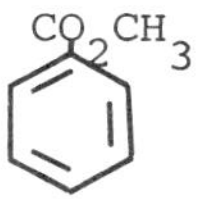 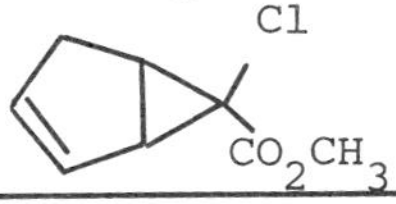	52
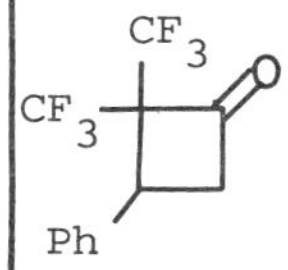	C_6H_6	Pyrex	--	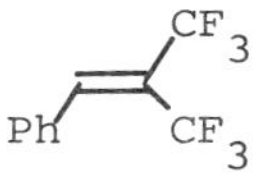	--	--	25

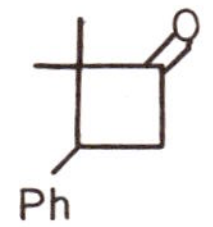	CH_3OH	Pyrex	--	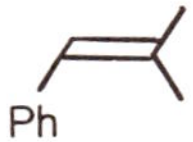 (15%)	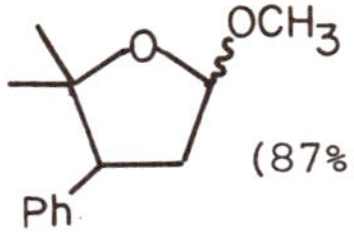 (87%)	--	25
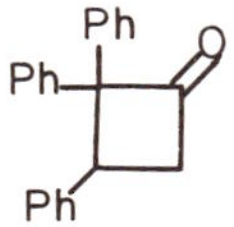	--	--	--	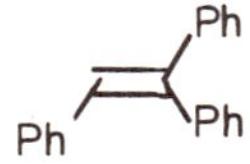 $H_2C{=}C{=}O$	--	--	55
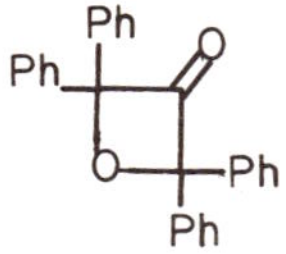	--	270 nm 77°K		--	--	--	56
	EtOH	313 nm	 (5-10%)	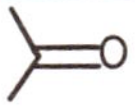 (95%) 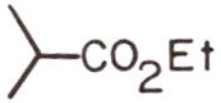 (90%)	--	--	45
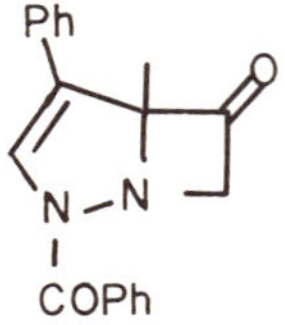	C_6H_6	Pyrex	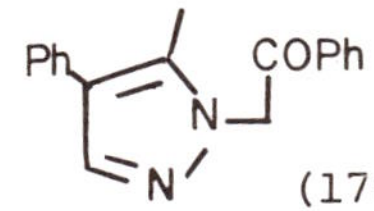 (17%) 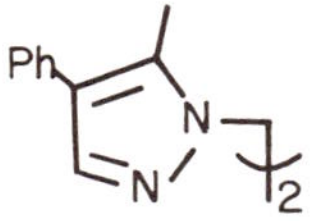 (7%)	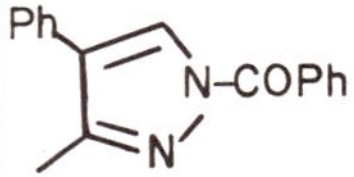 (8%)	--	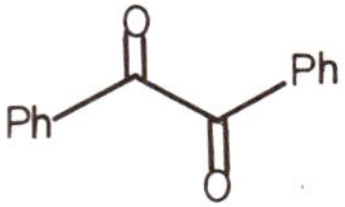	57

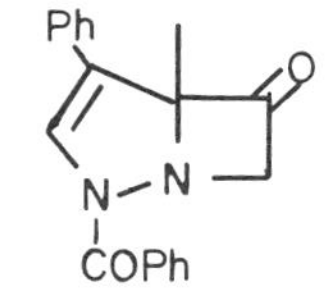	CH_3OH	Pyrex	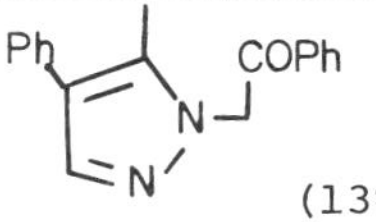 (13%) 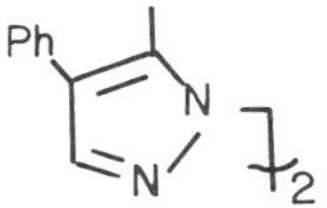 (5%) 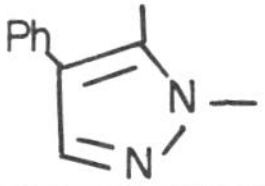 (11%)	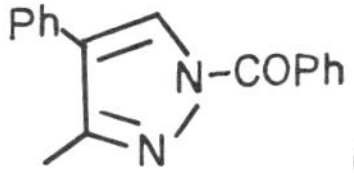 (18%)	--	--	57
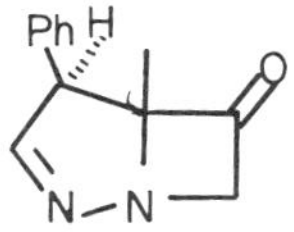	C_6H_6	Pyrex	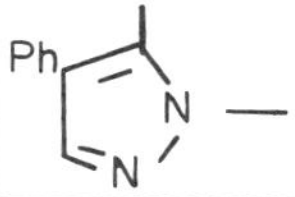 (54%)	--	--	--	57
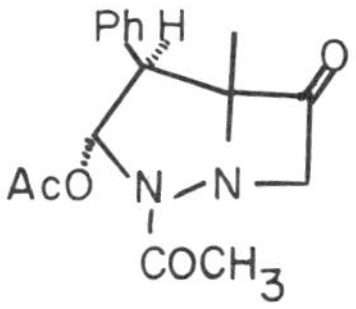	CH_2Cl_2	Pyrex	--	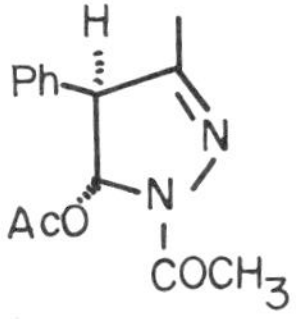 (-)	--	--	57
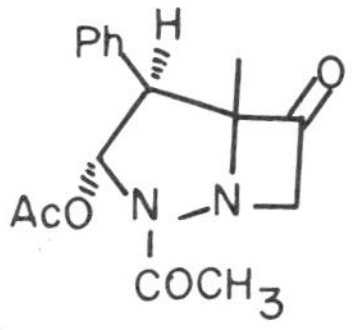	CH_3OH	Pyrex	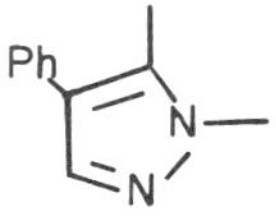	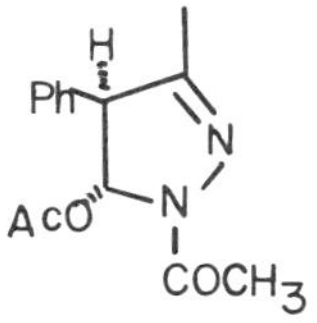	--	--	57

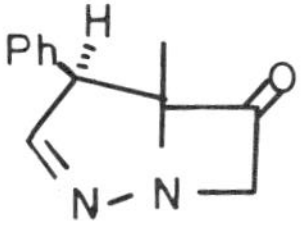	CH_3OH	Pyrex	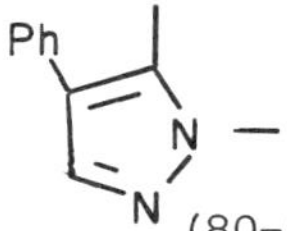 (80–90%)	--	--	--	57
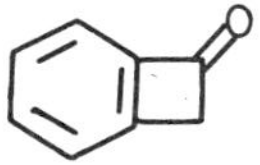	CH_3OH	Pyrex	--	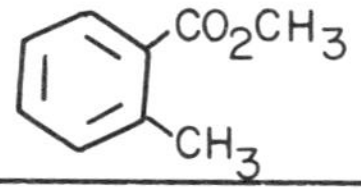	--	--	58
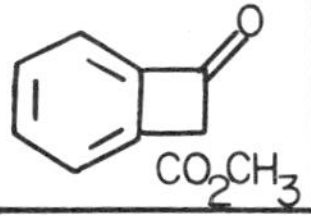	CH_3OH	Pyrex	--	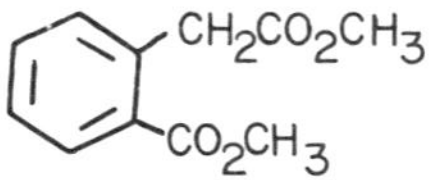	--	--	58
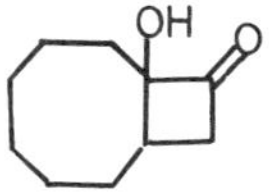	C_6H_6 (H_2O)	Pyrex	--	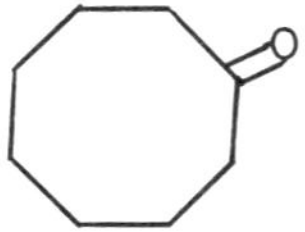 CH_3CO_2H	--	--	59
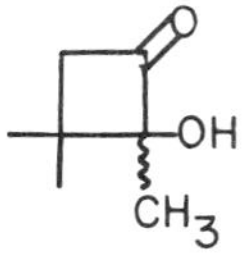	C_6H_6	Pyrex	--	--	--	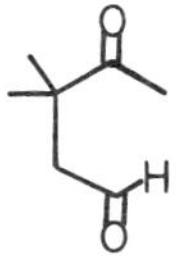 (36%) 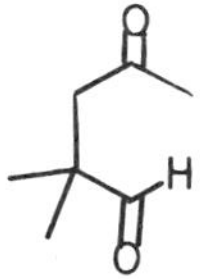 (29%)	44

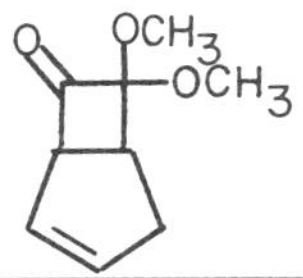	Et_2O	>300 nm	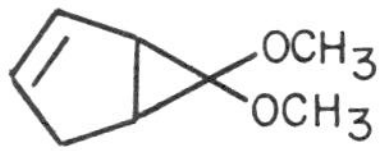	--	--	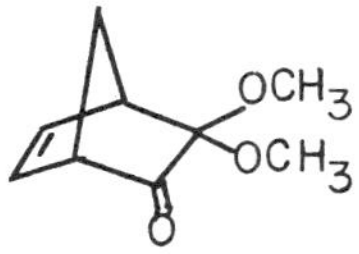	65
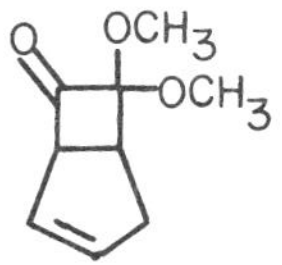	CH_3OH	>300 nm	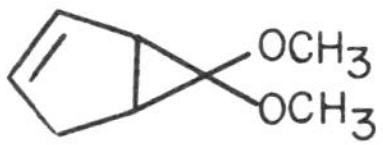	--	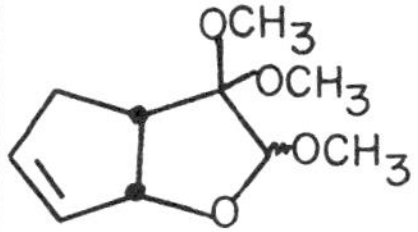	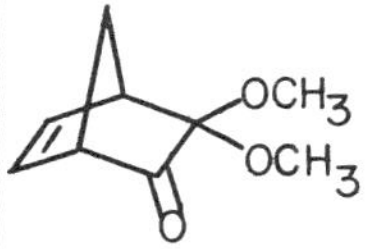	65
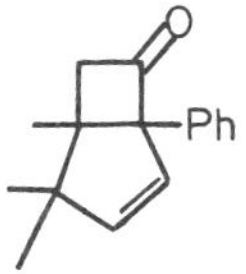	$c\text{-}C_6H_{12}$	Pyrex	--	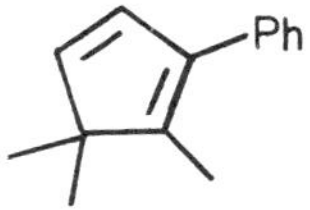	--	--	79
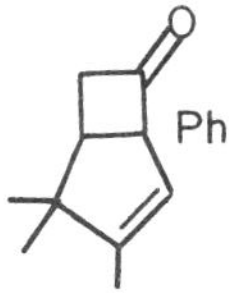	$c\text{-}C_6H_{12}$	Pyrex	--	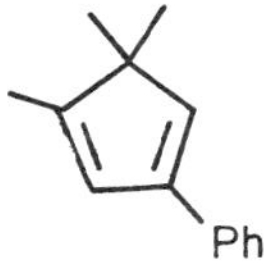	--	--	79
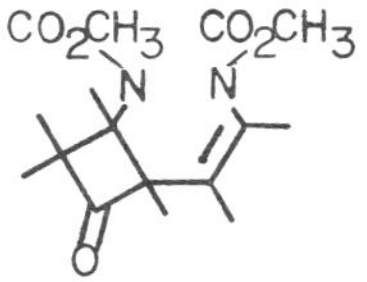	Et_2O	Pyrex	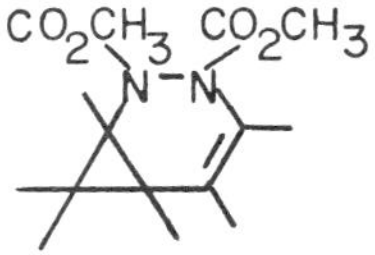	--	--	--	78

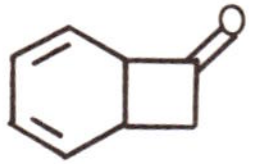	CH_3OH	Quartz	--	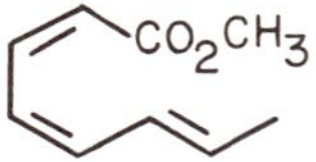	--	via 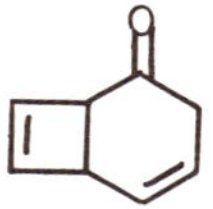 ?	87
	--	--	--	--	--	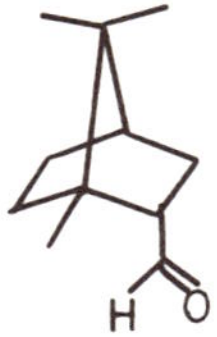	80
	$c\text{-}C_6H_{12}$	Vycor	--	--	--	 (18%)[1]	21b
	$c\text{-}C_6H_{12}$	Vycor	--	--	--	 (27%)[1]	21b
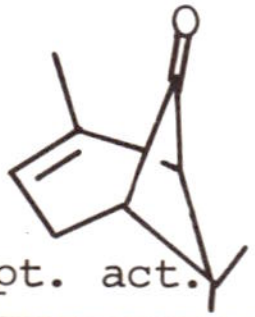 Opt. act.	$c\text{-}C_6H_{12}$	Pyrex	 (2%)  (13%)	--	--	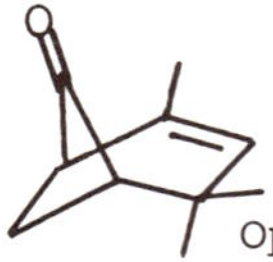 (52%)[1] Opt. act.	21a

 Opt. act.	c-C_6H_{12}	Pyrex	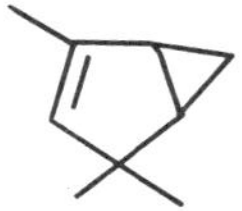 (4%) 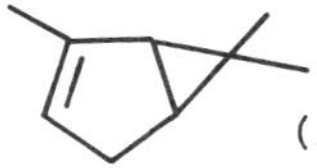 (13%)	--	--	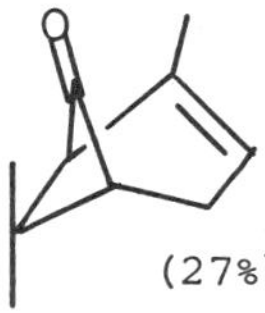 (27%)[1] Opt. act.	21a
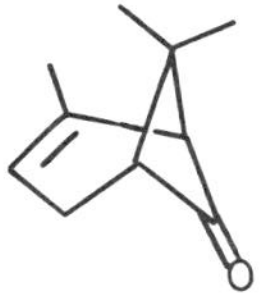	CH_3OH 30-35°C	--	--	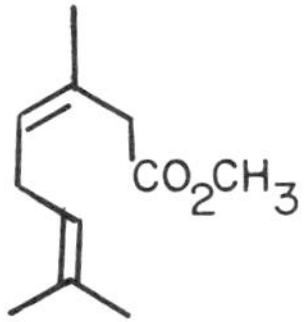	--	--	85
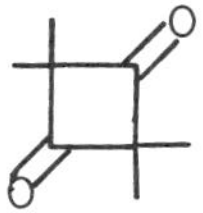	C_6H_6	366 nm	 CO (80%)	--	--	--	60
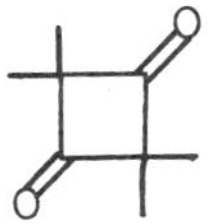	CH_3OH	254 or 313 nm	(60-70%) 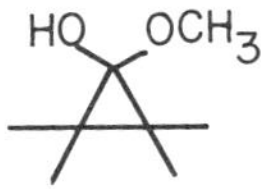	(40-30%) 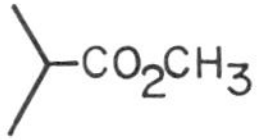	--	--	61

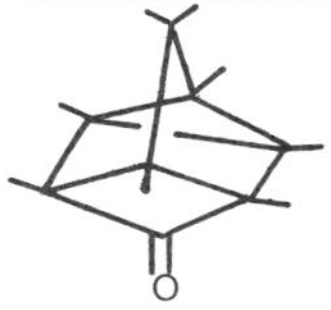	--	--		--	--	89
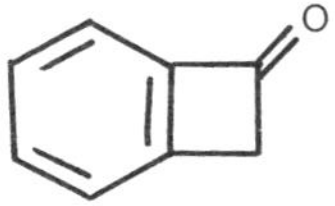	CH_3OH	--	--	--	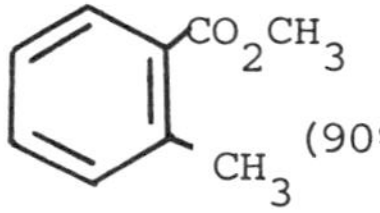 (90%) 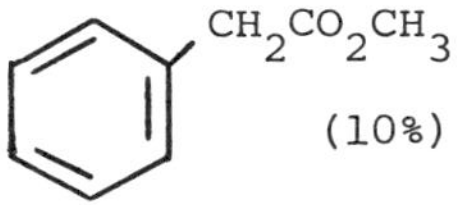 (10%)	90
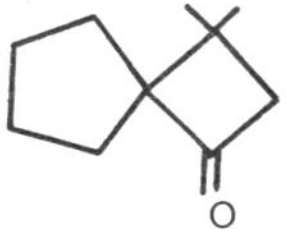	CH_3OH	--	--	--	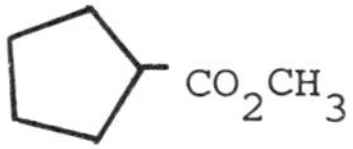	91

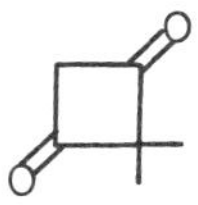	CH_2Cl_2 CH_3OH	--	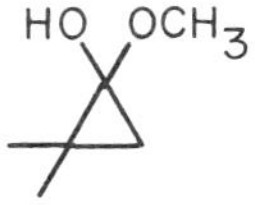	--	--	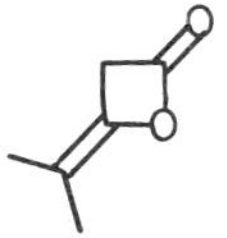 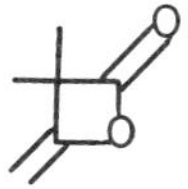 	88

[a]Excitation wavelength or filter when specified.
[b]Percentage yields, when available, are given in parentheses.
[c]Relative percentages. Mixture of 75% alkene to 25% cyclopropane.
[d]Relative percentages. Mixture of 70% alkene to 30% cyclopropane.
[e]Photolysis in cyclohexane afforded only polymeric material.
[f]Relative yield with respect to the other photoproducts significantly increased when irradiation is conducted at 254 nm using m-xylene as sensitizer.
[g]Under these same conditions, when cyclopentadiene is used as solvent, an analogous oxetane (40%) is produced.
[h]Deuterium atoms statistically scrambled among the trimethylene methane carbon atoms.
[i]Deuterium atoms statistically acrambled among methylene carbons in recovered starting material.
[j]Only acetal formed at low conversion. $\Phi = 0.14$.
[k]$\Phi = 0.1$ at low conversion. Formation sensitized by benzophenone, acetophenone, xanthone, or triphenylene.
[l]Photoequilibrium established with starting materials.

ADDENDUM

Since the original writing of this review two new annual surveys of carbonyl photochemistry have appeared (92,92) in addition to a review article (94). A communication describing the photochemistries of cicyclo[6.2.0]-decan-9-0nes [49c] and [49t] has appeared (95) as well as a full account of the photochemical ring expansion reaction of cyclobutanones and cyclopentanones (96). Denis and Conia (97) have recently reported that dione [154] gives lactone [155] and cyclopropyl ether [156] upon photolysis in methanol. The postulated mechanism is shown below. Interestingly, when photolysis of [154] was carried out in non-degassed hydrocarbon solvents, anhydride [157] was isolated.

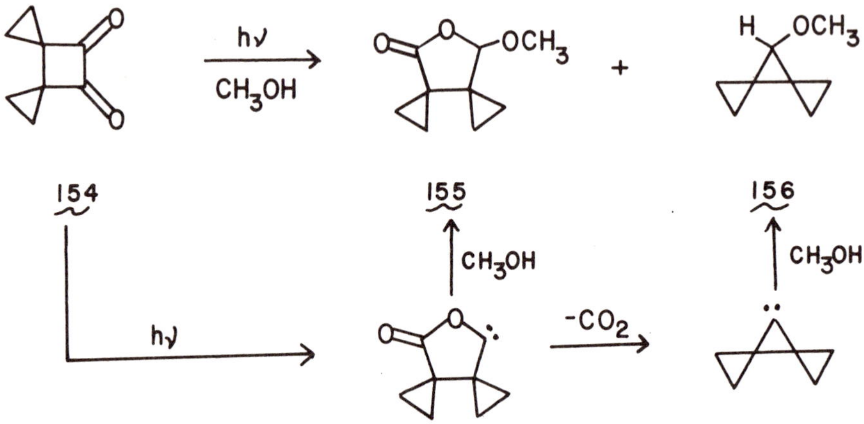

157

REFERENCES

1a. R. Srinivasan in Advances in Photochemistry, W. A. Noyes, Jr., G. S. Hammond, and J. N. Pitts, Jr., Eds., Vol. 1, Interscience, New York (1963), pp. 83-113.

1b. J. G. Calvert and J. N. Pitts, Jr., Photochemistry, Wiley, New York (1966), pp. 389-427.

1c. N. J. Turro in Energy Transfer and Organic Photochemistry, A. A. Lamola and N. J. Turro, Eds., Interscience, New York (1969), pp. 133-296.

1d. N. J. Turro in Annual Survey of Photochemistry, N. J. Turro, G. S. Hammond, J. N. Pitts, Jr., and D. Valentine, Jr., Eds., Vol. 1, Wiley-Interscience, New York (1969), pp. 77-91.

1e. N. J. Turro, D. M. Pond, and F. D. Lewis, Annual Survey of Photochemistry, N. J. Turro, G. S. Hammond, J. N. Pitts, Jr., and D. Valentine, Jr., Eds., Vol. 2, Wiley-Interscience, New York (1970), pp. 47-56.

1f. N. J. Turro, D. M. Pond, D. R. Morton, and J. C. Dalton in Annual Survey of Photochemistry, N. J. Turro and G. S. Hammond, Eds., Vol. 3, Wiley-Interscience, New York (1971), pp. 80-95.

1g. W. M. Horspool in Photochemistry, D. Bryce-Smith, Ed., Vol. 2, The Chemical Society, London (1971), pp. 299-354.

1h. J. C. Dalton and N. J. Turro in Annual Review of Physical Chemistry, H. Eyring, Ed., Vol. 21, Annual Reviews, Inc., Palo Alto, California (1970), pp. 499-560.

2. G. Quinkert, Pure Appl. Chem., 9, 607 (1964).

3a. G. Quinkert, J. Palmowski, H. P. Lorenz, W. W. Wiersdorff, and M. Finke, Angew. Chem. Internat. Edit., 10, 198 (1971).

3b. G. Quinkert, T. Tabata, E. A. J. Hickmann, and W. Dobrat, ibid., 10, 199 (1971).

4a. J. E. Starr and R. H. Eastman, J. Org. Chem., 31, 1393 (1966).

4b. R. S. Cooke and G. D. Lyon, J. Am. Chem. Soc., 93, 3840 (1971).

5. R. Srinivasan and S. E. Cremer, ibid., 87, 1647 (1965).

6. G. Quinkert, E. Blanke, and F. Homburg, Chem. Ber., 97, 1799 (1964).

7. G. O. Schenck and F. Schaller, ibid., 98, 2056 (1965).

8a. G. Quinkert, Angew. Chem. Internat. Edit., 1, 166 (1962).

8b. G. Quinkert, ibid., 4, 211 (1965).

8c. J. C. Dalton, K. Dawes, N. J. Turro, D. S. Weiss, J. A. Barltrop, and J. D. Coyle, J. Am. Chem. Soc., 93, 7213 (1971).

9. P. Yates, Pure Appl. Chem., 16, 93 (1968).

10a. C. C. Badcock, M. J. Perona, G. O. Pritchard, and B. Rickborn, J. Am. Chem. Soc., 91, 543 (1969).

10b. J. A. Barltrop and J. D. Coyle, Chem. Commun., 1081 (1969).

10c. N. C. Yang and R. H. K. Chen, J. Am. Chem. Soc., 93, 530 (1971).

10d. W. C. Agosta and W. L. Schreiber, ibid., 93, 3947 (1971).

11a. D. C. Moule, Can. J. Phys., 47, 1235 (1969).

11b. R. F. Whitlock and A. B. F. Duncan, J. Chem. Phys., 55, 218 (1971).

11c. W. D. Chandler and L. Goodman, J. Mol. Spectrosc., 35, 232 (1970).

11d. L. H. Scharpen and V. W. Laurie, J. Chem. Phys., 49, 221 (1968).

11e. W. D. Chandler and L. Goodman, J. Mol. Spectrosc., 36, 141 (1970).

11f. H. E. Howard-Lock and G. W. King, ibid., 36, 53 (1970).

12. H. U. Hostettler, Tetrahedron Lett., 687 (1965).

13. H. U. Hostettler, Helv. Chim. Acta, 49, 2417 (1966).

14. For a review, see E. K. C. Lee in Excited State Chemistry, J. N. Pitts, Jr., Ed., Gordon and Breach, New York (1970), pp. 59-91.

15a. N. J. Turro and R. M. Southam, Tetrahedron Lett., 545 (1967).

15b. D. R. Morton, E. Lee-Ruff, R. M. Southam, and N. J. Turro, J. Am. Chem. Soc., 92, 4349 (1970).

16. R. F. Klemm, Can. J. Chem., 48, 3320 (1970).

17. J. C. Hemminger, C. F. Rusbult, and E. K. C. Lee, J. Am. Chem. Soc., 93, 1867 (1971).

18. J. P. Wasacz and M. M. Joullie, Tetrahedron Lett., 2501 (1970).

19. P. Dowd, A. Gold, and K. Sachdev, J. Am. Chem. Soc., 92, 5724 (1970).

20. R. R. Hautala and N. J. Turro, ibid., 93, 5595 (1971), and references cited therein.

21a. W. F. Erman, ibid., 89, 3828 (1967).

21b. W. F. Erman and H. C. Kretschmar, ibid., 89, 3842 (1967).

22. G. Quinkert, G. Cimbollek, and G. Buhr, Tetrahedron Lett., 4573 (1966).

23. N. J. Turro and D. M. McDaniel, J. Am. Chem. Soc., 92, 5727 (1970).

24. D. M. McDaniel and N. J. Turro, Columbia University, unpublished results.

25. N. J. Turro and D. R. Morton, J. Am. Chem. Soc., 93, 2569 (1971).

26. J. P. Friend and B. P. Dailey, J. Chem. Phys., 29, 577 (1958).

27. R. G. Ford and R. A. Beaudet, ibid., 48, 4671 (1968).

28. F. M. K. Lam and B. P. Dailey, ibid., 49, 1588 (1968).

29. J. G. Pacifici and C. Diebert, J. Am. Chem. Soc., 91, 4595 (1969).

30. N. J. Turro, D. R. Morton, E. Hedaya, M. E. Kent, P. D'Angelo, and P. Schissel, Tetrahedron Lett., 2535 (1971).

31a. W. L. Mock, Ph.D. Thesis, Harvard University, Cambridge, Mass. (1964).

31b. J. R. Wiseman and H. F. Chan, J. Am. Chem. Soc., 92, 4749 (1970).

31c. C. R. Johnson, G. F. Katekar, R. F. Huxol, and E. R. Janiga, ibid., 93, 3771 (1971).

31d. B. M. Trost and M. J. Bogdanowicz, ibid., 93, 3773 (1971).

31e. J. R. Salaun and J. M. Conia, Chem. Commun., 1579 (1971).

32. For a review, see F. Weiss, Quart. Rev., 24, 278 (1970).

33. P. Dowd and K. Sachdev, J. Am. Chem. Soc., 89, 715 (1967).

34a. P. Dowd, A. Gold, and K. Sachdev, ibid., 92, 5725 (1970).

34b. P. Dowd, G. Senqupta, and K. Sachdev, ibid., 92, 5726 (1970).

35. D. R. Morton and N. J. Turro, Columbia University, unpublished results.

36. N. J. Turro, E. Lee-Ruff, D. R. Morton, and J. M. Conia, Tetrahedron Lett., 2991 (1969).

37. J. K. Crandall and D. R. Paulson, Indiana University, private communication.

38a. P. Yates and L. Kilmurry, Tetrahedron Lett., 1739 (1964).

38b. P. Yates and L. Kilmurry, J. Am. Chem. Soc., 88, 1563 (1966).

39. For a review, see W. Kirmse, Carbene Chemistry, 2nd Ed., Academic Press, New York (1971), pp. 47-51.

40a. W. Kirmse, Carbene Chemistry, Academic Press, New York (1964), pp. 203-212.

40b. R. Gleiter and R. Hoffmann, J. Am. Chem. Soc., 90, 5457 (1968).

40c. H. D. Hartzler, ibid., 92, 1412, 1413 (1970).

41. R. W. Hoffman and H. Hauser, Tetrahedron Lett., 197 (1964).

42. P. D. Bartlett and T. G. Traylor, J. Am. Chem. Soc., 84, 3408 (1962). See also reference (39), pp. 422-423.

43. A. G. Brook, H. W. Kucera, and R. Pearce, Can. J. Chem., 49, 1618 (1971).

44. For an exception, see W. H. Urry, J. C. Duggan, and M. S. Pai, J. Am. Chem. Soc., 92, 5785 (1970).

45. P. J. Wagner, C. A. Stout, S. Searles, and G. S. Hammond, ibid., 88, 1242 (1966).

46. N. J. Turro and D. M. McDaniel, Mol. Photochem., 2, 39 (1970).

47. J. C. Hemminger and E. K. C. Lee, J. Chem. Phys., 54, 1405 (1971

48. R. G. Shortridge, C. F. Rusbult, and E. K. C. Lee, J. Am. Chem. Soc., 93, 1863 (1971).

49. P. M. Rentzepis, Chem. Phys. Lett., 2, 117 (1968).

50a. W. C. Agosta and A. M. Foster, Chem. Commun., 433 (1971).

50b. W. C. Agosta, J. Amer. Chem. Soc., 94, 5777 (1972).

51. Reference (39), pp. 467-473.

52. B. E. Kaplan and L. T. Turner, 158th National Meeting of the American Chemical Society, September 1969, New York, ORGN-40.

53. N. J. Turro, J. C. Dalton, G. Farrington, M. Niemczyk, and D. M. Pond, J. Am. Chem. Soc., 92, 6978 (1970), and references cited therein.

54. D. C. England, ibid., 83, 2205 (1961).

55. A. S. Kende, Tetrahedron Lett., 2661 (1967).

56. T. Do-Minh, A. M. Trozzolo, and G. W. Griffin, J. Am. Chem. Soc., 92, 1402 (1970).

57. E. J. Volker and J. A. Moore, J. Org. Chem., 34, 3639 (1969).

58. M. P. Cava and R. J. Spangler, J. Am. Chem. Soc., 89, 4550 (1967).

59. W. H. Urry, D. J. Trecker, and D. A. Winey, Tetrahedron Lett., 609 (1962).

60. P. A. Leermakers, H. T. Thomas, L. D. Weis, and F. C. James, J. Am. Chem. Soc., 88, 5075 (1966).

61. N. J. Turro and T. Cole, Tetrahedron Lett., 3451 (1969).

62. G. Quinkert, B. Wegemund, and E. Blanke, ibid., 221 (1962).

63. H. Ona, H. Yamaguchi, and S. Masamune, J. Am. Chem. Soc., 92, 7495 (1970). See also P. B. Shevlin and A. P. Wolf, ibid., 92,406.

64. R. Moore, A. Mishra, and R. J. Crawford, Can. J. Chem., 46, 3305 (1968).

65. H. D. Scharf and W. Kuster, Chem. Ber., 104, 3016 (1971).

66. J. Ipaktschi, Tetrahedron Lett., 2153 (1969).

67. D. I. Schuster, M. Axelrod, and J. Auerbach, ibid., 1911 (1963).

68. D. I. Schuster, G. R. Underwood, and R. P. Knudsen, J. Am. Chem. Soc., 93, 4304 (1971).

69. R. B. Woodward and R. Hoffmann, The Conservation of Orbital Symmetry, Academic Press, New York (1970).

70. N. J. Turro, P. A. Leermakers, H. R. Wilson, D. C. Neckers, G. W. Byers, and G. F. Vesley, J. Am. Chem. Soc., 87, 2613 (1965).

71a. E. M. Kosower, J. Chem. Phys., 38, 2813 (1963).

71b. E. A. LaLancette and R. E. Benson, J. Am. Chem. Soc., 83, 4867 (1961).

72. N. J. Turro, W. B. Hammond, and P. A. Leermakers, ibid., 87, 2774 (1965).

73. R. C. Cookson, A. G. Edwards, J. Hudec, and M. Kingsland, Chem. Commun., 98 (1965).

74. H. A. J. Carless and E. K. C. Lee, J. Am. Chem. Soc., 92, 6683 (1970).

75. G. M. Breuer and E. K. C. Lee, J. Phys. Chem., 75, 989 (1971).

76. S. W. Benson and H. E. O'Neal, Kinetic Data on Gas Phase Unimolecular Reactions, National Bureau of Standards Publ., (1970), p. 589.

77a. R. D. Miller and V. Y. Abraitys, J. Am. Chem. Soc., 94, 663 (1972)

77b. R. D. Miller, Abstracts IV IUPAC Symposium Photochemistry, Baden-Baden, July 16-22, 1972, p. 172.

78. H. Hart, R. K. Murray, and G. D. Appleyard, Tetrahedron Lett., 4785 (1969).

79. D. E. Bays and R. C. Cookson, J. Chem. Soc., B, 226 (1967).

80. J. Doyle, Ph.D. Thesis, Oxford University, 1970, ref. 125.

81. H. A. J. Carless, J. Metcalfe, and E. K. C. Lee, J. Am. Chem. Soc., 94, 7221 (1972).

82. R. S. Givens and W. F. Oettle, ibid., 93, 3963 (1971).

83. R. S. Givens, W. F. Oettle, R. L. Coffin, and R. F. Carlson, ibid., 93, 3957 (1971).

84a. H. U. Hostettler, Tetrahedron Lett., 1941 (1965).

84b. R. C. Cookson, A. G. Edwards, J. Hudec, and M. Kingsland, Chem. Commun., 98 (1965).

85. J. J. Hurst and G. H. Whitman, J. Chem. Soc., 2864 (1964); Proc. Chem. Soc., 160 (1959).

86. H. E. Zimmerman and D. J. Sam, J. Am. Chem. Soc., 88, 4905 (1966).

87. G. Buchi and E. M. Burgess, ibid., 84, 3104 (1962).

88. N. J. Turro, R. B. Gagosian, S. S. Edelson, T. R. Darling, J. R. Williams, and W. B. Hammond, Trans. N. Y. Acad. Sci., 33, 396 (1971).

89. H. Hart, Pure Applied Chem., in press.

90. D. R. Arnold, E. Hedaya, V. Y. Marritt, L. A. Karnischky, and M. E. Kent, Tetrahedron Lett., 3917 (1972).

91. W. C. Agosta, Abstracts IV IUPAC Symposium Photochemistry, Baden-Baden, July 16-22, 1972, p. 1.

92. W. M. Horspool in Photochemistry, D. Bryce-Smith, Ed., Vol. 3, The Chemical Society, London (1972), pp. 367-421.

93. W. M. Horspool in Photochemistry, D. Bryce-Smith, Ed., Vol. 4, The Chemical Society, London (1973), pp. 453-512.

94. N.J. Turro, J. C. Dalton, K. Dawes, G. Farrington, R. Hautala, D. Morton, M. Niemczyk, and N. Schore, Accounts Chem. Res., 5, 92 (1972).

95. D.M. McDaniel and J. J. Turro, Tetrahedron Lett., 3035 (1972).

96. D. R. Morton and N. J. Turro, J. Am. Chem. Soc., 95, 3947 (1973).

97. J. M. Denis and J. M. Conia, Tetrahedron Lett., 461 (1973).

Phosphorescence-Microwave Multiple Resonance Spectroscopy

M. A. EL-SAYED
Chemistry Department
University of California
Los Angeles, California

CONTENTS

I. INTRODUCTION

A. Important Developments in Triplet-State Research

Active studies on the relationship between phosphorescence and the lowest triplet state started in the 1940s by G. N. Lewis and his group (1a-1f). In the 1950s and 1960s the electric dipole radiation of the phosphorescence was shown to result from spin-orbit interaction by demonstrating heavy atom effects on the emissior probability as well as from polarization studies (2,3). In the same period rapid advances in the ESR studies of the triplet state were made (3,4a,4b). Triplet-triplet (T-T) energy transfer (5), T-T absorption by flash photolysis (6), and T-T annihilation (7) were demonstrated. The field of triplet photochemistry started and was greatly advanced in the 1960s (8). At present the properties of the triplet state are being studied by a large number of scientists in different research disciplines, e.g., optical and magnetic resonance spectroscopists, theoretical chemists, and physical organic photochemists. Molecules reaching the lowest triplet state are very popular among researchers in all these fields. However, the same molecules have recently become very unpopular among scientists working in the dye laser field due to the optical losses these molecules introduce because of their triplet → triplet absorption.

During the past five years two research disciplines of optical spectroscopy and magnetic resonance have merged when it became evident that at low temperatures, microwave radiation of resonance frequencies with the zero-field (zf) transitions of the lowest triplet state could have observable effects on the phosphorescence intensity as well as the spectrum. Quantitative information can then be obtained from these phosphorescence-microwave multiple-resonance experiments from which the magnetic, the radiative, and the nonradiative as well as the structural properties of the triplet state can be determined.

B. Historical Development of Microwave-Optical Multiple-Resonance Techniques

Historically this field is an extension of the optical detection of magnetic resonance transitions of simple atoms and ions. Research in optical detection of radio-frequency transitions was first made in 1925 on gases by Fermi and Rasetti (9) as well as by Breit and Ellet (10). This was done by observing a change in the polarization of resonance fluorescence upon a change in frequency of an applied alternating magnetic field. The first determination of magnetic transitions in ionic solids was not made (11-13) until 1959. For a review of the field, the reader is referred to the book by Bernheim (14).

In 1967 the optical detection of the ESR transitions of the triplet state was made (15-17). This was done at 4.2°K, where the difference in the Boltzmann population of the zf levels is large enough so that microwave radiation can have a measurable effect on the emission intensity. The observation and assignment of the state of spin alignment, in which the population of the zf levels is not equal due to slow spin lattice relaxation processes, was first detected (18) in 1966 by observing ESR emission, rather than absorption lines. In actuality, manifestation of the state of spin alignment was first observed (19) in 1963 by observing the nonexponential behavior of the phosphorescence decay of naphthalene in durene. This behavior was not then understood. Attributing the observed nonexponentiality in the decay to the presence of a state of spin alignment was first done in 1967 (20). The first optical detection of the zero-field (zf) transitions was made in 1968 (21). Since then the field of studies on the effect of microwave radiation on the phosphorescence emission has become very active. The first application of phosphorescence-microwave double resonance (PMDR)* techniques (in zero-field) to the understanding of the optical spectroscopy of the triplet state was made (22) in 1969. In the same year the first optical detection of electron-nuclear double resonance (23,24) (ENDOR) in zero-field was made. A great many advances were accomplished in the zf studies during 1970, e.g., the use of the delay PMDR techniques (25) in determining the relative rates of the intersystem crossing (ISC) process to the different zf levels, the optical detection of electron-electron double resonance (26) (EEDOR), the determination of the relative ISC rates using steady-state saturation methods (27), the optical determination of the polarization of the zf magnetic transitions (28), and the optical detection (29) of the level anticrossing in the triplet state.

The following important reports on zf studies were made in 1971. The detection of phosphorescence modulation due to the coherent coupling of the triplet spin levels to a strong microwave field was made (30). Using PMDR techniques to determine the coherent length of triplet excitons was also reported (31). The use of pulsed excitation pulsed (or fast sweep) microwave techniques to determine the relative intersystem crossing rates was made (32). The detection of the state of spin alignment at 77°K using ESR and short-pulsed excitation was also made (33).

---

*PMDR distinguishes this field from the general field of optical microwave double resonance (OMDR) by being confined to the studies of the magnetic, radiative, nonradiative, and structural properties of the lowest triplet state of molecules that exhibit phosphorescence radiation. Titles such as optical detection of ESR transitions are not correct since the interest in this field goes beyond the magnetic properties of the triplet state.

Up to the date of writing this article, the following important reports have been made in 1972. Detection of zero-field magnetic signals at 77°K was recently reported (34 Triplet-triplet absorption-microwave double resonance was demonstrated (35). The use of PMDR methods to determine molecular geometry in the triplet state was indicated (36). Hole burning in the PMDR signal using EEDOR techniques has been demonstrated (37). Using PMDR methods, the zero-field transitions of triplet excimers were determined (38).

In the following sections the basic principles and equations of PMDR methods are discussed. This is followed by a description of the experimental methods. The application and results of PMDR spectroscopy to the different research disciplines of genuine interest to the author are summarized and discussed.

II. SPIN ALIGNMENT IN THE TRIPLET STATE

A. Production of State of Spin Alignment by Optical Pumping

For an orbitally nondegenerate triplet state, the spin angular momentum can be quantized in three orthogonal planes, corresponding to the spinning of the two unpaired parallel spins in three orthogonal planes. Due to the nonspherical electron distribution in molecules, the magnetic energy of the two spinning electrons in one plane may differ from that in the other. This gives rise to the zero-field splittings, which is of the order of a fraction of a wavenumber for the triplet states of organic molecules and a couple of wavenumbers in the O_2 ground triplet state. This magnetic energy is due to the magnetic dipolar repulsive interaction between the two parallel spins and to the spin-orbit magnetic interaction.

The contribution of the latter to the zf splittings of aromatic hydrocarbons is vanishingly small.

Three different optical methods can be used to excite the lowest triplet-state (T_1) molecules: singlet-singlet absorption, singlet-triplet absorption, and triplet-triplet energy transfer which follows absorption by the donor. After direct $S_o \rightarrow S_x$ excitation, molecules that do not photoreact or decompose in S_x return to the lowest singlet state, S_1, from which they either fluoresce or cross over to the triplet manifold by the intersystem crossing (ISC) process. The probability of the latter process depends on the degree of spin-orbit interactions, which are anisotropic in molecules. Thus molecules reaching the different zf levels do so via a specified spin-orbit interaction scheme which, in general, is different for the different zf levels. If pumping of the lowest triplet state takes place via singlet-triplet absorption, the pumping rate of the different zf levels depends on the radiative strength of the transition from the ground state to each zf level. The latter depends on the spin-orbit coupling strength of this transition to the radiative spin allowed transitions. If pumping is carried out by triplet-triplet energy transfer from a donor, the rate of pumping of a certain zf level of the acceptor depends on the population and the projections of the magnetic axes of the different zf levels of the donor onto those of the acceptor. This pumping process is also expected to be anisotropic in many mixed crystals due to the anisotropic spin-orbit perturbations in the donor (which determines the relative population density of the donor zf levels). From these considerations, it is obvious that the rates of pumping of the different zf levels are expected to be different, independent of the pumping mechanism.

The decay of the different zf levels of the lowest triplet state (from which phosphorescence takes place) to the ground singlet state depends also on the strength of the spin-orbit coupling and is thus expected to be, in general, different for the different zf levels. Since the pumping rates and decay rates are different for the different zf levels, it should be possible to prepare the lowest triplet state with its zf levels having unequal population, if the spin-lattice relaxation between these levels is not very fast. Under these conditions, the triplet state is said to be in a state of spin alignment.

If one defines the degree of spin alignment as a measure of the deviation from the equal population distribution for the different zf levels, then it is obvious that the SLR processes* should be absent (or very slow compared with the rates for the other processes) if the triplet state is to exist in a state of spin alignment. The actual value of the degree of spin alignment could be related to the population ratios of the zf levels.

In the absence of the SLR processes, the value of this ratio would depend on the mode of preparation of the triplet state. Three extreme cases could be realized. In the first case, the system is being continuously excited and decays continuously, i.e., the system is in a steady state. By equating the rates of pumping of any zf level with its rate of decay, one obtains the following ratios:

$$n_j/n_i = K_j k_i/K_i k_j \qquad (1)$$

---

*Unless the system is at very low temperatures so that $kT \ll \Delta E$, where ΔE is the average energy separation between the zf levels.

where n, k, and K are the population, decay constants, and pumping constants, respectively, of the zf levels i and j.

If after the system reaches the steady state, the excitation is cut off sharply, then the above ratio decays according to the following equation:

$$(n_j/n_i)_t = (K_j k_i/K_i k_j)e^{-(k_j - k_i)t} \quad (2)$$

Thus if the value of eq. (1) is not different from unity, i.e., the degree of spin alignment is small, larger values for the spin alignment could be created by allowing (25) the system, whose decay constants k_i and k_j are different, to decay.

If the K values are different, a large value of spin alignment could be obtained before the system reaches its steady state. This is accomplished experimentally by using pulsed excitation (32). The population ratio is simply equal to the ratio of the pumping rates, if the rate of pumping the zf level is much faster than its decay rate (a situation which is true at short times after short-pulsed excitation) (32), i.e.,

$$n_j/n_i = K_j/K_i \quad (3)$$

These are different situations by which the triplet state could be formed in a state of spin alignment. In order for the triplet state to absorb microwave radiation, $n_j - n_i \neq 0$ or $n_j/n_i \neq 1$. Thus a necessary, but not sufficient, condition for a successful PMDR experiment is that the triplet state should be in a state of spin alignment.

B. Detection of Spin Alignment in Zero-Field

An accurate method for detecting the state of spin alignment is by observing large effects of microwave field at resonance frequencies with the zf transitions, on the population of the corresponding zf levels. The observation of a stimulated microwave emission instead of absorption is better at proving the existence of the spin alignment. The observation of microwave emission, instead of absorption of ESR lines, gave us the first strong indication of the presence of the state of spin alignment (18

In zf, a number of optical methods can very easily detect the state of spin alignment. The large effects that microwaves have on the phosphorescence spectrum (22), intensity (21), lifetime (39), and polarization (40) give strong indications of the presence of spin-alignment state. Some of these will be discussed in detail below. The effect of increasing temperature, instead of applying microwave fields on the relative bands originating from the different zf levels, is also used to detect the presence of the state of spin alignment (41). At temperatures for which the SLR processes have rates comparable to the unimolecular decay processes, the decay curve becomes sensitive to temperature. The decay constant of the phosphorescence at temperatures for which the SLR are fast is related to the unimolecular decay constants k_i of the individual zf levels by the equation (42)

$$k = \sum_i k_i X_i \qquad (4)$$

where X_i is the fraction of the total population of the zf level i. This equation was first tested for quinoxaline (20) and later for pyrazine (43). The different decay components observed at low temperatures for pyrazine are found to have different polarization characteristics (44), indicating that all three zf levels in this molecule are at least partially radiative.

For systems whose emission is composed of progressions with false origins, different progressions could originate from different zf levels. If the population ratio of the different zf levels is changed by temperature, decay, pulsed excitation (time-resolved spectroscopy), or magnetic fields, the relative intensity of the different progressions is expected to change. The effects of the first and last perturbations have been demonstrated. Higher temperatures increase the rates of the SLR processes, which tends to decrease the degree of spin alignment. An applied magnetic field mixes the zf functions, thereby equalizing the radiative and pumping rates of the different (45) levels. An applied magnetic field also increases the average energy spacings between the levels and increases the SLR rates (20). If the application of a magnetic field, a microwave field at resonance frequencies, or a temperature increase affects the relative band intensities in the spectrum or changes the shape of the nonexponential decay, then the system must exist in a state of spin alignment prior to the application of the perturbation.

C. Basic Equations of the Effect of Microwave Radiation

If the system under study has its triplet state in a state of spin alignment, then the application of microwave radiation would change the degree of spin alignment. More important, microwave radiation of resonance frequencies should change the number of radiating molecules in a certain level, thus causing a change in the intensity of the emission from that level. The ratio of the intensity in the presence of resonant microwaves to that in the absence of the microwaves depends on the degree of spin alignment (which depends on the excitation mode), the difference in the radiative rate constant of the two levels being saturated, the manner by which the microwave is applied (pulsed, rapidly or slowly swept), and the microwave power used. Microwaves could be applied slowly, causing a new steady state to be established. On the other hand, microwaves could be swept rapidly as compared to the decay times of the levels. Depending on the power of the microwaves, the population of the two levels could be equalized (saturation) or even completely inverted (46a,46b) upon the fast passage of the microwaves. Slow sweep or the continuous exposure of the microwave can be applied only to a system in a steady state, i.e., when the degree of spin alignment is given by eq. (1). For simplicity, let us assume that the total emission intensity of the phosphorescence or the emission of a vibronic band whose intensity is being monitored, originates from the τ_i zf level. Furthermore, if the rate of sweep is very slow to ensure the satisfaction of the steady-state requirements, saturation of the zf levels is the maximum that can be attained when microwaves of sufficient power are used. If I^{ν} and I are the phosphorescence intensity with microwaves (that saturates the $\tau_i \leftrightarrow \tau_j$ transition) and without microwaves, respectively, then I^{ν}/I is given by (47)

$$I^{\nu}/I = n_i^{\nu}/n_i = [(K_i + K_j)/K_i] \cdot [k_i/(k_i + k_j)] \quad (5)$$

The sweeping of the microwaves across resonance can be used in all three extreme cases whose population ratios are given by eqs. (1) - (3). The sweep rate across resonance should be fast compared with the decay rate of the zf levels. Two types of effects can be discussed: (a) the microwaves have enough power to saturate the zf transition being swept (i.e., to equalize the population of the corresponding zf levels) or (b) the microwaves invert the population of the zf levels involved.

If the microwaves saturate the $\tau_i \leftrightarrow \tau_j$ transition, then the intensity ratio of the emission from zf level τ_i is given by

$$I^{\nu}/I = \left[\frac{n_i + n_j}{2}\right] /n_i = \frac{1}{2}\,(1 + n_j/n_i) \qquad (6)$$

Depending on the mode of excitation, n_j/n_i would be related to the rate constants involving the τ_i and τ_j levels as given by eqs. (1) - (3). Thus for a system which exists in a steady state before sweeping the microwaves, the following equation is applicable:

$$I^{\nu}/I = 1/2(1 + K_j k_i/K_i k_j) \qquad (7)$$

If the steady state is allowed to decay and the microwaves are swept at time t after excitation ceases, then

$$(I^{\nu}/I)_t = 1/2 \left[1 + (K_j k_i/K_i k_j) \quad e^{-(k_j - k_i)t}\right] \qquad (8)$$

On the other hand, if the state of spin alignment is formed by pulsed excitation, and the microwave radiation is swept shortly after excitation to saturate the $\tau_i \leftrightarrow \tau_j$ transition, then

$$I^{\nu}/I = 1/2(1 + K_j/K_i) \qquad (9)$$

Equations have also been derived (48,49) for systems in which the emission originates from two zf levels.

If the microwave radiation is swept in a very short time with enough power to completely invert the population of the τ_i and τ_j levels, then

$$I^{\nu}/I = n_j/n_i \tag{10}$$

Substituting eqs. (1) - (3) in eq. (10), one obtains

$$I^{\nu}/I = K_j k_i/K_i k_j \tag{11}$$

and

$$(I^{\nu}/I)_t = (K_j k_i/K_i k_j)e^{-(k_j - k_i)t} \tag{12}$$

and

$$I^{\nu}/I = K_j/K_i \tag{13}$$

for the steady-state, decay, and pulsed-excitation experiments, respectively.

It is clear from the above that the signal-to-noise ratio obtained in an optical detection method for a given system would greatly depend on the mode of excitation, mode of exposure to the microwaves, and mode of detection. The difference between the signals obtained for different molecules, using the same technique, would depend on the relative values of the rate constants of pumping and decay. In general, steady-state excitation methods [thus using eqs. (7) or (11)] are most useful when one of the zf levels being irradiated with microwaves has a larger K but smaller k than the other zf levels. If the two pumping rates are comparable, but the decay constants are very different, then the decay technique [thus eqs. (8) or (12)] is more appropriate to use. On the other hand, if the opposite is true, i.e. the decay constants are comparable but the pumping rates are different, the pulsed method [thus eqs. (9) or (13)] is the most useful one.

III. EXPERIMENTAL TECHNIQUES

For a recent detailed examination of the different experimental methods in this field, the reader is referred to references (49) and (50). A summary of these methods is given below.

There are several methods by which microwave resonance can be optically detected. The methods that are utilized in our laboratory, in order of increasing complexity of the attendant electronics, are (a) continuous-wave (cw) operation; (b) AM or FM modulation of the microwaves with lock-in detection at the modulation frequency (lock-in); (c) cut-off of the exciting light followed by a sweep of a microwave frequency region after a specified delay time (delay); and (d) sweep of microwave frequency region after subjecting the sample to a short excitation U.V. pulse (pulsed-excitation method). A computer of average transients (CAT) could be used to average the noise in all but method (b). Each of these methods incorporates certain advantages; which one is most advantageous for a given molecular system will depend on the rate constants for pumping and decay of the individual triplet sublevels as shown by the equations and discussion given in the previous section.

In order for optical detection of microwave resonances to be feasible for a given molecular system, four important conditions must be satisfied: (a) the triplet state in question must display luminescence; (b) some mechanism must exist for creating the triplet state in a state of spin alignment; (c) this population imbalance, once created, must persist for a time sufficient to allow interrogation of the system by the microwaves; and (d) the radiative rate constants of, or the polarization of the emission from, the two zf levels being pumped with microwaves should be different.

The first condition, though by no means universally satisfied, is not a serious limitation, since many molecules, both organic and inorganic, display phosphorescence when in glass or solid solution at reduced temperatures (77°K or lower). The second of these conditions is generally readily satisfied, inasmuch as intersystem crossing processes not only exist with significant yields for many molecular systems, but are also subject to spin-orbit symmetry selection rules which cause them preferentially to populate specific sublevels in the triplet manifold. When the intersystem crossing process is not anisotropic, other modes of pumping could be used, e.g., direct singlet-triplet absorption and triplet-triplet energy transfer from a donor or from the host crystal in which the molecule of interest is dissolved.

Satisfaction of the third condition above depends on the rates of the spin-lattice relaxation processes between the spin sublevels. These rates are highly temperature dependent and depend also on the environment within which the molecular system is placed. In order to maintain a steady-state triplet sublevel population imbalance, the rates of spin-lattice relaxation must be slower than the rates at which the sublevels decay to the ground state. To reduce the spin-lattice relaxation rates to this level requires temperatures of the order of 4.2°K or lower, in the proper solvent. Whether or not spin-lattice processes can be "frozen" at liquid helium temperatures may even depend on the solid phase of the crystals. For pyrazine in cyclohexane (51), for example, a slow cooling of the solution to below its freezing point prior to immersion in liquid helium gives a spin-polarized system, but rapid cooling yields a system in which spin-lattice relaxation rates are too fast even at 1.4°K for having spin polarization. The signal has been observed in EPA and sometimes in water as solvents, but in these cases power amplifiers are normally required.

In principle, optical detection of microwave resonances should also be possible, even in the presence of fast spin-lattice relaxation, e.g., at 77°K, using a pulsed light source to create an instantaneous triplet sublevel population imbalance and observing the effect of microwaves before the system has the opportunity to relax. Such an effect has been observed using ESR techniques (33). In zero-field optical detection of a fixed-frequency cavity has been observed (34) at 77°K with high microwave powers, the signal could be observed as long as the rate of microwave pumping overcomes the spin-lattice relaxation rates.

The experimental setup can be summarized as follows: the sample is inserted in a copper coil slow-wave helix structure which is attached to a stainless steel coaxial waveguide that is connected to microwave sweepers. The sweepers have output in the 0.06 watt range. Power amplifiers capable of producing power output of up to 20 watts are now available, and this assembly is immersed in liquid helium in a double-jacketed, bubble-free, quartz window Dewar assembly. Exciting light from a UV light source suitable for excitation of the phosphorescence emission of the sample is focused on the sample through suitable filters which block unwanted scatter light that would otherwise increase detection noise. The emitted light from the sample is collected by a lens system (optional), passed through a filter or monochromator, and detected by a phototube, whose output is amplified and displayed using suitable electronics.

The experiment then consists of changing the frequency output of the microwave generator, whereupon, at the frequency corresponding to a transition between magnetic sublevels of the sample, an intensity change in the phosphorescence emission will be observed, provided that the microwave power and the difference in population and the radiative rate constants between the triplet sublevels are sufficiently large. This basic scheme remains the same for the various modifications used in this field which represent the ways of enhancing the observed signal, either by increasing the selectivity of the detection network or by increasing the population difference at the instant of passage through the microwave resonance.

In a few special cases, the intensity change upon passage through a microwave transition may be observed with the naked eye. For pyrimidine in benzene, for example, the total steady-state emission intensity increases by about 20% upon passage through the 0.923 GHz (2E) transition. This change is extremely sensitive to the microwave sweep rate, and the resulting flashes of light can be readily seen with the naked eye. Such large changes are unusual; intensity variations of the order of a few percent are the norm.

In the next three sections a description is given of the different applications of PMDR spectroscopy to the different research areas involved in the study of the magnetic, optical, structural, and nonradiative properties of the triplet state.

IV. PMDR AND MAGNETIC STUDIES

A. Zero-Field Transitions

The first optical determination of zf transitions was done by Schmidt and van der Waals (21) for quinoxaline in durene at 1.4°K. Recently the measurement was extended (34) to 77°K, but a cavity was used, thus limiting the method to one frequency at a time. Until a variable-frequency cavity (e.g., helix) is used, the PMDR zf measurements at 77°K would remain impractical. The optical detection of zf transitions has also been reported for 2,3-dichloroquinoxaline (22), 8-chloroquinoline (52,53), quinoline (53), 1,2,4,5-sym-tetrachlorobenzene (54) in durene, as well as its x-traps (31) and a number of compounds of short-lived triplet states, e.g., the n,π^* states of the carbonyl compounds (55,56) and a number of halogenated benzenes (54,31,57). It was not possible to determine the zf parameters of these compounds by ESR, due to the short lifetime and thus the low steady-state concentration of the triplet state. Certainly the low steady-state concentration of the triplet state tends to decrease the PMDR signal, but this effect is compensated by the high emission intensity of these strongly radiative molecules.

Using PMDR methods, the broad excimer emission formed at defect sites in the lattice of hexachlorobenzene crystal at 1.6°K (by doping it with high-energy triplet molecules at 1.6°K) is shown (38) to have zf transitions. This provides strong evidence that it is a triplet excimer. For molecules with atoms having nuclear spin, transitions between hyperfine levels of different zf levels are also observed.

The nitrogen hyperfine structure observed in quinoxaline and its 2,3-dichloro derivatives is satisfactorily explained (23,24,53) as arising from the ^{14}N quadruple interaction and a second-order ^{14}N hyperfine interaction. These experiments yield, for the first time, a measurement of excited-state ^{14}N, ^{35}Cl, and ^{37}Cl coupling constants (23,24,53,58,59) for the excited triplet state. As has been observed in ESR, transfer hyperfine interaction resulting from the transfer of a small but finite electron spin-density from the excited triplet state of a guest to the nucleus of the host has also been optically detected (60) in zf, as was observed in ESR (61,62).

The detection of what is believed to be optically detected signals for the zf transitions of triplet excitons has been recently reported (31). By fitting the observed band shape to a theoretical expression based on a one-dimensional model (63), a minimum coherent length of 50-100 Å was obtained for the triplet exciton in sym-1,2,4,5-tetrachlorobenzene at 3.2°K.

A very elegant experiment has been done by van der Waals and his group recently (30) in which the phosphorescence modulation due to coherent coupling of the triplet spin levels with a strong microwave field of resonance frequencies was demonstrated. The system was quinoxaline-d_7 in durene, and the zf transition at 1.001 GHz was coupled to the microwave field. The experiment was done in the following manner. The excitation was cut off after the steady state was accomplished. A few seconds later, the radiative level was empty, yet the other level was still densely populated At this instant, powerful microwave pulses of 1.001 GHz frequency but of different pulse widths were applied in different experiments. The population of the radiative level, and thus the intensity of pulse of the phosphorescence produced, is shown to be a periodic function of the interaction time with the microwave field, i.e., a periodic function of the pulse width. In agreement with theory, the maximum intensity is found to occur when $\gamma H_1 t = \pi$ and the minimum occurs when $\gamma H_1 t = 2\pi$, where γ, H_1, and t are the gyromagnetic ratio, microwave magnetic field, and pulse width, respectively. This type of experiment was discussed in the same year by Harris (46a,46b).

The main advantages of the optical detection of zf transitions are (a) the need of using single crystals and magnets is eliminated; (b) zf transitions for very short-lived triplet states may be detected as long as the short lifetime is a result of an allowed radiative process; and (c) the cost of the experimental set-up is not as high as the ESR equipment. The disadvantages are (a) a liquid helium set-up is required, at least at the present time, (b) for molecules with a single radiative zf level, only two of the three zf levels are optically detected. The transition involving the dark zf levels will not be detected if the SLR is absent. Electron-electron double resonance techniques (31) (EEDOR), described later, could be used to determine the third transition. Other methods that would slightly increase the coupling of the levels, e.g., heating or application of small magnetic fields, might also be useful in detecting the third transition; and (c) the elimination of a direction indicator, e.g., the magnetic field in the ESR experiments, renders the PMDR method (as described above) to be incapable of determining the polarization and thus the assignment of the zf transitions.

An assignment of the zf transitions can be made using PMDR techniques by one of the following two methods. If the phosphorescence spectrum is analyzed and well understood, and its polarization is determined for symmetric-type molecules, the response of the different vibronic bands upon saturating the different zf transitions could give results from which the assignment of the different zf transitions may be inferred. On the other hand, the assignment could be made directly if the polarization of the magnetic zf transitions is determined. PMDR techniques have been extended (28,64) to make these measurements as given below.

B. Polarization of the zf Transitions from PMDR

The polarization of the zf transitions (28,64) is determined in the absence of a laboratory magnetic field by using a single mixed crystal oriented in the microwave helix or cavity with one of its crystal axes parallel to the axial axis of the helix. This method makes use of the fact that the magnetic field of the microwaves is not isotropic inside the helix. An equation is derived which relates the change in the optical signal to the power of the microwave radiation in resonance with the zf transition whose polarization is to be determined. At low microwave powers, the equation gives a straight-line relationship between $\frac{dI}{dP}$ and $\frac{1}{P}$, where I is the phosphorescence intensity from a certain zf level and P is the microwave power. From the ratio of the slope to the intercept, a quantity proportional to the microwave absorption cross section along the crystal axis employed is determined. Repeating the experiment with the crystal oriented differently, a quantity proportional to the absorption cross section along the other axis is determined. From these two values the polarization ratio of the magnetic zf transition in the molecular framework can be determined from a knowledge of the crystal structur of the host. It should be pointed out that the experimentally determined values of the polarization ratios for the zf transitions of 2,3-dichloroquinoxaline in durene (64) are in better agreement with the oriented gas model than those obtained for the phosphorescence emission of the same crystal. This difference might be due to the fact that depolarization effects are larger for the phosphorescence emission than for the microwave radiation.

This method was also used to determine the polarization of the zf transition of 1,2,4,5-sym-tetrachlorobenzene (54) in durene. This work points out that an assignment of all the zf transitions as well as the relative order of the zf levels can be made from the lifetimes of the zf levels, the spectroscopy, and the knowledge of the polarization of one crucial zf transition.

C. Optical Detection of Multiple-Resonance Magnetic Transitions

In magnetic resonance, electron-nuclear double resonance experiments have been performed (65-67). Using optical techniques, similar experiments can be performed on the triplet state in the absence of the laboratory magnetic field, but in the internal nonvanishing zero-field of the triplet state. Two types of these optical detection methods are described below.

1. Optical Detection of Electron-Nuclear Double Resonance (ENDOR) Transitions in Zero-Field

If a molecule contains an atom with a nucleus having nuclear spin $I > 1/2$ (e.g., N or Cl), the observed microwave spectrum will be more complex due to the electron-nuclear hyperfine interaction as well as nuclear quadrupole interactions (23,24). For each of the three electron-spin directions, there are a number of different nuclear spin quantization directions. For nitrogen,

I = 1; thus for each electron-spin level there are three different nuclear spin directions. If the rate of formation of the different zf states of the triplet state from S_1 is too fast as compared to the nuclear spin reorientation time in the magnetic field of the two unpaired electrons, then nuclear polarization will not be produced in the intersystem crossing process. Thus the steady-state population is expected to be equal for the different nuclear spin levels belonging to the same electron-spin level at very low temperatures in zero-field. If this is true, then the nuclear transitions in any zf level of the triplet state cannot be optically detected directly, since the corresponding levels would have equal populations and their transitions would not have net radiowave absorption. Even if there is a nuclear polarization in a certain electron-spin level, the change in the population density of these levels by radiofrequency would not change the total radiative power of the electron-spin level. This is due to the fact that the radiative probability of each nuclear spin level in the same electron spin level is the same.* One can make use of the difference in population of the different electron-spin (zf) levels to create a net population difference in the hyperfine (nuclear spin) level. If one saturates two electron-nuclear spin levels continuously with microwave radiation, then the nuclear levels being saturated would have different populations from the other nuclear levels in the same zf

*This might not be the case if the spin-orbit coupling in these molecules involves the orbital motion of the electron and the total spin angular momentum of the electron and nuclear spins. In this case, the intersystem crossing process might also be different for different nuclear spin levels belonging to the same electron-spin level.

manifold. If a radio-frequency sweeper is then used, net absorption would occur (and a corresponding change in the phosphorescence intensity would be observed) at frequencies corresponding to nuclear transitions between the nuclear spin levels being saturated with microwaves and other nuclear spin levels in the manifold belonging to the same electron-spin state. The equation relating the change in the phosphorescence intensity to the different pumping and decay constants has been derived for the steady-state experiment (47).

It is thus obvious that the energies required to change the direction of the nuclear spin in the field of the electron spin can be determined optically in zero-field. This is an optically detected n.m.r.-type experiment whereby the laboratory field is replaced by the field of the two unpaired spins (zf) of the triplet state in the molecular framework. The ENDOR frequencies are important in determining hyperfine and quadrupole parameters in the excited triplet state (23,24,53,57,58).

2. Optical Detection of the Electron-Electron Double Resonance (EEDOR) in Zero-Field

For molecules whose emission originates from only one zf level, e.g., symmetrical aromatic hydrocarbons and their N-heterocyclics, or when the population density of two zf levels is the same, optical methods could determine only two zf transitions at very low temperatures. In order to determine the third transition, an electron-electron double resonance experiment has to be performed (26). Let us assume the three electron-spin levels are τ_1, τ_2, and τ_3. If τ_1 is the only radiative level, or the level whose population density differs from the other two that have equal population, then the $\tau_2 \leftrightarrow \tau_3$ transition at low temperature would induce no change in the phosphorescence intensity upon microwave saturation. However, if the $\tau_1 \leftrightarrow \tau_2$ transition is saturated continuously with microwave radiation ν_1 and another microwave sweeper is swept, the population of the emitting τ_1 level changes, and a corresponding change in the phosphorescence intensity from I_{ν_1} to I_{ν_1,ν_2} is observed, when the frequency of the variable-frequency microwave ν_2 is equal to $(E_2 - E_3)/h$. An equation is derived (47) which relates $I_{\nu_1,\nu_2}/I_\nu$ to the pumping and decay constants of the radiating system in a steady-state experiment.

D. Hole Burning in the PMDR Signal

The optically detected microwave signals are much broader than the uncertainty width. One possible cause for the extra width is inhomogeneous broadening. Different molecules in the solid could have slightly different environment, causing the zero-field transition energy of the different molecules to have a distribution with a width that contributes to the observed width. If the difference in environment is not large enough to cause difference in the spin-orbit interactions, one would expect that the population ratio of the two zf levels being saturated with microwaves would be independent of the part of distribution being examined. The application of microwaves to the ensemble would give rise to the same change in the phosphorescence intensity for each molecule, independent of the part of distribution. The size of the signal at a certain frequency would be proportional to the fraction of molecules having transition energy corresponding to the applied frequency. If while sweeping the microwave, another microwave with a fixed frequency, ν_f, is used to equalize the population of a certain portion of the distribution, the PMDR signal obtained from the variable-frequency microwave source will not register signal at frequency ν_f. This is due to the fact that molecules with this frequency have equal zf population and do not interact with the variable-frequency microwave. Thus with this technique one could isolate a certain fraction of the distribution for examination. In fact, by modulating the power of the fixed-frequency microwaves and using phase-sensitive detection, the PMDR spectrum of the fraction of distribution having zf energy = $h\nu_f$ could be recorded (37).

This technique should be useful in resolving hyperfine structure in glasses, in which the signal is very broad. Energy transfer studies could also be used with these techniques. Energy transfer from one molecule in one part of the distribution to another molecule in another part should lead to "filling the holes." Studies along these lines are now in progress.

E. Optical Detection in a Magnetic Field: MODR

As was pointed out in the introduction, optical detection of the Zeeman transitions of the triplet state preceded the optical detection in zf. Since these former experiments resemble those in ionic crystals, researchers in this field called this technique MODR (microwave-optical double resonance). The assignment of the zf transitions as well as the relative order of the zf levels could be concluded also from the MODR techniques as in the PMDR technique. The first reported MODR experiment was made by Sharnoff (15), in which the $\Delta m = 2$ transition of the $C_{10}D_8$ triplet state in a biphenyl host, using amplitude modulation of the microwave power. A few months later Kwiram reported the optical detection of the $\Delta m = \pm 1$ for phenanthrene in biphenyl (16). The experiments were done at a fixed microwave frequency (8.88 GHz) but slowly sweeping the magnetic field through resonance. The output signal from a photomultiplier detector was stored in a computer of average transients (CAT). One hundred sweeps were required. In the same year, but four months later, the optically detected spectrum of quinoxaline in durene was reported (17). These workers compared two techniques. In one technique, the field was modulated at 12 Hz and phase-sensitive techniques were used for detection. In the other technique, the microwave was amplitude-modulated and phase-sensitive detection was again used.

These MODR techniques have the same advantages as PMDR methods in being able to determine zf transition of short-lived triplet states. They have the added advantage of being able to modulate either the field or the microwave and thus are able to obtain better signal-to-noise ratio. Because of this, it is not necessary that the SLR processes be absent for the successful optical detection using the MODR methods. In these cases, however, the temperature has to be low enough (e.g., 4.2°K) so that a detectable Boltzmann population difference can be maintained. PMDR is more convenient than MODR for determining the zf energies since single crystals of known structures are not required for the former technique. A detailed examination of the different factors affecting the sensitivity of MODR has been given by Sharnoff (68). An equally extensive examination of the different experimental techniques used in the PMDR field has also recently been made (49,50).

MODR techniques have been used to obtain the magnetic properties of new short-lived systems, e.g., pyrazine (69,70), benzophenone (71,72), the oxidation product of durene (carbonyl compound) (73), and cyclopentanone (68).

The absolute order of the zf levels in 1,2,4,5-sym-tetrachlorobenzene in durene was determined recently by the effect of increasing the temperature on the MODR signal (74).

In addition to optically detecting and assigning zf transitions, MODR has also been used by Veeman and van der Waals (29) to show level anticrossing and cross-relaxation effects between the Zeeman levels of the lowest triplet state of benzophenone (29). If the magnetic field is exactly aligned along one of the magnetic principal axes of a triplet molecule, the field mixes the function corresponding to the other two orthogonal magnetic axes to give the $m = \pm 1$ states. The state which corresponds to the magnetic axis along the field is that correlated with the $m = 0$ high-field state. At fields when the $m = 0$ and $m = +1$ (or $m = -1$) states have equal energies, or have energy separation smaller than the radiative line-width, a coherent interference takes place. Since the two states have different symmetry properties, their levels cross one another as the magnetic field changes its value. If the magnetic field is not exactly aligned with one of the molecular magnetic principal axes, the $m = 0$ and $m = +1$ (or $m = -1$) would no longer be orthogonal to one another, but would be mixed. Each state would have a small part of the other state. This would result in a repulsion between their energy levels, and the phenomenon of level anticrossing, rather than crossing, would be observed. This is found to be the case for the benzophenone (29) crystal due to the difficulty of exact alignment of the magnetic field with the principal magnetic axes. Furthermore, at high exciting light intensities, the phenomena of cross-relaxation between the electron-spin directions of the Zeeman state of different adjacent equivalent triplet molecules have been demonstrated (29). This possibility was previously discussed by Gorfer (75).

The field of MODR has not progressed at the same rate as the PMDR field. The number of papers contributed to the PMDR field during the past few years is more than five times those appearing in the MODR field. One importan contributing factor is the convenience of using powder, rather than single crystals, in the PMDR method. This eliminates the difficulty of aligning the single crystal in the magnetic field. Another equally important contributing factor is the fact that the measurements done using PMDR are characteristic of the molecule without the external perturbation of the field. A third factor is a result of the convenience in using sweepers and helices rather than magnets and tuned cavities. These factors made PMDR methods attractive to researchers not only in the magnetic field, but also in other fields, e.g., optical spectroscopy and the field of nonradiative relaxation. For systems of low symmetry and whose phosphorescence spectrum is not fully understood, the assignment of their field transitions is best made from the MODR studies.

V. PMDR IN OPTICAL AND STRUCTURAL SPECTROSCOPY

The earlier experiments done on optical detection of the Zeeman transitions were actually focused on determining the zero-field origin of the total intensity of the phosphorescence emission (15-17). The previous workers did not use a spectrometer as a part of their detection system. Fortunately most of the phosphorescence intensity of the systems studied by MODR originates from a single zf level. No optical spectroscopy has been done so far with the MODR methods in spite of the very important information that can be obtained from it. This is probabl due to the fact that researchers in the MODR field came to it from the ESR field. It is certain that such studies will soon be made. The importance of PMDR methods in phosphorescence spectroscopy was first realized (22) in 1969, and its usefulness has since been extended. In the four sections below, the different applications of PMDR to this field will be briefly discussed.

A. Zf Origin of Emission Bands

It has been shown that PMDR is useful in identifying the zf origin(s) of the different vibronic bands in the phosphorescence spectrum (22,39,40,76). If a vibronic band originates from a single zf level and if the zf transitions are known and assigned, then the response of the intensity of this band to the microwave saturation of the different zf transitions could determine its zf origin. On the other hand, the band could be emitted from two zf levels. If the molecule has reasonable symmetry (not too high and not too low), the radiation emitted from each zf level would have different polarization characteristics. In these systems, microwave saturation of the zf transitions connecting the two radiative zf levels would have little effect on the intensity of the band but could have large effects on the polarization characteristics. This is due to the fact that microwave saturation has the effect of transferring the excess molecules from one level to the other. If the radiative rate constants of the two levels are comparable, the excess molecules emit radiation at the same rate but with different polarization. It is thus convenient to conclude the zf origin of the vibronic bands from the following simple rules:

a. For bands of single origin, saturation of two zf

transitions should change its intensity, but not its polarization. Saturation of the third transition, or the one connecting the levels that do not contribute to the intensity of the band, should not have any effect on the intensity or the polarization of the band.

b. For bands of mixed origin, with two zf levels emitting in molecules of reasonable symmetry, saturating all three zf transitions could change the intensity; the one having the least effect is the one that should cause large changes in the polarization of the band. The latter transition connects the zf origins of the band.

If the symmetry increases or decreases drastically, the spin-lattice relaxation process is fast, or the optical pumping is not successful in producing large spin alignment, the above rules would not be as definite. A convenient method by which these experiments are done is to use phase-sensitive detection. The microwave frequency that is capable of saturating a certain zf transition is modulated, and the phosphorescence intensity change is detected at the same modulation frequency while the optical spectrum is slowly scanned. The spectrum obtained is termed (40) the amPMDR (amplitude-modulated PMDR) spectrum. It is better to take this spectrum when the solute is in a single crystal of the host and an analyzer parallel to one of the crystal faces is used with its transmission maximum along one of the crystal axes. The spectrum is also recorded with the analyzer axis rotated 90°. The spectrum obtained is termed (40) the polarized amPMDR spectrum. In this spectrum, bands that originate from one single level should show only a change in intensity upon changing the direction of the polarizer, while bands of mixed origin could show additional change in phase (or phase angle).

It should be pointed out that once the zf origin of the different bands is determined (in particular the true and false origins of the spectrum), a complete description could be given for the spin-orbit perturbations that give the lowest triplet state its radiative properties. A very extensive work was conducted by Tinti and El-Sayed (39) in which different perturbations—e.g., heating, applying a magnetic field, as well as saturation of the zf transitions with microwave radiation-were used to determine the property of the individual zf levels. The effect of these perturbations not only on the phosphorescence spectrum but also on the observed decays and polarizations has been examined. Limits on the importance of the different spin-orbit interactions are obtained. This spectroscopic work represents the type of experiments that can be done and the kind of information that can be obtained from PMDR and other methods.

B. AmPMDR and Spectral Analysis

One very powerful utilization of the amPMDR studies is in differentiating (77) bands that belong to the molecule of interest from those belonging to impurities or emission from different sites. In many instances, one obtains bands that are separated from each other by lattice vibration frequencies. However, they could very well be due to emission from different sites of the guest molecule under examination in the lattice. These emissions complicate the analysis of the emission, in particular in the region of the spectrum away from the 0,0 band (true origin). The fact that sites (or impurities) do have different zf energies that could be resolved in the microwave region makes amPMDR spectroscopy an attractive technique that could be used to obtain the spectrum of each site (separately). It could also be used to obtain the spectrum of the molecules being studied without any impurity bands. Furthermore, progressions originating from different zf origins could be obtained in different spectra or in the same spectrum with a different sign for the intensity changes observed when different zf transitions are saturated. This would identify false origins, assist in assigning the symmetry of the observed vibrations, and greatly assist in making the spectral analysis.

C. PMDR and Assignment of Triplet-State Symmetry

The spacial symmetry of the emitting state can be uniquely determined for symmetric molecules from a knowledge of the zf origin of the 0,0 band of the phosphorescence emission and its spin-orbit origin as determined from the polarization of the emission (78). On the other hand, if the zf origin and the symmetry of the emitting state are known, the spin-orbit interaction responsible for the emission could be identified.

In spite of the extensive optical research that has been carried out on the spectroscopy of the phosphorescence of benzene, no definitive and direct proof of the assignment of the emitting state could be given (3) from the properties of the observed emission. The reason for this is that the studies were made at 77°K, at which temperature the structure of benzene is D_{6h}. In this high symmetry, the emission is symmetry forbidden as well as spin forbidden. The 0,0 band is thus not observed. The vibronically induced part of the spectrum, which constitutes the total emission intensity, can be successfully described by spin-orbit schemes equally well for the $^3B_{1u}$, as for the $^3B_{2u}$ state. C_6H_6 in C_6D_6 at

1.6°K is found by ESR workers (79) to be of D_{2h} symmetry in the triplet state. The emission is thus symmetry allowed, and the 0,0 band is allowed by internal direct spin-orbit perturbation. The determination of the zf origin of this band could identify not only the special symmetry of the state but also the spin-orbit scheme responsible for the radiation. The emitting state is found (78) to correlate with the $^3B_{1u}$ state in D_{6h}, in agreement with previous identifications. Furthermore, in agreement with the observed polarization of the 0,0 band of larger condensed aromatic hydrocarbons—e.g., naphthalene and phenanthrene—the direct spin-orbit interaction between the $T_{\pi,\pi*}$ state and the $\sigma,\pi*$ singlet state is found to be responsible for the 0,0 band intensity of benzene as well (78). More recently the symmetry as well as the relative order of the zf levels has been determined (54) for 1,2,4,5 symmetric tetrachlorobenzene. This is done from a knowledge of the spin-orbit coupling routes (both direct and vibronically induced), the polarization of the magnetic zf transitions, and the lifetimes of the individual zf levels. Furthermore, it was shown that PMDR results could distinguish vibronic coupling in the triplet from those in the singlet manifold for the vibronically induced portion of the spectrum.

D. PMDR and Triplet-State Geometry [80]

Planar aromatic molecules possess a plane of reflection symmetry containing their molecular plane. They may thus belong to C_s, C_{2v}, D_{2h}, D_{3h}, and D_{6h} point groups. In all but the C_s point group, the short and long in-plane molecular axes coincide with principal magnetic axes. In these molecules, these axes also correspond to axes that contain atoms (a-axis) or bisect bonds (b-axis). Let us thus select a set of three orthogonal axes—a, b and N—where a and b are the in-plane axes and N is normal to the molecular plane. The spin functions would then be τ_a, τ_b, and τ_N and transform like rotations along a, b, and N-axis, respectively.

Let us confine our discussion to observations on the lowest triplet state of aromatic hydrocarbons, the ^3L_a state. τ_N should have the lowest zf energy if spin-spin, but not the spin-orbit interactions, is the dominant cause for the zero-field splittings (as is the case for aromatic hydrocarbons). For molecules in which the molecular dimension along the b-axis is longer than that along the a-axis, e.g., naphthalene (3), quinoxaline (3), anthracene (3), and the quinoidal structure of benzene (81a), τ_b is found to be higher in energy than τ_a. On the other hand, if the molecule is longer along a than along b, e.g., phenanthrene (3) and the antiquinoidal structure of benzene (79), τ_a is found to have higher energy than τ_b. This system is thus convenient in predicting the sign of the zero-field parameter E. It can also assist in making conclusions concerning the geometrical distortion of highly symmetric molecules, as will be shown below.

The phosphorescence of the ^3L_a and ^3L_b states is forbidden by symmetry (as well as spin) for D_{3h}- and D_{6h}-type molecules. Their 0,0 band should thus be missing. Its appearance indicates deviation to lower symmetries. The distortion of these molecules in either of these states could be a result of Pseudo-Jahn-Teller interaction resulting from the interaction of the ^3L_a or ^3L_b state with the nearby doubly degenerate triplet state. This is an in-plane distortion and could lead either to contraction along the a-axis, giving rise to the quinoidal form of benzene, or to an elongation along the a-axis, giving rise to the antiquinoidal form of benzene.

It is a well-established fact that direct spin-orbit interaction between the T_{π,π^*} and S_{σ,π^*} states is to be blamed for most, if not all, the intensity of the true origin (the 0,0 band) of the phosphorescence of aromatic compounds whose lowest triplet state is of the π,π^* type (2,3). For this to be true, the symmetry of the product

$$<\Gamma_{S_{\sigma,\pi^*}} x\Gamma_{H_{so}} x\Gamma_{T_{\pi,\pi^*}} x\Gamma_{\tau_i} x\Gamma_{x_s}>$$

must contain the totally symmetric representation, where $\Gamma_{H_{so}}$ and Γ_{x_s} (symmetry representation of the singlet spin function) belong to the totally symmetric representation. Thus, $\Gamma_{S_{\sigma,\pi^*}} x\Gamma_{T_{\pi,\pi^*}}$ must be of the same symmetry as Γ_{τ_i}, the symmetry representation for one of the three spin levels, τ_i.

All molecules have a σ_{ab} reflection symmetry plane. In addition, C_{2v}-type molecules possess σ_{bN}, and molecules of higher than C_{2v} symmetry possess both σ_{bN} and σ_{aN}. The lower triplet π,π^* states of these molecules are of the ^3L_a and ^3L_b types in Platt's notation. The first state has the excited π-electron distributed on the atoms, i.e., its nodal planes are across bonds; whereas the latter state has its excited π-electron distributed between atoms with its nodal planes containing the atoms.

In Table I, the symmetry properties of the L_a, L_b, S_{σ,π^*}, τ_a, τ_b, and τ_N to the three reflection planes are summarized. From these properties one can conclude that the zf level that gives rise to radiation to the 0,0 band is the τ_b level for the ^3L_a state and the τ_a level for the ^3L_b state.

PMDR can now be used to determine the origin of the 0,0 band. The lowest triplet state of the molecules selected is the ^3L_a state, even for benzene in C_6D_6, which

TABLE I
Symmetry Properties of Electronic Excited States and Spin Levels to Reflection Planes in Planar Aromatic Molecules

Function / Reflection plane[a]	σ_{ab}	σ_{aN}	σ_{bN}
$L_a(\pi,\pi^*)$	+	+	-
$L_b(\pi,\pi^*)$	+	-	+
$S_{(\sigma,\pi^*)}$	-	+	+
τ_a	-	-	+
τ_b	-	+	-
τ_N	+	-	-

[a] a and b are in-plane axes: a goes through atoms and b bisects bonds. N is an axis normal to the molecular plane.

PMDR has recently shown that its 0,0 band originates from the τ_b level (78). If the intensity of the 0,0 band changes upon using microwave in resonance with the highest energy transition, then τ_b is highest in energy, indicating contraction along the a-axis. On the other hand, if saturating the middle transition changes the 0,0 band intensity, then τ_b is lower in energy than τ_a. This suggests a distortion in which an elongation along the a-axis has taken place. If the intensity of the 0,0 band changes when both transitions are saturated, then a distortion to a symmetry below C_{2v} has taken place. This could be due to crystal field effects, since the site symmetry is lower than C_{2v}, or due to nonplanar distortion in crowded aromatic compounds, e.g., 2,3-dibromonaphthalene (36).

Results on benzene have shown that it has an anti-quinoidal structure (79) in C_6D_6 but a quinoidal structure in cyclohexane (81a) at 1.6°K. The structure of hexachlorobenzene (81b), trichlorobenzene (82a), and triphenylene (82b) are also found to depend on the host crystal. It is thus obvious that, if the molecule has a tendency to distort, it does so in a manner that is "acceptable" to the host lattice.

VI. PMDR AND THE NONRADIATIVE PROPERTIES OF THE TRIPLET STATE

A. Intramolecular Processes

The first discussion of the mechanism of the intersystem crossing process in N-heterocyclics (83,84) was made in 1961. This work suggested that the strong phosphorescence properties and weak fluorescence characteristics of these and aromatic carbonyls are due to the strong spin-orbit interaction between n,π^* and π,π^* singlet and triplet states. This would suggest that if the lowest excited singlet state is of the n,π^* type, the most probable intersystem crossing is to a triplet state of the π,π^* type, and vice versa.

The first experimental confirmation of the above proposal was made from the studies involving the detection of the state of spin alignment (20,85) from the optical decay at very low temperatures. The results showed that the strong radiative level is populated fastest by the intersystem crossing process. From a knowledge of the symmetry of the spin radiative level, it was easily concluded that in quinoxaline, the most favored intersystem crossing process is $S_{n,\pi^*} \rightsquigarrow T_{\pi,\pi^*}$. The first quantitative measurements of the relative intersystem crossing probability to the different zf levels, and thus the relative importance of the different mechanisms, were made (86) in 1968. This was done by decomposing the decay curve of pyraxine at 1.4°K upon direct singlet-singlet excitation, i.e., the pumping rate constants K_i and K_j are proportional to the intersystem crossing rate constants. The three zf levels of the lowest triplet state of this molecule were shown (44) to be radiative. At very low temperatures, when the spin-lattice relaxation is assumed to be absent, the steady-state initial intensity ratios of the different decay components of the decay curve are given by

$$I_i^o/I_j^o = k_i^r\, n_k/k_j^r\, n_j \approx k_i n_i/k_j n_j = K_i/K_j \tag{14}$$

In this expression, the observed decay constants are assumed to be radiative, i.e., the quantum yield of the phosphorescence is assumed to be unity. Later other expressions were given using PMDR methods, from which the ratios of the radiative constants may be determined (27).

The important equations used in the different methods of determining the ratios of the intersystem crossing rates are given in Section II.C by eqs. (5), (7-9), and (11-13) if pumping is done by direct excitation from the ground state to the excited singlet state transitions. In this case, the K's are the intersystem crossing rate constants. A detailed description of the different experimental techniques used at the present time, together with equations derived for molecules whose emission results from two zf levels, is given in a recent review of the subject (87). In this section we will give only a brief description of these PMDR methods. In all but the pulsed-excitation method, these assumptions are made: (a) the spin-lattice relaxation is absent and (b) the microwave radiation either saturates or inverts the population of the zf levels being coupled. The pulsed-excitation method makes the second assumption, as well as the assumption that pumping the different zf levels occurs from the same excited singlet state, presumably the lowest one, in a time shorter than the decay and spin-lattice relaxation times.

1. Microwave-Induced Delayed Phosphorescence (MIDP) Method

In addition to its usefulness in determining the K_i/K_j ratios, this technique (29) is powerful in resolving the values of the decay constants of the different zf levels, even if they are of comparable magnitude. The technique has been applied to obtain K_i/K_j for the case when one zf level is strongly radiative and the other two levels are long-lived. The principle involved depends on the fact that if level i has a very short radiative lifetime as compared to j, then at a time $t \gtrsim 5/k_i$ after the exciting light is sharply cut off, the population of level i decays to an almost zero value, while that of level j is still appreciable. If a microwave pulse of frequency ν_{ij} and of sufficient power to saturate the $\tau_i \leftrightarrow \tau_j$ transition is now applied, 1/2 $n_j(t)$ will now be in level i that gives a pulse of phosphorescence signal whose leading edge, h(t), is proportional to $n_j(t)k_i^r$. Starting with the same steady-state population for each level, prior to cutting off the excitation, these pulses can be applied with different delay times. The time dependence of h(t) reflects the time dependence of $n_j(t)$. A plot of $\ln(h_{(t)})$ vs. delay time should give a straight line with a slope = τ_j. The decay of any one pulse in this case is determined by k_i. In the special case for systems having one strongly radiative level, i, the intensity at t = 0 is given by

$$I_o = Cn_i^o k_i^r \tag{15}$$

At any time,

$$h(t) = (1/2)Cn_j(t)k_i^r = (1/2)Cn_j^o e^{-k_j t} \cdot k_i^r \tag{16}$$

$$\therefore I_o/h(t) = 2n_i^o/(n_j^o e^{-k_j t}) = \frac{2K_i/k_i}{(K_j/k_j)e^{-k_j t}} = 2(K_i/K_j)(k_j/k_i)e^{+k_j t} \tag{17}$$

Thus from the initial intensity and the leading edge of the microwave-induced phosphorescence signal resulting from the microwave pulse given at a delay time t after cutting off the excitation, the ratio K_i/K_j can be determined.

Equations have recently been derived (87) which extend the MIDP method to systems whose monitored emission originates from two zf levels.

2. c.w. Method

In some systems, a detectable state of spin alignment can exist in the steady-state mode of excitation and decay. The application of microwave radiation of resonance frequencies could change the observed phosphorescence intensity from a single zf level (47) or the relative intensity from any two zf levels (27,87). For the case of an emission from a single zf level, eq. (5) has been used previously (39,47) to determine the intersystem crossing rate ratios. The c.w. technique seems to be reliable. However, the fact that in a number of systems, the degree of spin alignment is not large under steady-state conditions might limit its usefulness. A better signal-to-noise ratio could be obtained if the mode of excitation and decay remains under steady-state conditions but the microwave is swept rapidly to equalize or to invert the population of the levels. In this case, eq. (7) is used, if the microwave saturates the zf transition being swept, and eq. (11) is used if the microwave radiation leads to a complete population inversion. This technique has not as yet been demonstrated.

3. Pulsed-Excitation Method

In this method (32), the system studied is pumped and irradiated with microwaves at a faster rate than its decay rate. In this case, the ratio of the intersystem crossing rates is measured by the ratio of the population of the zf field. The latter can be measured from the intensity ratio before and after the fast microwave sweep, as given in eq. (9) or (13) depending on whether the microwave radiation saturates or completely inverts the population of the levels. These equations assume that the pumping of the zf level occurs from the same excited singlet state, presumably the lowest excited singlet state. This technique does not require the spin-lattice relaxation to be frozen, as long as the pulsed excitation and sweep of the microwaves occur in a time short compared with the spin-lattice relaxation times. This can only occur if (a) the intersystem crossing process is faster than the spin-lattice relaxation processes, (b) a very short excitation pulse is used, and (c) fast microwave sweep or very short microwave pulse is employed. This, in principle, could be used at temperatures higher than liquid He temperatures. In order to increase the micro-wave sweep rate and still be able to saturate or invert the zf population, high microwave powers will be required.

4. $S_i \twoheadrightarrow T_i$ Intersystem Crossing in Aromatic Compounds

In addition to the results obtained by the optical

methods in zf, a number of important results concerning the relative intersystem crossing processes in a number of aromatic hydrocarbons have been obtained using ESR methods (88). The results of these workers were obtained without selective excitation of the guest. It is thus expected that the ratio of the pumping of the zf levels obtained in these experiments is a result of a combination of host-guest energy transfer as well as intramolecular intersystem crossing processes following direct absorption of the guest molecules.

The important results obtained so far concerning the relative intersystem crossing processes can be summarized as follows:

a. In N-heterocyclics, the zf level of the lowest triplet state which is the most favored is the one which could be involved in an intersystem crossing process involving direct spin-orbit coupling between singlet and triplet states of different electronic type, e.g., $S_{n,\pi*} \rightsquigarrow T_{\pi,\pi*}$ or $S_{\pi,\pi*} \rightsquigarrow T_{n,\pi*}$. This is in agreement with previous proposals concerning the intersystem crossing process in these systems (82,83).

b. In aromatic hydrocarbons, both ESR results (88) and PMDR results (89,82b) showed that the zf level favored in the intersystem crossing process is the one which is coupled to the lowest singlet state by spin-orbit coupling in a nonplanar configuration. In recent PMDR experiments on benzene, naphthalene, phenanthrene, triphenylene, and coronene, the effect of out-of-plane distortion is found (82b) to enhance the intersystem crossing process by a factor greater than 5. One cannot conclude at the moment whether the nonplanarity is a result of Herzberg-Teller-type interaction or simple nonplanar equilibrium configuration for some of the excited states involved in the intersystem crossing process.

It should be pointed out that the above two types of results obtained for aromatic hydrocarbons and N-heterocyclics are consistent with the following conclusion: the strongly radiative zf level is also one of the most favored zf levels in the intersystem crossing process. This conclusion is not hard to explain. It merely restates the fact that the zf level which interacts strongly with the singlet manifold via spin-orbit interaction would be favored in both the nonradiative and radiative processes involving a change in the spin quantum number. In aromatic hydrocarbons, strong spin-orbit coupling involves the lowest triplet π,π^* state and the high-energy σ,π^* (or π,σ^*) singlet states. On the other hand, in N-heterocyclics, the coupling is between the π,π^* and n,π^* states. It is thus clear that in N-heterocyclics, the triplet manifold is mixed with the lowest singlet state by spin-orbit interaction and no need for vibronic or geometrical changes to cross over from the lowest singlet state to the triplet manifold. In aromatic hydrocarbons, the triplet states are mixed with singlet σ,π^* states, whereas the lowest singlet state is of the π,π^* type. In order for the σ,π^* and π,π^* singlet states to mix, a nontotally symmetric vibration or geometrical distortion that would eliminate the plane of symmetry is required. This is in agreement with the experimental conclusions.

B. Intermolecular Triplet-Triplet Energy Transfer

The state of spin alignment in the lowest triplet state can be prepared (47) by a number of other methods in addition to its preparation by singlet-singlet absorption, e.g., by ground-state singlet→excited triplet-state absorption, excitation by electron scattering, and donor-acceptor triplet-triplet energy transfer. In these cases, the relative pumping rates are no longer related to the relative intersystem crossing rates. The equations derived in Section II.C are still applicable if K_i and K_j are rates of pumping. For excitation using the $S_o \nrightarrow T_{\tau_i}$ and $S_o \nrightarrow T_{\tau_j}$, the K's are the radiative rate constants (or absorption cross sections) for these two transitions. Pumping using triplet-triplet energy transfer is discussed below.

1. Spin Memory in Triplet-Triplet Energy Transfer

Magnetic forces are not responsible for T-T energy transfer. It is thus expected that the unpaired spins do not suffer any direction changes upon being exchanged in the T-T energy transfer process, i.e., the spin direction is conserved in spin space in the triplet-triplet energy transfer process. Spin space is defined in terms of the magnetic principal axis of the donor and acceptor. The exchange interaction [leading to the energy transfer between aromatic molecules (90)] is usually larger than the dipolar or spin-orbit interactions (leading to the zf splitting). This indicates that the energy transfer frequency is faster than electric spin reorientation time. This means that the spin Hamiltonian matrix should be diagonalized with respect to a set of magnetic orthogonal axes common to both the donor and acceptor axes. The new set of magnetic axes of the donor-acceptor pair should then be determined from the relative orientation of the donor and acceptor magnetic axes as well as their zf parameters. The simplest case (that was discussed previously) (47,91) is that when the zf splittings of the donor and acceptor are the same and the magnetic axes of the donor are parallel to those of the acceptor. In this case, an exact "printing" of the spin population of the donor is found in the acceptor. Thus, if the two unpaired electrons are spinning in the molecular plane of an aromatic molecular donor, energy transfer to an aromatic acceptor triplet would result in having the two unpaired spins being in the molecular plane of the acceptor.

Another situation arises when the zf parameters are the same but the magnetic axes of the donor are not parallel to those of the acceptor. In this case, a new set of orthogonal magnetic axes system for donor-acceptor pair is obtained by rotating the individual magnetic axes. If the donor spins are polarized in a certain plane, the spin polarization of the acceptor triplet state can be obtained by simple projection of the donor polarization axis onto the different magnetic axes of the acceptor. The rate of populating a certain zf level, τ_i, of the acceptor is proportional to the square of the direction cosine of the angle between the magnetic axis corresponding to the spin polarization of the donor and that of the zf level, τ_i of the acceptor. Thus a high degree of spin polarization in the donor could give rise to a small degree of spin alignment in the acceptor if the magnetic axes of the two molecules are not parallel or their zf parameters are widely different.

An experiment (91) has been carried out using a system satisfying the first condition. Quinoxaline is used as a donor. The acceptor is naphthalene x-traps, created by dissolving quinoxaline in the naphthalene crystal. The naphthalene located next to the quinoxaline molecules have their triplet state lower than the rest of the molecules in the crystal, including the neighboring quinoxaline molecules. Exciting quinoxaline at 3660 Å (at which wavelength naphthalene does not absorb) leads to the population of τ_L (where L is the in-plane molecular long axis) of the lowest triplet state of quinoxaline. Since the quinoxaline molecules are located next to naphthalene x-traps, they transfer their energy to these molecules. If the excitation is carried out for the quinoxaline in predeuterated naphthalene, quinoxaline phosphorescence is observed instead, since the lowest triplet state of quinoxaline is lower than that of the $C_{10}D_8$ x-traps. The results are as follows: exciting quinoxaline in $C_{10}D_8$ produces quinoxaline molecules in the τ_L zf level, while in $C_{10}H_8$, T-T energy transfer from quinoxaline molecules to the $C_{10}H_8$ x-trap leads to populating the τ_L level of naphthalene x-traps. The crystal structure of naphthalene has the long axes of the two molecules in the unit cell parallel to each other. Combining this and the fact that the zf parameters of naphthalene and quinoxaline are similar shows that the observed results above strongly support the expected conservation of spin direction in T-T transfer. More recently similar experiments have been carried out by other workers using conventional ESR (92) and MODR (93) methods.

2. Mechanism of Intermolecular Energy Transfer

Recently (94) the PMDR methods have been extended to determine the mechanism by which the triplet state of a dissolved guest is populated by intermolecular energy transfer when the host is optically excited to its high-energy excited singlet state. Two mechanisms are possible (a) singlet-singlet energy transfer leads to an excited singlet guest molecule which populates its own triplet by an intersystem crossing process from S_1, or (b) the host singlet excitation is converted to triplet excitation by intersystem crossing in the host. This process is then followed by triplet-triplet energy transfer to the guest molecule. The PMDR relative signals, and thus the guest molecule relative zf populations, should have the same values as those obtained by direct excitation of the excited singlet state of the guest molecule if mechanism (a) is more efficient than mechanism (b). The relative signal intensities obtained by pumping via mechanism (b) are determined by the relative intersystem crossing rate constants in the host, the projections of the magnetic principal axes of the host onto that of the guest, as well as the zf parameters of the guest and the host molecules. The results indicate that if aromatic hydrocarbons or some of their chloro derivatives are used as hosts, mechanism (a) seems to be predominant, while in bromo derivative hosts, mechanism (b) is important.

CLOSING REMARKS

PMDR spectroscopy is only four years "young", yet its applications in many fields have been pursued very actively (see Table II). It enabled us to see the triplet

TABLE II
Summary of Relevant Reviews

	Topic	Reference No.
1.	Triplet state, premultiple resonance era	2
2.	Optical pumping in atomic and ionic system	14
3.	MODR, experimental	68
4.	Spin alignment and spin levels	84
5.	PMDR, general	98
6.	Optical pumping of the triplet state, steady-state equations	47
7.	The spectroscopy of the individual zf levels	39
8.	Experimental methods in PMDR	49,50
9.	PMDR and the intersystem crossing process	87

state as being composed of three levels. It has also shown us that the orbitally nondegenerate triplet state, because of the fact that it has three levels, gives more information about molecules and molecular processes than the nondegenerate singlet state, that has only one level. In a time when picosecond-spectroscopy (95) has determined the absolute values of the rate constants of the non-radiative processes, PMDR is giving us the mechanisms of these processes. The combination of both the magnetic and the radiative properties is a very effective combination in determining the dynamic and the static properties of the triplet state. It is a fact that because we lost the spherical symmetry of the atom that we are able to do away from using magnetic field in these zero-field double-resonance experiments. It is the anisotropy of spin-spin and spin-orbit interactions that gives rise to (a) the zero-field splittings, (b) the state of spin alignment, and (c) the observed difference in the radiative properties of the zf levels. The failure to satisfy any of these conditions would lead to an unsuccessful PMDR experiment. The satisfaction of these three conditions is necessary but not sufficient to observe the PMDR signal. The spin-lattice relaxation (SLR) processes between the spin levels have to be slow to enable the microwave radiation to change the population of, and thus the emission intensity from, the radiative zf levels. Due to the temperature dependence of the SLR processes, PMDR has to be carried out at low temperatures ($< 1.6°K$). An understanding of the SLR processes in molecular systems is lacking. Thus while the PMDR signal in some systems is large (a change of phosphorescence intensity of 300% has been observed), in many others the signal requires averaging methods, high microwave powers, and a large number of sweeps.

In addition to its usefulness in the many research fields discussed in this article, PMDR could be extended to understand heat conductivity in molecular crystals at low temperatures. Due to the low efficiency of heat conduction at low temperatures, heating effects (96) due to intramolecular nonradiative processes have been detected from the decrease of the observed PMDR signal upon increasing the intensity or the frequency of the exciting light. PMDR could also be used in examining the spin direction of emitting triplet molecules formed in the very few cases of photochemical decompositions or reactions that might occur at low temperatures. From the results a deep understanding of the nature of the photochemical process could be accomplished since in these (very few) solid-state cases everything is fixed not only in our three-dimensional space but also in spin space. Observing a change in the delayed fluorescence intensity upon the microwave saturation of the zf transitions of the interacting triplet molecules is still to come. These studies are now under examination (97).

ACKNOWLEDGMENTS

The author wishes to acknowledge, with pleasure, his collaborators. This collaboration was not only fruitful but it was also extremely enjoyable. I wish to thank Messrs. Chodak, Leung, and Lin and Miss Wilkerson; Drs. Chen, Gwaiz, Hall, Kalman, and Owens; and Professors Moomaw, Olmsted, and Tinti. The financial support of the U.S. Atomic Energy Commission is greatly appreciated.

REFERENCES

1a. G. Lewis, D. Lipkin, and T. Magel, J. Am. Chem. Soc., 63, 3005 (1961).

1b. A. Terenin, Acta Physico Chim. USSR, 18, 210 (1943); Zhur. Fiz. Khim., 18, 1 (1944).

1c. G. Lewis and M. Kasha, J. Am. Chem. Soc., 66, 2100 (1944).

1d. G. Lewis and M. Calvin, J. Am. Chem. Soc., 67, 1232 (1945).

1e. G. Lewis, M. Calvin, and M. Kasha, J. Chem. Phys., 17, 804 (1949).

1f. S. Weissman and D. Lipkin, J. Am. Chem. Soc., 64, 1916 (1942).

2. S. K. Lower and M. El-Sayed, Chem. Reviews, 66, 199 (1966).

3. S. P. McGlynn, T. Azumi, and M. Kinoshita, Molecular Spectroscopy of the Triplet State, Prentice-Hall, Englewood Cliffs, New Jersey (1969).

4a. C. A. Hutchison, Jr. and B. W. Mangum, ibid, 29, 952 (1958).

4b. C. A. Hutchison, Jr. and B. W. Mangum, ibid, 34, 908 (1960).

5. A. Terenin and V. Ermolaev, Dokl. Akad. Nauk SSSR, 85, 547 (1952)

6. G. Porter and M. Windsor, J. Chem. Phys., 21, 2088 (1953).

7. C. Parker, Advance Photochem., 2, 305 (1964).

8. J. G. Calvert and J. N. Pitts, Jr., Photochemistry, Wiley, New York (1966).

9. E. Fermi and F. Rasetti, Nature, 115, 764 (1925).

10. G. Breit and A. Ellet, Phys. Rev., 25, 888 (1925).

11. F. Varsangi, D. C. Wood and A. L. Schawlow, Phys. Rev. Letters, 3, 544 (1959).

12. S. Geschwind, R. Collins, and A. L. Schawlow, ibid, 3, 545 (1959).

13. J. Brossel, S. Geschwind, and A. Schawlow, ibid, 3, 549 (1959).

14. For a review see: Bernheim, R. A., Optical Pumping: An Introduction, Benjamin, New York (1965).

15. M. Sharnoff, J. Chem. Phys., 46, 3263 (1967).

16. A. L. Kwiram, Chem. Phys. Letters, 1, 272 (1967).

17. J. Schmidt, I. A. Hesselmann, M. S. de Groot, and J. van der Waals, ibid, 1, 434 (1967).

18. M. Schwoerer and H. C. Wolf, in Proc. of the XIVth Collogue Ampere, 1966 ed., R. Blinc, Ed., North Holland, Amsterdam (1967), p. 87.

19. A. W. Hornig and J. S. Hyde, Mol. Phys., 6, 33 (1963).

20. M. S. de Groot, I. A. Hesselmann, and J. H. van der Waals, Mol. Phys., 12, 259 (1967).

21. J. Schmidt and J. H. van der Waals, Chem. Phys. Letters, 2, 640 (1968).

22. D. S. Tinti, M. A. El-Sayed, A. H. Maki, and C. B. Harris, Chem. Phys. Letters, 3, 343 (1969).

23. I. Y. Chan, J. Schmidt, and J. H. van der Waals, Chem. Phys. Letters, 4, 269 (1969).

24. C. B. Harris, D. S. Tinti, M. A. El-Sayed, and A. H. Maki, Chem. Phys. Letters, 4, 409 (1969).

25. D. A. Antheunis, J. Schmidt, and J. H. van der Waals, ibid, 6, 255 (1970).

26. T. S. Kuan, D. S. Tinti, and M. A. El-Sayed, Chem. Phys. Letters, 4, 507 (1970).

27. M. A. El-Sayed, J. Chem. Phys., 52, 6438 (1970).

28. M. A. El-Sayed and O. F. Kalman, J. Chem. Phys., 52, 4903 (1970).

29. W. S. Veeman and J. H. van der Waals, Chem. Phys. Letters, 7, 65 (1970).

30. J. Schmidt, W. G. Van Dorp, and J. H. van der Waals, ibid, 8, 345 (1971).

31. A. H. Francis and C. B. Harris, Chem. Phys. Letters, 9, 188 (1971).

32. M. A. El-Sayed and J. Olmsted, Chem. Phys. Letters, 11, 568 (1971).

33. H. Levanon and S. Weissman, J. Am. Chem. Soc., 93, 4309 (1971).

34. A. Shain, M. Sharnoff, and M. Stombler, J. Chem. Phys., 56, 2475 (1972).

35. R. H. Clarke and J. M. Hayes, J. Chem. Phys., 57, 569 (1972).

36. M. A. El-Sayed, M. Leung, and C. T. Lin, Chem. Phys. Letters, 14, 329 (1972).

37. M. Leung and M. A. El-Sayed, Chem. Phys. Letters, 16, 454 (1972).

38. C. T. Lin, A. A. Gwaiz, and M. A. El-Sayed, J. Amer. Chem. Soc., 94, 8234 (1972).

39. D. S. Tinti and M. A. El-Sayed, J. Chem. Phys., 54, 2529 (1971).

40. M. A. El-Sayed, D. V. Owens, and D. S. Tinti, Chem. Phys. Letters, 6, 395 (1970).

41. M. A. El-Sayed, D. S. Tinti, and D. V. Owens, Chem. Phys. Letters, 3, 339 (1969).

42. T. Azumi, C. M. O'Donnell, and S. P. McGlynn, J. Chem. Phys., 45, 2735 (1966).

43. L. Hall, A. Armstrong, W. Moomaw, and M. A. El-Sayed, J. Chem. Phys., 48, 1395 (1968).

44. M. A. El-Sayed, W. R. Moomaw, and D. S. Tinti, J. Chem. Phys., 50, 1888 (1969).

45. L. H. Hall, D. V. Owens, and M. A. El-Sayed, Mol. Phys., 20, 1025 (1971).

46a. C. S. Harris, Proc. of 5th Molecular Crystal Symposium, Philadelphia, Pennsylvania (1970).

46b. C. S. Harris, J. Chem. Phys., 54, 972 (1971).

47. M. A. El-Sayed, J. Chem. Phys.,54, 680 (1971).

48. M. A. El-Sayed, unpublished results.

49. T. H. Cheng and N. Hirota, J. Chem. Phys., 56, 5019 (1972).

50. J. Olmsted and M. A. El-Sayed, "Experimental Methods in Phosphorescence-Microwave Double Resonance," to be published in The Creation and Detection of the Excited State, Dekker, New York (1972).

51. L. Hall, Ph.D. dissertation, "Spin-Lattice Relaxation and Zeeman Effects in the Triplet State of Pyrazine d_4," UCLA (1971).

52. D. Owens, M. A. El-Sayed, and S. Ziegler, J. Chem. Phys., 52, 4315 (1970).

53. M. J. Buckley and C. B. Harris, Chem. Phys. Letters, 5, 205 (1970).

54. C. R. Chen and M. A. El-Sayed, Chem. Phys. Letters, 10, 307 (1971).

55. T. H. Cheng and N. Hirota, Chem. Phys. Letters, 13, 194 and 14, 415 (1972).

56. A. Nishimura and D. S. Tinti, ibid, 13, 278 (1972).

57. A. H. Francis and C. B. Harris, J. Chem. Phys., 57, 1050 (1972); and references therein.

58. M. J. Buckley and C. B. Harris, Chem. Phys. Letters, 4, 591 (1970).

59. J. Schmidt and J. H. van der Waals, ibid, 3, 546 (1969).

60. M. D. Fayer, C. B. Harris, and D. A. Yuen, J. Chem. Phys., 53, 4719 (1970).

61. C. A. Hutchison, Jr. and G. A. Pearson, ibid, 47, 520 (1967); and C. A. Hutchison and B. E. Kohler, ibid, 51, 3327 (1969).

62. R. C. McCalley and A. L. Kwiram, Phys. Rev. Letters, 24, 1279 (1970).

63. A. H. Francis and C. B. Harris, Chem. Phys. Letters, 9, 181 (1971).

64. O. F. Kalman and M. A. El-Sayed, J. Chem. Phys., 54, 4414 (1971).

65. For basic reference see: A. Abragam, The Principles of Nuclear Magnetism, The Clarendon Press, Oxford (1961).

66. J. A. Pople, W. G. Schneider, and H. J. Bernstein, High Resolution Nuclear Magnetic Resonance, McGraw-Hill, New York (1959).

67. C. P. Slichter, Principles of Magnetic Resonance, Harper and Row, New York (1963).

68. M. Sharnoff, Mol. Cryst., 9, 265 (1969).

69. M. Sharnoff, Chem. Phys. Letters, 2, 498 (1968).

70. L. J. Cheng and A. L. Kwiram, ibid, 4, 457 (1969).

71. M. Sharnoff, J. Chem. Phys., 51, 451 (1969).

72. I. Y. Chan and J. Schmidt, Symposium on "Magnetic Optical Effects," Faraday Soc. No. 3 (1969).

73. M. Sharnoff, Mol. Cryst., 5, 297 (1969).

74. C. R. Chen, J. A. Mucha, and D. W. Pratt, Chem. Phys. Letters, 15, 73 (1972).

75. C. J. Gorfer, in Polarisation, Matiere et rayonnement, Societe Francais de Physique, Paris (1969), p. 259.

76. D. S. Tinti and M. A. El-Sayed, Presentation at the International Conference on Molecular Luminescence, Newark, Delaware, August 1969.

77. D. S. Tinti and M. A. El-Sayed, Organic Scintillators and Liquid Scintillation Counting, Academic Press, New York (1971), p. 563.

78. A. A. Gwaiz, M. A. El-Sayed, and D. S. Tinti, Chem. Phys. Letters, 9, 454 (1971).

79. M. S. deGroot, I. A. Hesselmann, and J. H. van der Waals, Mol. Phys., 13, 583 (1967).

80. M. A. El-Sayed, "Proceedings of the International Conference on Luminescence, August 1972," J. Luminescence, in press.

81a. M. A. El-Sayed, W. R. Moomaw, and J. B. Chodak, J. Chem. Phys., 57, 4061 (1972).

81b. M. A. El-Sayed, A. A. Gwaiz, and C. T. Lin, Chem. Phys. Letters, 16, 281 (1972).

82a. C. T. Lin and M. A. El-Sayed, unpublished results.

82b. W. R. Moomaw, M. A. El-Sayed, and J. Chodak, unpublished results

83. M. A. El-Sayed, J. Chem. Phys., 36, 573 (1962).

84. M. A. El-Sayed, J. Chem. Phys., 38, 2834 (1963).

85. J. H. van der Waals and M. S. de Groot, The Triplet State, A. B. Zahlan, Ed., Cambridge University Press, London (1967), p. 101.

86. M. A. El-Sayed, in the Proceedings, International Conference on Molecular Luminescence (Loyola University, Chicago, Illinois), E. Lim, Ed., Benjamin, New York (1968), p. 715.

87. M. A. El-Sayed, Advances in Electronic Excitation and Relaxation, E. Lim, Ed., Vol. 1, Academic Press, New York (1972).

88. H. Sixl and M. Schwoerer, Z. Naturforsch, 25a, 1383 (1970).

89. M. A. El-Sayed and C. R. Chen, Chem. Phys. Letters, 10, 313 (1971).

90. D. L. Dexter, J. Chem. Phys., 21, 836 (1953).

91. M. A. El-Sayed, D. S. Tinti, and E. M. Yee, J. Chem. Phys., 51, 5721 (1969).

92. R. H. Clarke, Chem. Phys. Letters, 6, 413 (1970).

93. M. Sharnoff and E. B. Iturbe, Phys. Rev. Letters, 27, 576 (1971).

94. A. Wilkerson, C. T. Lin, and M. A. El-Sayed, Chem. Phys. Letters, 17, 175 (1972).

95. For a recent review: P. M. Rentzepis and G. E. Busch, IEEE Trans. of Instrumentation and Measurement, IM21, 488 (1972).

96. J. Chodak and M. A. El-Sayed, unpublished results.

97. A. Wilkerson and M. A. El-Sayed, in progress.

98. M. A. El-Sayed, Accts. Chem. Res., 4, 23 (1971).

Photochemistry of the Troposphere

HIRAM LEVY II
Smithsonian Institution
Astrophysical Observatory
Cambridge, Massachusetts

CONTENTS

INTRODUCTION

The troposphere, a uniformly mixed region where the temperature on average decreases with height, extends upward from the earth's surface to the stratosphere, a region of relatively constant temperature. The height of the upper boundary of the troposphere, the tropopause, ranges from 8 km at the poles to 18 km at the low latitudes and is a complex function of the radiation balance of the lower atmosphere. While the exact position of the tropopause is not always easily determined, it is sufficient for our needs to define the troposphere as a region of strong vertical mixing whose temperature decreases with height, whose lower bound is the earth's surface, and whose upper bound is a region of relatively constant temperature and weak vertical mixing.

Serving as a link between the biosphere and the stratosphere, the troposphere is intimately involved in the chemistry of both. The biosphere serves as a source (outgassing, decomposition, and bacterial and microbiological activity) and as a sink (dry and wet deposition, plant and animal gaseous uptake, and surface reactions) for important constituents of the troposphere, while transport across the tropopause serves as an important sink and source for both tropospheric and stratospheric species.

The complicated photochemistry of local, regional, and global air pollution must be considered in the context of the existing tropospheric chemistry. Those determining the allowable levels of pollutant concentrations must take into consideration the existing chemistry and minor constituent concentrations of the normal troposphere.

The temperature and density structure of the troposphere, along with the concentrations of major constituents, are well documented; and altitude profiles have been measured over a wide range of seasons and latitudes for the minor species water, carbon dioxide, and ozone. A few profiles are available for carbon monoxide, nitrous oxide, methane, and molecular hydrogen, while only surface or low-altitude measurements have been made for nitric oxide, nitrogen dioxide, ammonia, sulfur dioxide, hydrogen sulfide, and nonmethane hydrocarbons. No direct measurements of nitric acid and formaldehyde are available, though indirect information does exist. The concentrations of a number of other important species, such as peroxides and oxy and peroxy radicals, have never been determined. Therefore, while considerable information concerning trace constituent concentrations is available, the picture is far from complete.

Starting with the work of Bates and Witherspoon (12), there have been a number of studies of tropospheric chemistry. The book by Leighton (149), while primarily a text on photochemical air pollution, contains a great deal of information concerning the troposphere, and one by Junge (128) reviews in detail the composition and chemical reactions of the troposphere as they were understood in 1962. The proceedings of the three International Symposia on Trace Gases and Radioactivity (110-112) provide extensive information on measurements of trace constituent concentrations, along with discussions of possible chemical reactions. A series of review articles by Robinson and Robbins (214-216) summarize the knowledge of tropospheric concentrations, sinks, and sources of minor constituents through 1970, while a review by Cadle and Allen (28) does the same for tropospheric photochemistry. At that time, radicals and metastable species, known to be important both in the stratospheric photochemistry [see, e.g., Nicolet (180), Leovy (150), and Hunt (104)] and in the photochemistry of smog [see, e.g., Leighton (149) and Altshuller and Bufalini (4)], were not thought to be present in sufficient concentrations to be important in the troposphere.

On the basis of ratios of C^{14} and C^{12} present in carbon dioxide, Weinstock (250) estimated a carbon monoxide lifetime of 0.1 year. This was more than an order of magnitude less than previous estimates of Bates and Witherspoon (12) and Robinson and Robbins (214), which were based on calculations of the anthropogenic source of carbon monoxide. Weinstock (250) suggested that if a sufficient concentration of hydroxyl radical were availabl the oxidation of carbon monoxide by hydroxyl radical, first proposed by Bates and Witherspoon (12) for the stratosphere, would provide the rapid loss mechanism for carbon monoxide that appeared necessary. By extension of previous stratospheric models of Hunt (104), Leovy (150 Nicolet (180), and others, Levy (152) demonstrated that a large source of hydroxyl radical, the oxidation of water by metastable atomic oxygen, which was itself produced by the photolysis of ozone, existed in the troposphere and that a chain reaction involving the hydroxyl and hydroperoxyl radicals would rapidly oxidize both carbon monoxide and methane. It was then pointed out that all the loss paths for the formaldehyde produced in the methar oxidation led to the production of carbon monoxide [McConnell, McElroy, and Wofsy (171) and Levy (153)]. Similar chain mechanisms were shown to provide troposphere sources for molecular hydrogen, nitric acid, hydrogen and methyl peroxide, and formaldehyde (153). The photochemistry of the troposphere, involving radicals, metastables, and other reactive species, has been developed in detail (153,154) and applied to the general question of carbon monoxide [Weinstock and Niki (251)] and methane [Ehhalt (52)] in the troposphere and to the effects of the troposphere on the stratosphere [Wofsy, McConnell, and McElroy (256); Nicolet (182); and Crutzen (43)].

The present review offers a unified model of the photochemistry of the troposphere and discusses the effects of variations in altitude, latitude, season, and time of day. The general questions of the sinks and sources for minor tropospheric species will be discussed in terms of their atmospheric reactions.

Experimental measurements of the tropospheric density and temperature structure, along with a brief discussion of the major transport process and experimental estimates of related mixing rates, are presented in Section I. The available experimental data on minor constituent concentrations, together with current estimates of their sinks, sources, and lifetimes, will be summarized in Section II. While these first two sections are intended to provide a detailed summary of current experimental information regarding the physical and chemical properties of the troposphere, they are not meant to be an exhaustive review of the literature. Therefore, some measurements are not directly cited.

In Section III the solar-radiation intensity and its transmission through the troposphere are discussed. The solar flux is tabulated as a function of wavelength, and its dependence on azimuthal angle and total ozone column density is discussed.

The photochemical reactions of those species present in the troposphere are examined in detail in Section IV, and reaction schemes are developed.

In Section V, numerical models are developed and calculations performed. In Section VI the results are discussed in terms of number-density profiles; seasonal, latitudinal, and diurnal variations; and photochemical sources and sinks.

In the final section, VII, the effects of uncertainties in reaction rate and absorption coefficients on the numerical predictions are examined, important areas for further theoretical and experimental research are recommended, and the results of Sections IV, V, and VI are summarized.

I. PHYSICAL PROPERTIES OF THE TROPOSPHERE

The photochemistry of the troposphere is influenced by both the physical structure and the transport mechanism of that region. This section briefly reviews the important features.

A. Physical Structure

The troposphere is a uniformly mixed region that closely obeys the equation of state for an ideal gas,

$$P = \frac{\rho RT}{m}$$

where P is the atmospheric pressure, ρ is the atmospheric density, T is the kinetic temperature, R is the universal gas constant, and m is the mean molecular weight with an essentially constant value of 28.966 g $mole^{-1}$ for dry air.

It has on the average a high lapse rate Γ,

$$\Gamma = - \frac{\partial T}{\partial Z}$$

while its upper boundary, the tropopause, has a low or zero lapse rate.

1. Tropospheric Temperature Profiles

The temperature profile strongly influences those reactions whose rate coefficients have large activation energies. As will be shown in Sections IV, V, and VI, a number of reaction paths, while dominant in the lower troposphere, lose their importance with increasing altitude as the temperature drops sharply. Particularly affected are the altitude profiles of the hydroxyl radical, formaldehyde, and nitric oxide number densities.

Temperature profiles have been measured for many years throughout the troposphere and are well summarized (41,243,244). The representative profiles for various seasons and latitudes in Table I are taken from the U.S. Standard Atmosphere Supplements (1966).

Average surface temperatures in the summer range from 300°K at the low latitudes to 280°K near the poles; and in the winter, from 300°K at the low latitude to 250°K at the pole. The variation of surface temperatures with season for a particular location increases with latitude, while the variation with latitude is greater in the winter than in the summer. Overall, the variation of the surface temperature over the earth seldom exceeds 30°K, except for the very low average temperatures at the winter pole. The variation of temperature with altitude is much greater, ranging from 60°K to 100°K, except for the winter polar region, which shows a very low surface temperature and only a 30°K reduction at the tropopause. Tropopause temperatures range from 195°K near the equator to 220°K near the poles, and at any given altitude the temperature may vary by as much as 30°K, depending on the latitude and the season. The general

TABLE I
Tropospheric Temperature Profiles

Altitude (km)	15° (Annual) (°K)	30° (Winter) (°K)	30° (Summer) (°K)	45° (Winter) (°K)	45° (Summer) (°K)	75°N (Winter) (°K)	75°N (Summer) (°K)
0	300	287	301	272	294	249	278
1	294	284	294	269	290	252	276
2	288	281	288	265	285	251	273
3	283	275	283	262	279	245	269
4	277	268	277	256	273	240	262
5	271	262	272	250	267	234	256
6	264	255	266	244	261	229	249
7	257	249	259	238	255	223	242
8	250	242	252	232	248	218	236
9	243	236	245	226	242		236
10	237	229	238	220	234		
11	230	223	231		229		
12	223	216	223		222		
13	217		217		217		
14	210		210				
15	203		203				
16	197						
17	195						

features of the tropospheric temperature structure are the following:

a. For a given altitude, the temperature may vary by as much as 30°K as a function of the season and latitude. This will result in a significant, though not overwhelming, variation in the rate coefficients for those reactions with large activation energies. In general the dominant photochemical reaction paths remain independent of season and latitude, though production and loss rates may vary significantly.

b. For a given latitude and season, the temperature may vary by as much as 100°K with altitude, and this does result in variations in the dominant photochemical reaction paths owing to the loss at lower temperatures of many reactions with high activation energies.

c. The exception to a and b is the winter polar regions, where the average surface temperature is very cold, 250°K, and the tropopause is quite low, 8 km. In this region most reactions with large activation energies are unimportant throughout the troposphere.

2. Tropospheric Density Profiles

Since the atmosphere obeys the ideal gas law, the magnitude of the variations in density will show a dependence on altitude, latitude, and season similar to that shown by temperature. Unlike temperature, which enters into the rate expression in an exponential and therefore may have a very large effect in a particular production or loss rate, the density enters only as a single product either as the density of a third body or as that of some reactant that is uniformly mixed. In general throughout the troposphere the density is high enough to make three-body reactions—many of which can be treated as effective two-body reactions—very effective.

The major variation in density is with altitude, which is to be expected. Surface densities range from 2.4×10^{19} to 3.0×10^{19} molecules cm^{-3}, while densities at the tropopause range from 3.2×10^{18} to 9.3×10^{18} molecules cm^{-3}. The much larger range at the tropopause is due to the wide range of tropopause heights. The actual variation with season and latitude for a given altitude is similar to that shown at the surface. Representative density profiles for the troposphere are shown in Table II.

The general conclusion to be drawn from Sections I.A. 1 and 2 is that the major effect of the physical structure of the troposphere on the photochemistry is to cause the actual model to be altitude dependent. Except for the winter solar regions, the variations in season and latitude do not greatly affect the model, although they will affect the quantitative results.

B. Transport Process

Given the problems of theoretical complexity, numerical instability, and lengthy computation, the development of a realistic three-dimensional model of tropospheric motions is very difficult [see Manabe and Hunt (163) and Mahlman and Manabe (162)]. The considerable research in the general area of transport processes in the lower atmosphere has been reviewed by Reiter (204-206).

Although a unified treatment of transport and photochemistry has not yet been developed for the lower atmosphere, simplified photochemical models have been included in detailed transport calculations [Reiter (205)], and aeronomers [see Colegrove, Hanson, and Johnson (39,40)] have attempted to account for the effects of vertical transport by including a one-dimensional treatment of eddy diffusion [Lettau (151)] in their photochemical models.

TABLE II
Tropospheric Density Profiles

Altitude (km)	15° (Annual) (molecules cm^{-3})	30° (Winter) (molecules cm^{-3})	30° (Summer) (molecules cm^{-3})	45° (Winter) (molecules cm^{-3})	45° (Summer) (molecules cm^{-3})	75°N (Winter) (molecules cm^{-3})	75°N (Summer) (molecules cm^{-3})
0	2.4^{19}[a]	2.6^{19}	2.4^{19}	2.7^{19}	2.5^{19}	3.0^{19}	2.6^{19}
1	2.3^{19}	2.3^{19}	2.2^{19}	2.4^{19}	2.3^{19}	2.6^{19}	2.4^{19}
2	2.0^{19}	2.1^{19}	2.0^{19}	2.2^{19}	2.0^{19}	2.2^{19}	2.1^{19}
3	1.8^{19}	1.9^{19}	1.8^{19}	1.9^{19}	1.9^{19}	2.0^{19}	1.9^{19}
4	1.7^{19}	1.7^{19}	1.7^{19}	1.7^{19}	1.7^{19}	1.8^{19}	1.7^{19}
5	1.5^{19}	1.5^{19}	1.5^{19}	1.6^{19}	1.5^{19}	1.6^{19}	1.5^{19}
6	1.4^{19}	1.4^{19}	1.3^{19}	1.4^{19}	1.4^{19}	1.4^{19}	1.4^{19}
7	1.2^{19}	1.2^{19}	1.2^{19}	1.2^{19}	1.2^{19}	1.2^{19}	1.2^{19}
8	1.1^{19}	1.1^{19}	1.1^{19}	1.1^{19}	1.1^{19}	1.1^{19}	1.1^{19}
9	9.8^{18}	9.8^{18}	9.8^{18}	9.7^{18}	9.8^{18}		9.7^{18}
10	8.8^{18}	8.7^{18}	8.8^{18}	8.5^{18}	8.7^{18}		
11	7.8^{18}	7.7^{18}	7.8^{18}		7.7^{18}		
12	6.9^{18}	6.8^{18}	6.9^{18}		6.8^{18}		
13	6.1^{18}		6.1^{18}		6.0^{18}		
14	5.4^{18}		5.4^{18}				
15	4.7^{18}		4.7^{18}				
16	4.1^{18}						
17	3.5^{18}						

[a]The superscript designates the power of 10 by which each number is to be multiplied.

The large fluctuations with altitude that are frequently observed in tropospheric measurements of ozone and water number densities suggest that vertical eddy diffusion with a constant diffusion coefficient is not adequate.

Fortunately most constituents in the troposphere fall into two classes: reactive species with photochemical lifetimes ranging from minutes to a few hours whose photochemically determined number densities can adjust to atmospheric motions; and much less reactive species with photochemical lifetimes ranging from months to years and with tropospheric number densities represented by a single mixing ratio.

There are, however, a number of species that have either photochemical lifetimes of an intermediate range or altitude profiles controlled by transport processes. Nitric acid and the peroxides are good examples of the former case, and their altitude profiles, when calculated from a purely photochemical model, may not be quantitativel correct. In the latter case, water and ozone being excellent examples, it is necessary to use measured altitude profiles that remain fixed and independent of the photochemical model.

While detailed transport calculations of trace constituent distributions in the troposphere are not generally available, a number of simple box models have been fitted to measurements of seasonal fluctuations in carbon dioxide and ozone and tropospheric distributions of fission debris. These box models may not accurately represent the physical processes responsible for tropospheric motions and mixing, but the parameters used in the model do provide reasonable estimates of macroscopic mixing rates in the troposphere.

The results from a number of parametric models have

been summarized in recent papers [Pressman and Warneck (196) Bolin and Bischof (23), and Lal and Rama (147)]. Estimated rates for vertical, longitudinal, intrahemispherical, and interhemispherical mixing, as well as transport across the tropopause and rainout, are reviewed and their relevance to photochemistry discussed.

1. Vertical Mixing

Bolin and Bischof (23) found a 25- to 30-day phase lag between the seasonal variations of carbon dioxide number densities at the tropopause and those at the surface layer. Assuming that vertical eddy diffusion is responsible for the transport from the ground to the tropopause, they derived for the vertical eddy-diffusion coefficient K_z a value of 2×10^5 cm^2 sec^{-1}. This value is in harmony with earlier estimates that Jacobi and Andre (113) determined by fitting measured profiles of R_n^{222} and R_n^{220} along with their products Pb^{210} and Po^{210} to a one-dimensional eddy-diffusion model ($K_z = 1$ to 5×10^5 cm^2 sec^{-1}) and with the estimates of Junge (127) and Kroening and Ney (141) based on one-dimensional eddy-diffusion models of the tropospheric ozone distribution, $K_z = 7 \times 10^4$ cm^2 sec^{-1} and 0.1 to 1×10^5 cm^3 sec^{-1}, respectively.

Both Jacobi and Andre and Kroening and Ney found that it was necessary for K_z, while remaining constant for altitudes above a few hundred meters, to decrease rapidly near the ground if the models for fission debris and ozone are to agree with measurement.

In general, the troposphere has a vertical mixing time of a month or less, and the altitude profiles for species with photochemical lifetimes of one month or more are each well described by a single mixing ratio.

2. Longitudinal Mixing

It is generally assumed that a horizontal eddy-diffusion model of longitudinal transport requires an eddy-diffusion coefficient $K_H = 10^{10}$ to 10^{11} $cm^2\ sec^{-1}$ [Lettau (151)]. Taking the lower value, Junge (127) found a longitudinal mixing time ranging from 15 to 60 days, depending on the values of other model parameters. Bolin and Keeling (24) found $K_H = 3 \times 10^{10}$ $xm^2\ sec^{-1}$ in an analysis of the worldwide distribution of carbon dioxide.

The fact that K_H is much larger than K_z is the usual justification given for the treatment of eddy diffusion in photochemical models as vertical diffusion only.

3. Intrahemispherical Mixing

Using a four-cell model, Lal and Rama (147) found that the dispersion of fission debris injected into the atmosphere by Russian nuclear tests was best explained by a mixing time of no more than 3 weeks for each cell and an overall intrahemisphere time ranging from 15 days in the winter to 3 months in the fall. Junge (127) estimated from Be^7 distributions in the troposphere that the intrahemisphere mixing time, both latitudinal and longitudinal, was approximately one month. It is generally assumed that those species with photochemical or residence lifetimes of a few months or more are well mixed throughout the hemisphere.

4. Interhemispherical Mixing

Lal and Rama (147) estimated a mixing time of 1.2 yr based on the dispersion of fission debris, while Nydal (186) suggested a mixing time of 1 yr based on similar data. Junge (129) concluded from the carbon dioxide data of Keeling (137) that the exchange time was greater than 0.9 yr, while Fergusson (61) estimated a time greater than 1.8 yr; and Bishop, Delafield, Eggleton, Peabody, and Taylor (21), from the distribution of tritiated methane, estimated 3 to 4 yrs. From these data we see that transport across the equator is much slower than the other tropospheric mixing processes. Only those species with tropospheric lifetimes of one or more years or with a uniform distribution of sinks and sources can be represented by a single mixing ratio throughout the troposphere.

5. Troposphere-Stratosphere Exchange

From a study of fission debris, Pressman and Warneck (196) summarized a number of estimates of this rate that ranged from 1 to 2 yr and recommended 1.5 ± 0.5 as a good average. Based on the carbon monoxide profile in the lower stratosphere, Junge, Seiler, and Warneck (133) estimated an exchange rate of a few years. From an estimated exchange rate of 1.5 yr and an assumed depth of the lower stratosphere that mixes with the troposphere, Pressman and Warneck derived a flux through the tropopause of 8×10^{16} molecules cm^{-2} sec^{-1}. This is in reasonable agreement with the value calculated by Reiter, Glasser, and Mahlman (207) of 3×10^{16} molecules cm^{-2} sec^{-1} and the estimate of Junge (127) based on ozone transfer. Smith (231) found a higher value, 6×10^{17} molecules cm^{-2} sec^{-1}, using tritium as a tracer, but he also found an exact balance between upward and downward fluxes. It should be noted that transport through the tropopause is both latitude dependent [Martell (165); Gudiksen, Fairhall, and Reed (88); and Fabian, Libby, and Palmer (60)] with the greatest contribution coming at midlatitudes [Reiter, et al. (207); Danielsen (45); and Smith (231)] and seasonal dependent, as determined from the ozone model of Junge and Czeplak (130).

6. Rainout and Washout

Junge (128) estimated that tropospheric aerosols had a lifetime of approximately 10 days, although the rate of removal decreased rapidly with increasing altitude. On the basis of the distribution of fission products, Jacobi and Andre (113) found a longer time of 20 to 50 days, which is consistent with the lifetime calculated by Bates and Hays (11) for odd nitrogen. However, Francis, Chester and Haskin (66) found that dust containing fission products had been aged and that analysis of filtered rainwater gave a much shorter lifetime of approximately 10 days. Recently Poet, Moore, and Martell (195) estimated a mean atmospheric residence time of about four days for particles in the lower troposphere and about one week for particles in precipitation.

It is thought that rainout and washout may also be important removal mechanisms for soluble tropospheric gases with photochemical lifetimes longer than a few days. From the data for aerosols and fission debris, we can estimate a lifetime of at least one week, and possibly longer. It would also appear that this process is significant in only the first 5 km of the troposphere [Junge (128)].

In summary, we find that vertical mixing and intrahemispherical mixing, both longitudinal and latitudinal, have a time scale of approximately one month, while interhemispherical mixing and mixing through the tropopause hav a time scale of one year or more. Rainout and washout provide lifetimes of one week or more for aerosols and soluble gases in the lower troposphere.

II. CHEMICAL COMPOSITION

A number of terms used to describe the relative

quantities of various species are defined below:

The number density of A, n(A), is the number of molecules or atoms of A found in a cubic centimeter of air.

The volume mixing ratio of A, r(A), equals n(A)/n(M), where n(M) is the total number density of air.

The volume percentage composition of A equals 100 x r(A).

Two other abbreviations, parts per million by volume (ppm) and parts per billion by volume (ppb), will also be used. All comparative measures will be made according to volume.

A. Major Species

The principal constituents of the dry troposphere are, by volume percentage, molecular nitrogen 78.1%, molecular oxygen 20.9%, and argon 0.9% [Glueckauf (80)]. Argon has no effect on the photochemistry, while molecular nitrogen serves as a nonreactive partner in three-body and quenching reactions. Molecular oxygen, besides having a nonreactive role, is involved in a number of secondary photochemical reactions such as the formation of ozone and peroxy radicals.

B. Water Vapor

Water vapor is one of the earliest known atmospheric tracers and meteorological variables [Junge (128)]; under the influence of the tropospheric temperature profile, it shows a sharp decrease in its mixing ratio with increasing altitude. At a given altitude, the relative humidity may fluctuate between 100% and almost 0% [see, e.g., Mastenbrook (166,167) and Mastenbrook and Purdy (168,169)]. This highly variable quantity may have a mixing ratio as high as 4×10^{-2} in the surface atmosphere, while it drops to 10^{-4} to 10^{-5} in the tropopause before reaching its minimum of 2 to 5 x 10^{-6} at the bottom of the stratosphere [Mastenbrook (167) and Sissenwine, Grantham, and Salmela (229)].

The distribution of water vapor in the troposphere is controlled by general climatic conditions rather than by atmospheric photochemistry. On the basis of rainfall measurements, Junge (128) estimated a 10-day average residence time for water vapor.

Representative altitude profiles for a number of latitudes and seasons (U.S. Standard Atmosphere Supplements 1966) are presented in Table III.

While there appears to be a decrease in relative humidity with altitude, as well as a latitude-dependent variation in the middle troposphere [Manabe, Smagorinsky, and Strickler (164) and Rashke (198)], the day-to-day fluctuations observed by Mastenbrook and Purdy (168,169) are comparable.

C. Ozone

Ozone is a highly variable minor constituent of the troposphere; its altitude profiles and global distribution are thought to be controlled by large-scale transport processes rather than by photochemical reactions.

Regener (201) first proposed that tropospheric ozone originates in the stratosphere and is destroyed at the ground. Using a simple box model and assuming a periodic rate of injection of ozone through the tropopause, Junge (127) correlated the ozone maximum of the lower troposphere in spring and early summer with the total ozone maximum in winter and estimated a tropospheric lifetime of two to three months. Danielsen, Bleck, Shedlovsky, Wartburg, Haagenson, and Pollock (46), Hunt (105), and Manabe and

TABLE III
Water Number-Density Profiles

Altitude (km)	15° (Annual) (molecules cm^{-3})	30° (Winter) (molecules cm^{-3})	30° (Summer) (molecules cm^{-3})	45° (Winter) (molecules cm^{-3})	45° (Summer) (molecules cm^{-3})	75°N (Winter) (molecules cm^{-3})	75°N (Summer) (molecules cm^{-3})
0	6.5^{17}[a]	3.2^{17}	7.3^{17}	1.2^{17}	4.6^{17}	2.1^{16}	1.9^{17}
1	4.6^{17}	2.3^{17}	4.0^{17}	8.6^{16}	3.2^{17}	2.2^{16}	1.5^{17}
2	3.2^{17}	1.4^{17}	2.6^{17}	6.0^{16}	2.0^{17}	1.8^{16}	1.1^{17}
3	1.6^{17}	8.4^{16}	1.9^{17}	4.0^{16}	1.1^{17}	1.0^{16}	7.4^{16}
4	7.5^{16}	4.0^{16}	1.1^{17}	2.3^{16}	6.5^{16}	5.8^{15}	4.0^{16}
6	3.0^{16}	1.3^{16}	4.0^{16}	7.5^{15}	2.0^{16}	1.8^{15}	1.2^{16}
8	8.4^{15}	4.2^{15}	1.3^{16}	1.7^{15}	7.1^{15}	2.8^{14}	2.8^{15}
10	1.7^{15}	1.2^{15}	2.9^{15}	4.3^{14}	2.2^{15}		6.4^{14}

[a] The superscript designates the power of 10 by which each number is to be multiplied.

Hunt (163) have performed detailed transport calculations using ozone as a tracer and found that a tracer produced at low latitudes in the stratosphere is transported downward toward the poles. It then enters the troposphere through folds or breaks in the tropopause at middle and high latitudes and is transported downward back toward lower latitudes, where it is removed at the ground.

Regener (202) first measured an ozone flux at the surface; the flux has been confirmed by later experiments of Aldaz (3) and Gabally (68). The currently accepted view of sources and sinks of ozone is that it enters the troposphere from the stratosphere, principally at middle and high latitudes, and is destroyed at the earth's surface. Kroening and Ney (141) have suggested that sheets of high-density ozone found in the troposphere are due to direct transport of stable lamina ("ozone rivers") from the stratosphere.

There is, however, some evidence for alternative sinks and sources. Kelly and McTaggart-Cowan (139) found at Point Barrow, Alaska, that while there is a total oxidant gradient toward the ground during the growing season, the gradient reverses after the growing season. They suggested that decaying plant matter is the source of oxidant, which is presumably, though not necessarily, ozone. Ripperton (209) found that he could produce ozone by radiating samples of surface air, and Ripperton, Jeffries, and Worth (210) found that ozone densities in the surface layer would increase during periods of temperature inversion, when downward transport is not likely. Ripperton and Vukovich (213) suggested the existence of an atmospheric sink for ozone, possible reactions with NO_2. Recently Crutzen (43) suggested that reactions involving peroxy radicals may be an indirect source of ozone in the lower troposphere and that the photolysis of NO_3 may be an indirect sink. While his mechanism involves some reactions whose rates are not well known, preliminary estimates suggest that the photochemical source and sink are comparable to the fluxes estimated both for surface destruction and for stratospheric injection. This possibility will be discussed in more detail in later sections.

Measurements made before 1963 are summarized by Junge (128). Hering and Borden (96-98) have reported on measurements made with an extensive network of ozonesondes that utilize a sensitive detection technique developed by Regener (203). Dutsch (51) presented data from a European network that uses the Brewer Mast Detector, and Bojkov (22) reported on data taken in the southern hemisphere. It would appear [Dutsch (51)] that the distribution and transport mechanisms are somewhat different for the two hemispheres. In general the tropospheric ozone mixing ratio, while highly variable, ranges from 0.1 to 0.01 x 10^{-6}, with a definite maximum in the spring and early summer and with a general trend of increasing mixing ratio with altitude. Representative ozone number-density and total column-density profiles for several summer and winter latitudes [Hering and Borden (98)] are shown in Table IV. It should be noted that the surface layer is particularly variable owing to climatic effects on the loss rate at the ground and to the possibility of local production in polluted air masses.

D. Sulfur Compounds

In recent reviews, Kellogg, Cadle, Allen, Lazrus, and Martell (138) and Robinson and Robbins (214) discussed both atmospheric sources and sinks for sulfur compounds and a general atmospheric sulfur cycle. This section

TABLE IV
Ozone Number-Density and Column-Density Profiles

Altitude (km)	9°W (Winter)		9°W (Summer)		30°N (Winter)		30°N (Summer)		45°N (Winter)		45°N (Summer)		76°5N (Winter)		76°5N (Summer)	
	$n(O_3)$	$\int_z n(O_3)$	$n(O_3)$	$\int_z n(O_3)$	$n(O_3)$	$\int_z n(O_3)$	$n(O_3)$	$\int_z n(O_3)$	$n(O_3)$	$\int_z n(O_3)$	$n(O_3)$	$\int_z n(O_3)$	$n(O_3)$	$\int_z n(O_3)$	$n(O_3)$	$\int_z n(O_3)$
0	4.5^{11}[a]	6.8^{18}	3.0^{11}	6.8^{18}	6.3^{11}	7.3^{18}	8.8^{11}	8.1^{18}	5.8^{11}	9.5^{18}	6.4^{11}	1.00^{19}	6.3^{11}	1.22^{19}	4.9^{11}	9.5^{18}
1	3.8^{11}	6.8^{18}	3.0^{11}	6.8^{18}	5.8^{11}	7.2^{18}	8.3^{11}	8.0^{18}	5.2^{11}	9.4^{18}	6.2^{11}	9.9^{18}	5.7^{11}	1.21^{19}	4.9^{11}	9.5^{18}
2	3.8^{11}	6.7^{18}	3.0^{11}	6.7^{18}	5.8^{11}	7.2^{18}	8.3^{11}	7.9^{18}	5.2^{11}	9.4^{18}	6.2^{11}	9.9^{18}	5.7^{11}	1.21^{19}	4.9^{11}	9.4^{18}
3	3.4^{11}	6.7^{18}	3.2^{11}	6.7^{18}	5.2^{11}	7.1^{18}	8.1^{11}	7.9^{18}	4.9^{11}	9.3^{18}	6.1^{11}	9.8^{18}	5.4^{11}	1.20^{19}	5.4^{11}	9.4^{18}
4	3.0^{11}	6.7^{18}	3.4^{11}	6.7^{18}	4.7^{11}	7.1^{18}	8.0^{11}	7.8^{18}	4.8^{11}	9.3^{18}	5.9^{11}	9.8^{18}	5.1^{11}	1.20^{19}	5.8^{11}	9.3^{18}
5	2.9^{11}	6.6^{18}	3.5^{11}	6.6^{18}	4.3^{11}	7.0^{18}	7.6^{11}	7.7^{18}	4.7^{11}	9.2^{18}	5.6^{11}	9.7^{18}	5.4^{11}	1.19^{19}	6.0^{11}	9.2^{18}
6	2.8^{11}	6.6^{18}	3.7^{11}	6.6^{18}	3.9^{11}	7.0^{18}	7.3^{11}	7.6^{18}	4.6^{11}	9.2^{18}	5.3^{11}	9.6^{18}	5.8^{11}	1.19^{19}	6.2^{11}	9.2^{18}
7	2.7^{11}	6.6^{18}	3.6^{11}	6.6^{18}	3.9^{11}	7.0^{18}	7.1^{11}	7.5^{18}	5.4^{11}	9.2^{18}	5.7^{11}	9.6^{18}	8.5^{11}	1.18^{19}	8.2^{11}	9.1^{18}
8	2.5^{11}	6.5^{18}	3.3^{11}	6.5^{18}	3.9^{11}	6.9^{18}	7.0^{11}	7.5^{18}	6.1^{11}	9.1^{18}	6.1^{11}	9.5^{18}	1.1^{11}	1.17^{19}	1.0^{12}	9.0^{18}
9	2.4^{11}	6.5^{18}	3.2^{11}	6.5^{18}	4.3^{11}	6.9^{18}	6.6^{11}	7.4^{18}	1.0^{12}	9.0^{18}	7.2^{11}	9.5^{18}			1.8^{12}	8.9^{18}
10	2.3^{11}	6.5^{18}	3.0^{11}	6.5^{18}	4.7^{11}	6.8^{18}	6.3^{11}	7.3^{18}	1.4^{12}	8.9^{18}	8.3^{11}	9.4^{18}				
11	2.2^{11}	6.5^{18}	3.0^{11}	6.4^{18}	5.5^{11}	6.8^{18}	6.2^{11}	7.3^{18}			9.8^{11}	9.3^{18}				
12	2.2^{11}	6.4^{18}	2.9^{11}	6.4^{18}	6.3^{11}	6.7^{18}	6.1^{11}	7.2^{18}			1.1^{12}	9.2^{18}				
13	2.2^{11}	6.4^{18}	3.0^{11}	6.4^{18}			6.6^{11}	7.2^{18}			1.3^{12}	9.1^{18}				
14	2.2^{11}	6.4^{18}	3.1^{11}	6.4^{18}			7.2^{11}	7.1^{18}								
15	2.3^{11}	6.4^{18}	4.0^{11}	6.3^{18}			1.0^{12}	7.0^{18}								
16	2.5^{11}	6.4^{18}	5.0^{11}	6.3^{18}												
17	5.0^{11}	6.3^{18}	9.4^{11}	6.2^{18}												

[a] The superscript designates the power of 10 by which each number is to be multiplied.

will be based to a large extent on their work, as well as on the earlier work of Junge (128) and Eriksson (59).

To summarize the atmospheric sulfur cycle as it is currently understood, the major natural source is the production of H_2S from biological decay, while SO_2 from the combustion of coal and petroleum is a comparable anthropogenic source. No significant natural sources of SO_2 are known, though volcanic activity may contribute a small amount. While some SO_2 may be absorbed by plants or washed out by rain, most of it, as well as most of the atmospheric H_2S, is believed to be oxidized to sulfuric acid and sulfate salts. The major sink for atmospheric sulfur is then the rainout and dry deposition of the sulfates and sulfuric acid. While sea spray is also a source of sulfate aerosols, most of it is thought to return directly to the sea, leaving little, if any, net production. Some experimental evidence, while not conclusive, suggests that the sulfate aerosols over land originate from SO_2 and H_2S rather than from sea salt.

Unfortunately as Kellogg et al., Robinson and Robbins, Junge, and Eriksson have all pointed out, most of the atmospheric measurements have been made in polluted areas of the United States and Europe, so not much is known about normal background concentrations of sulfur compounds and their global distribution. Therefore the atmospheric cycle is somewhat speculative, as are also estimates of individual sinks, sources, and concentrations.

1. Sulfur Dioxide (SO_2)

The major known direct source of SO_2 is the combustion of coal and petroleum, although the refining of petroleum and nonferrous metals is also significant. Recent estimates of these sources range from 100 x 10^6

tons yr^{-1} [Kellogg et al. (138)] to 150 x 10^6 tons yr^{-1} [Robinson and Robbins (216)]. The major indirect source is the oxidation of atmospheric hydrogen sulfide, for which no measured value is available.

While SO_2 may be directly removed by rain [Beilke and Georgii (19)], plant absorption [Katz and Ledingham (134)], and direct deposition [Chamberlain (38)], the major sink is thought to be oxidation to sulfate. Though photoxidation in pure water appears to very slow [Robinson and Robbins (214)], the presence of metallic catalysts or dissolved ammonia produces very fast rates [Junge and Ryan (132)]. The photochemical oxidation of SO_2 in the presence of NO_2 and hydrocarbons has also been shown to produce considerable sulfuric acid aerosols [Schuck and Doyle (221)].

In summary, it appears [cf. Robinson and Robbins (214)] that photochemical reactions involving NO_2 and hydrocarbons and absorption by alkaline water drops with or without metal catalysts are the major mechanisms for oxidizing SO_2. The complex nature of these removal mechanisms will, however, make their inclusion in any atmospheric photochemical model difficult. This will be discussed in Section V.

While most measurements of SO_2 number densities have been made in polluted urban areas, some background measurements in the upper troposphere and in isolated areas are available. Georgii and Jost (77) found densities less than 1 ppb at 5 km or higher, and Georgii (76) found less than 0.5 ppb in the upper troposphere over Nebraska. Surface measurements in Hawaii were about 0.3 ppb, and in Florida, about 1 ppb [Junge (128)]. Similar values ranging from 13 to 1 ppb were measured by Cadle, Pate, and Lodge (31), and Cadle, Fischer, Frank, and Lodge (29) in Antarctica and the Panama Canal Zone. Over a wide range of Central America, Kuhme (143) found no SO_2 above his detection limit of 0.3 ppb. From these data Robinson and Robbins (214) recommended an average tropospheric concentration of 0.2 ppb for SO_2. They also obtained a residence time for SO_2 of four to two days, which is in agreement with other estimates [Junge (126,128)] and Georgii and Jost (77) from regional measurements of source strengths and ambient number densities. Weber (249) did find a very short lifetime of one hour for SO_2 in the highly polluted environment of a dispersin stack plume. It would appear that while the ambient level of SO_2 is quite low away from pollution sources, it can be much higher and quite variable in industrial and populated regions.

2. Hydrogen Sulfide (H_2S)

It is assumed that the major source of H_2S is decaying organic material [Kellogg et al. (138) and Robinson and Robbins (214)], although there are no data to support this conclusion. Earlier, Eriksson (59) and Junge (128) suggested a large ocean source, but Ostlund and Alexander (188) pointed out that the surface layer of the ocean is too oxidizing to allow the emission of H_2S, and Fonselius (64) found that the same is true for the stagnant basins of the Baltic Sea. Mud and tidal flats may, however, be a source. Robinson and Robbins (214) felt it necessary to retain a sizable ocean source to balance their model, although this assumption is certainly questionable. Their estimate of the land source was determined by balancing the total sulfur cycle, just as the early workers did. Currently there are no actual data concerning the natural sources of H_2S. Air pollution is not thought to be a significant source.

The major sink for H_2S appears to be its oxidation to SO_2. Using the reaction rate measured by Cadle and Ledford (30) for the heterogeneous oxidation of H_2S by O_3 and an estimate of particulate surface area based on measurements reported by Junge (128), Robinson and Robbins (214) found a 2-hr lifetime for H_2S in the surface atmosphere and a 28-hr lifetime in the background troposphere with its light particulate load.

The earlier measurements of H_2S number densities summarized by Junge (128) are now thought to be in error. Smith, Jenkins, and Cunningworth (230) reported a value of 0.1 to 0.3 ppb in rural areas of Wales, while Lodge (158 cited values generally less than 0.1 ppb in the Boulder, Colorado, area, except for stagnant days, when the values may reach 0.25 ppb. Robinson and Robbins (214) recommended an average tropospheric value of 0.2 ppb, though as they point out there are not sufficient data to make an accurate estimate.

3. Sulfate Aerosols ($SO_4^=$)

The major source of sulfate aerosols appears to be the oxidation of atmospheric H_2S and SO_2. While some is also produced by sea spray, most of the sea-salt aerosol returns directly to the ocean, resulting in only a small net production.

The major sinks for sulfate aerosol, both sulfate salts and sulfuric acid, are rainout and dry deposition. From limited data both Kellogg et al. (138) and Robinson and Robbins (214) estimated 270 x 10^6 tons yr^{-1} of $SO_4^=$ deposited in the northern hemisphere. No data are available for the southern hemisphere. Junge (128) estimated a residence time of 20 to 30 days for such aerosols, while Robinson and Robbins suggested 8 days.

Sulfate aerosols are the major form of atmospheric sulfur; Robinson and Robbins recommended an average tropospheric concentration of 2 μg m^{-3}.

4. Dimethyl Sulfide (Me_2S)

Lovelock, Maggs, and Rasmussen (161) have observed the emission of Me_2S by microorganisms in the soil and water and suggested that they may provide a large natural source of atmospheric sulfur, particularly in the ocean. No measurements of the source strength or atmospheric density are currently available. It is most likely that Me_2S is converted into either dimethyl sulfoxide or SO_2, but no detailed reaction mechanisms have been proposed.

5. Global Distribution

Robinson and Robbins (216) estimated that two-thirds of the atmospheric sulfur is produced in the northern hemisphere owing to the much larger land mass and to industrial production there. Assuming that the loss rates for sulfate aerosol and oxidation rates for SO_2 and H_2S are similar in the two hemispheres, a north-south asymmetry would be expected in the upper troposphere values for SO_2 and H_2S number densities, as well as for the sulfate aerosol concentration. The apparent lack of such asymmetry suggests either the existence of a large natural source, possibly the production of Me_2S, or a much longer lifetime for the sulfur compounds in the dry upper troposphere with its very light particulate load. On the basis of previous experience with carbon monoxide, it would not be surprising to find another large global source of sulfur. Further measurements in both hemispheres are certainly needed, particularly for H_2S.

E. Nitrogen Compounds

In a series of articles, Robinson and Robbins (214-216) have reviewed the sinks, sources, and concentrations of tropospheric nitrogen compounds, while a survey of the early literature is provided by Junge (128).

Robinson and Robbins have also developed a circulation model for atmospheric nitrogen; owing to a lack of accurate data, it is somewhat speculative and possibly inaccurate as a quantitative model. However, on the basis of their review and that of Bates and Hays (11), we can construct a qualitative model with three relatively separate cycles.

For the major atmospheric oxide of nitrogen—nitrous oxide—the source is biological activity at the surface, and the sink is transport into the stratosphere, where it is destroyed by photodissociation and reaction with $O(^1D)$. There are no important photochemical reactions for nitrous oxide in the troposphere.

For the other two naturally occuring oxides—nitric oxide and nitrogen dioxide—as well as their related acids, nitrous and nitric, the known source is pollution, which produces mainly nitric oxide; but biological activity is also thought to produce nitric oxide. During the day, the two oxides rapidly establish photostationary-state number densities, while at night, nitric oxide is converted to nitrogen dioxide. The sink for these oxides is the conversion of nitrogen dioxide to nitric acid, which is then either rained out as nitric acid and nitrate salts or removed by the dry deposition of nitrate aerosols. These species are deeply involved in the tropospheric photochemistry.

The third cycle is for ammonia, which is produced by biological activity in the soil and removed by the rainout and dry deposition of ammonium salt aerosols. It now appears that ammonia may play an important role in gas-phase tropospheric chemistry as a large atmospheric source of nitrogen oxides [Lodge and Pate (160); Stuhl and Niki (238); McConnell (170); and Crutzen (43)].

1. Nitrous Oxide (N_2O)

Adel (2) was the first to identify N_2O in the atmosphere, and he suggested a microbiological source. Then Arnold (6) found that it was produced by bacterial action on ammonium and nitrate salts. The ocean has been suggested as both a source and a sink [Robinson and Robbins (214)] and there are some indications of a biological sink [Burris (27)]. Bates and Witherspoon (12) suggested the oxidation of molecular nitrogen by ozone as a source, but experiments by Goody and Walshaw (82) showed that it is small.

Goody and Walshaw calculated a removal rate for N_2O based on its transport into the stratosphere and resultant photodissociation; this value is in agreement with estimates of its production rate from the data of Arnold (6). More detailed calculations by Bates and Hays (11) reduced the removal rate by almost a factor of 30, and this much slower rate has been confirmed by the calculations of McElroy and McConnell (171). Robinson and Robbins (214-216) have suggested both a soil sink and absorption by vegetation as means of balancing the large production estimate. In view of the approximate nature of these evaluations, it would seem just as reasonable to lower the net soil production rate. In either case, the tropospheric residence time is a number of years and N_2O can be considered well mixed. Summarizing a number of measurements, Robinson and Robbins (214) recommended a single tropospheric mixing ratio of 2.5 x

10^{-7} for N_2O, in harmony with more recent results of Junge and Hahn (131), Goody (81), and Schutz, Junge, Beck, and Albrecht (223). Schutz et al. did find some possible north-south asymmetry and seasonal dependence, though the differences were small, and both they and Junge and Hahn suggested a possible ocean source.

2. Nitric Oxide and Nitrogen Dioxide (NO and NO_2)

Robinson and Robbins (216) estimated the combustion source of NO and NO_2 to be 16 x 10^6 tons (N) yr^{-1}, and there is a qualitative indication of NO production by bacteria. A number of upper atmospheric reactions—photodissociation of N_2 [Bates and Hays (11)], oxidation of N_2O by metastable atomic oxygen [Nicolet (181) and McElroy and McConnell (173)], and reactions involving atomic nitrogen and oxygen [Nicolet (179)]—do not appear to be significant in the troposphere. Robinson and Robbin decision to balance the estimated sink, rainout, and dry deposition of nitric acid and nitrate salts [Eriksson (58) by an unknown bacterial source is reasonable given the lack of quantitative data, but, considering the large amount of NH_3 present in the atmosphere, its oxidation may be the major source. This possibility will be discussed in detail in Section IV.B.

Using measurements by Junge (125), Lodge, MacDonald and Vihman (159), Lodge and Pate (160), Ripperton, Worth, and Kornreich (212), Hamilton, Worth, and Ripperton (89), O'Connor (187), and Fischer, Lodge, Wartburg, and Pate (62 Robinson and Robbins (216) recommended number densities of 2 ppb for NO and 4 ppb for NO_2 over land areas between 65°N and 65°S, and 0.2 ppb for NO and 0.5 ppb for NO_2 over the remaining land and all oceans. It should be note that these values are for noontime at the surface. The ratio of NO to NO_2 will change both with the time of day and with altitude. Assuming that the sole sink is conversion to nitrates and their resulting deposition, we find a residence time of 21 days for NO and NO_2; if dry deposition of NO_2 is included, the lifetime drops to 9 days. In either case, the residence times appear to be short enough to prevent complete mixing. The findings of large number densities near the equator appears to support the existence of a large natural source, since combustion is concentrated in the midlatitudes.

3. Nitric Acid (HNO_3)

The source of HNO_3 is the photochemical oxidation of NO_2, and the sinks are rainout and dry deposition of HNO_3 and its related nitrate salts. While HNO_3 has been observed in a smog environment [Price and Stephens (197)], it has not been found in the normal atmosphere. Levy (153,154) has predicted that HNO_3 should exist in the troposphere, with a number density as high as 30 ppb. Eriksson (58) has determined the latitude dependence of nitrate precipitation; to some extent, it follows the behavior suggested by Robinson and Robbins (216) for NO and NO_2 number densities. Based on the estimated concentrations for nitrates of 0.2 $\mu g\ m^{-3}$ [Junge (128) and Lodge et al. (159)], the precipitation rate of 80 x 10^6 tons (N) yr^{-1} [Eriksson (58)] gives a residence time for nitrate aerosols of four days. While this time is quite short for an aerosol, it it likely that a considerable portion of the rained-out nitrate comes from direct washout of tropospheric HNO_3, with the result that the nitrate aerosol loss rate is overestimated.

4. Ammonia (NH_3)

The main source of NH_3 is thought to be the biological breakdown of amino acids in organic material

[Junge (128)], though the soil, depending on its pH, can be either a sink or a source. Emission sources are not important for NH_3. While there are no quantitative measurements of the source strength, Eriksson (58) gave measurements of a major loss rate, the rainout of ammonium salts, 170 x 10^6 tons (N) yr^{-1}. Another proposed sink [Lodge and Pate (160), Stuhl and Niki (238), McConnell (170), and Crutzen (43)] is the oxidation of NH_3 to oxides of nitrogen. This reaction will be discussed in greater detail in Section IV.B.

Robinson and Robbins (214) and Junge (128) reported a number of values for NH_3 number densities that range between 1 and 20 ppb, with a maximum at low latitudes and during the summer season. Lodge and Pate (160) have measured some values that exceeded 20 ppb. Murcray and Murcray (178), using a technique whose lower limit was 0.1 ppb, reported no detection of NH_3 at the tropopause. Robinson and Robbins (216) suggested an average tropospheric density of 6 ppb, but we can expect this value to vary.

Using Eriksson's (58) estimates of rainout and adding 25% for dry deposition, we find a residence time of approximately one month for atmospheric NH_3. The inclusion of gaseous deposition as a loss mechanism [Robinson and Robbins (216)] gives a residence time of one week. It is believed that the seasonal and altitude variations in the NH_3 number densities are due to variations in its biological source at the ground and its short residence time in the troposphere.

F. Carbon Compounds

1. Carbon Dioxide (CO_2)

Carbon dioxide, the major source of carbon-containing molecules in the troposphere, is not known to have any photochemical reactions in that region. Its number density is controlled by three basic cycles, which have been discussed in some detail by Junge (128) and are summarized here.

The first is the rapid circulation of CO_2 between the surface air and the earth's surface as well as the surface layer of the ocean. Plant respiration and bacterial decay are the important sink and source, respectively.

A slower exchange takes place between atmospheric CO_2 and the large reservoir of CO_2 dissolved in the ocean as a whole.

The last cycle is the weathering and precipitation of carbonates in the ocean and lakes and the formation of fossil fuels in the earth.

While sharp fluctuations are observed in the CO_2 number densities of the surface layer owing to variations in plant and bacterial activity, it is well mixed throughout the rest of the troposphere, with the mixing ratio currently established as 320 ppm [Bolin and Bischof (23)]. A small seasonal variation in the CO_2 level, believed due to seasonal fluctuations in the main sources, has been observed [Bolin and Bischof (23) and Junge and Czeplak (130)].

Although man is responsible for less than 2% of the global production of CO_2, the slow increase of 0.7 ppm yr^{-1} observed in the CO_2 level [Pales and Keeling (189) and Bolin and Bischof (23)] has been attributed to man's activity [cf. Junge (128) and Robinson and Robbins (214)]. Because of the much larger natural sources, the possibility of small variations in them as well as the possible effects of a small long-term temperature change on the ocean-atmosphere CO_2 balance should also be considered.

2. Carbon Monoxide (CO)

Migeotte and Neven (175) first observed atmospheric CO and found a mixing ratio of 0.2 to 0.3 ppm. In other data summarized by Junge (128), the mixing ratio in the troposphere had a range of 0.1 to 0.2 ppm.

Robinson and Robbins (214) concluded from a number of measurements made in Greenland, Point Barrow (Alaska), and the South Pacific that CO averaged 0.1 ppm in the northern hemisphere and 0.04 in the southern hemisphere, though values as high as 0.2 ppm of CO, now believed due to transported industrial air from the U.S., were observed in Greenland. On the basis of further measurements, Robinson and Robbins (216) reported a north-south asymmetry but raised the average values to 0.14 ppm and 0.06 ppm, with a global average of 0.1 ppm.

Seiler and Junge (226) cited measurements in the upper troposphere ranging from 0.1 to 0.15 ppm in both hemispheres, with an average value of 0.13 ppm and rather constant surface values of 0.1 to 0.12 ppm at the equator and 10°S. They found no diurnal variation, and none of their measurements dropped below 0.09 ppm. Sharply increased values approaching 0.2 ppm were reported for the air over the North Atlantic, which they attribute to an ocean source of CO.

Seiler (225) recently reported measurements that show a relatively constant value of 0.1 to 0.15 ppm between the midlatitudes, with a drop to 0.05 ppm at high southern latitudes and an increase of 0.2 ppm in the high northern latitudes.

From these data it is clear that the CO mixing rate varies locally and possibly regionally, with a maximum range of 0.05 to 0.2 ppm. The average tropospheric value appears to have the range 0.1 to 0.15 ppm.

It had been assumed [cf. Junge (128) and Robinson and Robbins (214)] that combustion was the major source of CO. On the basis of this assumption and an estimated CO tropospheric mixing ratio of 0.1 ppm, Bates and Witherspoon (12) calculated a CO tropospheric lifetime of up to 4 yr. Later calculations by Robinson and Robbins (214) found a lifetime of 2.7 yr. Weinstock (250) then derived from radiocarbon data a lifetime for CO that was more than 10 times shorter than the earlier estimates. Such a short lifetime required much larger sources and sinks than had been previously suggested [Bates and Witherspoon (12), Robinson and Robbins (214), and Seiler and Junge (226)]. Weinstock suggested that the oxidations of CO by hydroxyl radicals, first proposed by Bates and Witherspoon for the stratosphere, would provide the necessary sink if a sufficient source of radicals could be found. Levy (152) presented a simple photochemical model of the troposphere that provides both the necessary hydroxyl radicals and, by their destruction of methane, a large source of formaldehyde. It was then pointed out that the loss paths for formaldehyde all lead to the production of CO [McConnell et al. (171) and Levy (153)]. Calculations by McConnell et al. and by Levy (155) have shown that the oxidation of methane is a much larger source of CO than is combustion.

Swinnerton, Linnebom and Cheek (239) and Seiler and Junge (226) reported that the ocean, which they found to be supersaturated with CO, might be an important source. It is also possible that the oxidation of nonmethane hydrocarbons, particularly those whose production occurs over land, may be another huge source. Since, as we will see later, methane is thought to have a uniform distribution throughout the troposphere, some large sources of CO with a north-south asymmetry are needed. Combustion is not large enough to produce the apparent effects, and estimates of the ocean source [Swinnerton, Linnebom, and Lamontagne (240)] also appear too small.

The major sink would appear to be the oxidation of CO by hydroxyl radicals. The calculated sink for this process gives a lifetime [Levy (152,154)] in harmony with that predicted from radiocarbon data [Weinstock (250) and Weinstock and Niki (251)]. Ingersoll and Inman (107) and Inman, Ingersoll, and Levy (108) have shown that soil bacteria can destroy CO and suggested that it may be an important sink. Pressman and Warneck (196) calculated the sink due to the transport of CO into the stratosphere, where it is then oxidized to CO_2, and found that it was small.

3. Methane (CH_4)

Early measurements of tropospheric methane, first identified by Migeotte (174), are summarized by Junge (128 who concluded that CH_4 is well mixed in the troposphere, with a mixing ratio of 1.5 ppm. Cavanagh, Schadt, and Robinson (37) found a CH_4 mixing ratio of 1.6 ppm in the surface air of Point Barrow. Bainbridge and Heidt (8) measured two altitude profiles for CH_4 that showed a relatively constant mixing ratio in the range 1.4 to 1.6 ppm up to the tropopause. Ehhalt (53) reported a number of tropospheric measurements of CH_4 that, while giving an average mixing rate of 1.4 ppm, do show variations on the order of $\pm 20\%$ with altitude and from day to day. He did not, however, observe a significant seasonal variation.

It is generally accepted that the major source of methane is anaerobic decay of organic matter, though natural-gas emissions and combustion provide some methane. Recently, using the earlier work of Koyama (142) on paddy fields and other minor sources and of Robinson and Robbins (214), Ehhalt (52) estimated that the total column production rate for CH_4 ranged between 1.5 and 4×10^{11} molecules cm^{-2} sec^{-1}.

Bainbridge and Heidt (8) estimated from the rate of decrease of the CH_4 mixing ratio above the tropopause that at most 10% of the tropospheric CH_4 was lost by transport into the stratosphere and suggested that a large tropospheric sink was necessary. Ehhalt and Heidt (53) found that the calculated sink for methane due to transport into the stratosphere was much too small.

Levy (152) suggested that the oxidation of CH_4 by hydroxyl radicals would provide the necessary sink for methane. Calculations by Wofsy et al. (256) gave a lifetime of 7 yr for methane, while Levy (155), using average daily hydroxyl radical concentrations previously calculated for summer midlatitude [Levy (154)], found a column loss rate of 4.5×10^{11} molecules cm^{-2} sec^{-1} and a CH_4 lifetime of 2 yr. It would appear that the major sink for tropospheric CH_4 is its oxidation by hydroxyl radicals.

4. Nonmethane Hydrocarbons

Cavanagh et al. (37) made gas-chromatographic measurements of the organic content of the surface atmosphere at Point Barrow and found fractional part-per-billion levels of ethane, ethylene, butane, pentane, acetaldehyde, acetone, methanol, ethanol, benzene, and two unknown components, while n-butanol was found in the 10 ppb range. It is not known whether the butanol was present owing to contamination from an anthropogenic source or whether it was naturally present in the biosphere.

The release of volatile organic compounds by the foliage of green plants is well known. Went (252) estimated a total emission of 175 x 10^6 tons; Rasmussen and Went (200) found emissions ranging from 13.5 x 10^6 to 432 x 10^6 tons, depending on the means of measuring the particular foliage emission rate; and Ripperton, Worth, and Jeffries (211) obtained emissions 2 to 10 times larger than any previous estimate. Rasmussen (199) has concluded that, while the identity of the emissions is well known, a quantitative estimate of the worldwide terpene emission rate is not yet possible. It would appear that the natural emissions are much larger than that estimated for man's activities, 27 x 10^6 tons yr^{-1}. Ripperton et al. have suggested that the reaction of ozone with terpenes provides an important, if not dominant, sink for both compounds in the troposphere. While large terpene mixing ratios (ppm or more) have been measured locally in isolated areas [Ripperton et al. (211)], no global estimate is available.

Dhar and Ram (50) found formaldehyde in rain water and estimated a tropospheric mixing ratio of 0.7 ppb, while Cauer (36) measured a mean value of 0.4 ppb. Lodge and Pate (160) obtained an average value of 1.1 ppb for the total aliphatic aldehydes in surface air in the tropics. Levy (152) proposed the formation of formaldehyde via the tropospheric oxidation of methane and calculated (155) an upper limit of 1 ppb for the mixing ratio, with an altitude profile for a summer midlatitude decreasing from 0.6 ppb at the ground to less than 0.01 ppb in the upper troposphere, where methane oxidation is very slow (154).

G. Molecular Hydrogen (H_2)

Molecular hydrogen is assumed to be well mixed in the troposphere, with a mixing ratio of 0.4 to 0.6 ppm [Junge (128) and Scholz, Ehhalt, Heidt, and Martell (219)]. Koyama (142) found that swamps and paddies are very small natural sources. Levy (153) proposed both an atmospheric source (photodissociation of formaldehyde) and an atmospheric sink (oxidation by hydroxyl radical). From daily average number densities for the hydroxyl radical and a daily average hydrogen production rate, Levy (155) calculated an H_2 mixing ratio of 0.7 ppm and a tropospheric residence time of 2 yr. From this agreement it appears that both the major source and the major sink for H_2 are homogeneous gas-phase reactions in the troposphere.

III. SOLAR FLUX AND ITS TRANSMISSION

Detailed discussions of the transmission of solar radiation through the earth's lower atmosphere have been given by Leighton (149) and Elterman (54-56).

A. Incident Solar Radiation

Ackerman (1) has compiled values for the solar flux incident at the top of the atmosphere from the measurements by Detwiler, Garrett, Purcell, and Tousey (49) and Tousey (242) over the wavelength range 2299 to 3625 Å and the measurements by Johnson (115) for longer wavelengths. These flux data are presented in Table V for wavelength intervals starting at 2857 Å and extending to 4025 Å.

TABLE V
Solar-Radiation Data

Wavelength (Å)	Solar Flux (photons $cm^{-2}sec^{-1}$)	$\sigma_{O_3}(\lambda)$ (cm^{-2})	$K_{Rayleigh}$	$K_{aerosol}$
2857-2899	2.46^{14}[a]	1.92^{-18}	1.5	0.42
2899-2941	3.90^{14}	1.14^{-18}		
2941-2985	3.99^{14}	6.60^{-19}		
2985-3030	3.86^{14}	3.69^{-19}	1.22	0.41
3030-3077	5.08^{14}	1.97^{-19}		
3100 ± 25	5.92^{14}	1.05^{-19}		
3150 ± 25	6.05^{14}	5.23^{-20}		
3200 ± 25	6.94^{14}	2.91^{-20}	0.93	0.40
3250 ± 25	8.12^{14}	1.50^{-20}		
3200 ± 25	9.71^{14}	7.78^{-21}		
3350 ± 25	8.97^{14}	3.72^{-21}		
3400 ± 25	9.44^{14}	1.71^{-21}	0.72	0.38
3450 ± 25	1.01^{15}	7.46^{-22}		
3500 ± 25	1.03^{15}	2.66^{-22}		
3550 ± 25	1.03^{15}	1.09^{-22}		
3600 ± 25	1.04^{15}	5.49^{-23}	0.56	0.38
3650 ± 25	1.18^{15}	--		
3700 ± 25	1.23^{15}	--		
3750 ± 25	1.24^{15}	--		
3800 ± 25	1.17^{15}	--	0.45	0.36
3850 ± 25	1.11^{15}	--		
3900 ± 25	1.09^{15}	--		
3950 ± 25	1.19^{15}	--		
4000 ± 25	1.54^{15}	--	0.36	0.32

[a]The superscript designates the power of 10 by which each number is to be multiplied.

B. Atmospheric Transmission

The two important atmospheric processes are molecular absorption and molecular and aerosol scattering. Absorption by aerosols, both liquid and particulate, is not thought to be important.

1. Molecular Absorption

The absorption in the upper atmosphere by molecular oxygen in its Schumann-Runge and Herzberg continua and by ozone in the diffuse Hartley band reduces all radiation with a wavelength less than 2850 Å to an insignificant level in the troposphere. It has been suggested that an atmospheric window may exist in the region of 2000 Å between the major ozone and molecular oxygen absorptions. Calculations of transmission through the Schumann-Runge bands suggested that radiation in a narrow wavelength interval about 2000 Å might be transmitted into the lower stratosphere [Kockarts (140)], but our preliminary calculations have found no significant flux into the upper troposphere.

The major effect of molecular absorption on the radiation that does penetrate into the troposphere is absorption by ozone in the tail of the Hartley band, 2850 to 3250 Å. Absorption cross sections for this process have been compiled by Ackerman (1) from the measurements by Watanabe, Inn, and Zelikoff (247), Inn and Tanaka (109), and Vigroux (246) and are presented in Table V.

As far as the attenuation of solar radiation is concerned, absorption by NO_2 and SO_2 is not significant in the troposphere, because of their very small number densities. The very weak absorption by O_3 in the Chappuis bands is also insignificant.

2. Molecular (Rayleigh) and Aerosol Scattering

Rayleigh scattering is important throughout the troposphere, while aerosol scattering is important only in the lower troposphere, where the aerosol layer is dense. While there is a dust layer in the lower stratosphere [cf. Junge (128)], it does not cause a significant attenuation of incident radiation. Rayleigh scattering is also unimportant outside the troposphere. Elterman (54) has tabulated Rayleigh-scattering coefficients, using Penndorf's (192) calculation of the index of refraction of a standard atmosphere and Gucker and Basu's (87) value for the depolarization factor. He presented both the Rayleigh and the aerosol extinction values over the altitude range o to 50 km for a clear day (55) and for varying degrees of haze (56). The aerosol extinction coefficients were determined from a large number of atmospheric turbidity measurements (55).

Values of both Rayleigh and aerosol extinction at the surface for a clear day are given in Table V. For the detailed altitude behavior of these coefficients and their integrated extinction, the reader should refer to the reports by Elterman (55,56).

C. Solar Flux in the Troposphere

While radiation absorbed by O_3 is removed from the tropospheric solar flux, the scattered radiation only changes direction. Rayleigh scattering in general scatters equally in the forward and the backward directions, while particulate scattering increases the percentage in the forward direction with increasing particle size. At the same time, the amount of multiple scattering increases with aerosol size. It is reasonable to estimate that half the scattered radiation continues in the forward direction While a beam of radiation is greatly attenuated in a nonabsorbing atmosphere, the total radiation, direct plus diffuse, reaching a horizontal surface is much less reduced. This effect has been observed [cf. Leighton (149), p. 22] in surface measurements of ultraviolet radiation, where the diffuse sky radiation is equal to or greater than the direct.

It is necessary to consider the differing path lengths of direct and sky radiation as the overhead angle, zenith angle χ, varies with time. As the sun sets and the solar zenith angle increases, the amount of absorbing and scattering atmosphere traversed by the solar flux increases, with a resulting increase in the attenuation. On the other hand, the effective path length for absorption is increased. The net result is that the effects of Rayleigh and aerosol scattering are greatly reduced.

Leighton (149), who discussed the above points in detail, has derived the following formula for the incident flux at a given wavelength, altitude, and solar zenith angle χ:

$$q(z,\chi,\lambda) = q(\infty,\lambda) \exp [-\sec(\chi)\ K_{O_3}(z,\lambda)]$$

$$\left[\exp [-\sec (\chi)\ K_S(z,\lambda)] + \{1 - \exp[-\sec (\chi) K_S(z,\lambda)]\}\cos (\chi)\right] \tag{1}$$

where $K_S(z,\lambda)$ is the sum of the Rayleigh and aerosol extinction, and $K_{O_3}(z,\lambda)$, the extinction due to absorption by atmospheric O_3, is given by

$$K_{O_3}(z,\lambda) = \sigma_{O_3}(\lambda) \int_z^\infty n(O_3,h)\ dh \tag{2}$$

The ozone absorption cross sections are listed in Table V, and the integrated column densities of O_3 above an altitude z, $\int_z^\infty n(O_3,h)\,dh$, are given in Table IV for a variety of seasons and latitudes.

D. Altitude, Latitude, and Seasonal Effects

The total solar flux at a given altitude and its wavelength distributions are very important photochemical parameters in the troposphere. Of particular importance is the solar flux in the region 2850 to 3250 Å. This ultraviolet radiation, indirectly the major source of radicals in the troposphere, is very sensitive to fluctuations in the total ozone column density.

While most of the atmospheric ozone lies above the tropopause, there is enough in the troposphere to cause the ultraviolet flux to increase significantly with increasing height. At the same time, the decrease of the atmospheric column density and particulate load with increasing altitude leads to a reduction in atmospheric scattering of visible and ultraviolet radiation. The net effect is an increase in solar flux with increasing altitude, with the largest increase coming in the region affected by ozone absorption, 2850 to 3250 Å.

As shown in Table IV, for a given altitude in the troposphere there are significant seasonal and latitudinal fluctuations in the total column of ozone, and these lead to similar fluctuations in the ultraviolet flux. The maximum ozone column and minimum transmitted flux occur in late winter and early spring, while the minimum column and maximum flux occur in the fall. This effect is most pronounced at high latitudes, and much less so near the equator [cf. Junge (128)]. The seasonal and latitudinal variation in Rayleigh and aerosol scattering is much less and is not important to this discussion.

The range of the solar zenith angle and the number of

hours of sunlight also vary with latitude and season. The smaller the solar zenith angle is, the less atmosphere (particularly ozone) the total solar flux passes through before reaching the troposphere, and therefore the greater the incident flux. The solar zenith angle has its minimum in the summer and maximum in the winter. The variation of solar zenith angle with latitude for a given season is less important than the seasonal variation.

A detailed discussion of the seasonal and latitudinal effects on solar flux in the troposphere is given by Leighton (149).

IV. PHOTOCHEMISTRY

In recent years, there have been a number of reviews of reaction-rate data relevant to the photochemistry of the troposphere [Schofield (218), Heicklen (93), Johnston (116), Kaufman (135), Lloyd (157), and Wilson (254)]. Also available are the High Temperature Reaction Rate Data series compiled by Baulch, Drysdale, and Lloyd (13-15); Baulch, Drysdale, Horne, and Lloyd (16); and Baulch, Drysdale, and Horne (17,18) and the Chemical Kinetics Data Survey of the National Bureau of Standards compiled by Hampson, Brown, Garvin, Herron, Huie, McKinley, and Tsang (90); Hampson, Braun, Garvin, Herron, Huie, Kurylo, Laufer, Okabe, and Tsang (91); and Garvin and Gevantman (72).

The reactions and rate coefficients used in later calculations in this chapter are listed in Table VI, and the noontime altitude profiles for the photodissociation rates for the same calculations, in Table VII.

The following units will be used in this discussion.

1. Number density (molecule cm^{-3}),
2. First-order reaction (sec^{-1}),
3. Second-order reaction (cm^3 $molecule^{-1}$ sec^{-1}),
4. Third-order reaction (cm^6 $molecule^{-2}$ sec^{-1}).

TABLE VI
Reaction-Rate Data

	Reaction	Rate Coefficient	Reference
R1	$SO + O_2 + h\nu \rightarrow SO_3 + O$	Slow	78
R2	$SO + O + M \rightarrow SO_3 + M$	Not important	See Section IV.A
R3	$SO_2 + h\nu \rightarrow {}^1SO_2^*$	Not known	227
R4	${}^1SO_2^* + M \rightarrow {}^3SO_2^* + M$	Not known	227
R5	$RO_2 + SO_2 \rightarrow RO + SO_3$	Not known	70
R6	$HO_2 + SO_2 \rightarrow OH + SO_3$	$K6 = 1.5 \times 10^{-15} - 8 \times 10^{-16}$	191a
R7	$SO_3 + H_2O \rightarrow H_2SO_4$	Fast but most likely heterogeneous	138
R8	$H_2SO_4 + 2NH_3 \rightarrow (NH_4)_2SO_4$	Fast	32
R9	$H_2SO_4 + NH_3 \rightarrow (NH_4)HSO_4$	Fast	32
R10	$H_2S + O_3 \xrightarrow{\text{surface}} H_2O + SO_2$	Fast	30
R11	$H_2S + O \rightarrow OH + HS$	Not important	see Section IV.A
R12	$H_2S + OH \rightarrow H_2O + HS$	$K12 = 5 \text{ to } 10 \times 10^{-12}$	238
R13	$N_2 + O_3 \rightarrow N_2O + O_2$	$K13 = 5 \times 10^{-28}$	82
R14	$O^*(^3P) + N_2 \rightarrow N_2O$	Not known	170
R15a	$N_2O + h\nu \rightarrow N_2 + O(^1D)$ $\lambda < 3370$ Å	$J15a = 3 \text{ to } 4 \times 10^{-10}$	11
R15b	$N_2O + h\nu \rightarrow N_2 + O(^1S)$ $\lambda < 2100$ Å	Small in troposphere	11
R15c	$N_2O + h\nu \rightarrow NO + N(^4S)$ $\lambda < 2500$ Å	Small in troposphere	11
R16a	$N_2O + O(^1D) \rightarrow NO + NO$	1.2×10^{-10}	173
R16b	$N_2O + O(^1D) \rightarrow N_2 + O_2$	$.6 \times 10^{-10}$	173

TABLE VI
(cont.)

	Reaction	Rate Coefficient	Reference
R17	$NH_3 + HNO_3 \rightarrow NH_4NO_3$	Fast in presence of surface	184
R18	$NH_3 + OH \rightarrow NH_2 + H_2O$	$K18 = 1.5 \times 10^{-13}$	236
R19	$NO + O_3 \rightarrow NO_2 + O_2$	$K19 = 9 \times 10^{-13} \exp(-1200/T)$	91
R20	$NO + HO_2 \rightarrow NO_2 + OH$	$K20 = 5 \times 10^{-13}$	46b
R21	$NO_2 + h\nu \rightarrow NO + O$	$J21 = 1 \times 10^{-2}$ exp[-0.38 sec( )]	149
R22	$NO_2 + O_3 \rightarrow NO_3 + O_2$	$K22 = 1 \times 10^{-11}$ exp(-3500/T)	121
R23a	$NO_3 + h\nu \rightarrow NO + O_2$	$J23a = 5 \times 10^{-2}$	43 and 119

Units of rate coefficients:
First-order reactions: sec^{-1}
Second-order reactions: $cm^3\ molecule^{-1} sec^{-1}$
Third-order reactions: $cm^6\ molecule^{-2} sec^{-1}$

Fast reactions for which an activation energy is not known are assumed to have no temperature dependence.

TABLE VI
(cont.)

Reaction		Rate Coefficient	Reference
R23b	$NO_3 + h\nu \rightarrow NO_2 + O$	$J23b = 0$	119
R24	$NO_3 + NO \rightarrow NO_2 + NO_2$	$K24 = 1 \times 10^{-11}$	117
R25	$NO_3 + NO_2 \overset{M}{\rightarrow} N_2O_5$	$K25 = 3 \times 10^{-12}$	117
R26	$N_2O_5 \rightarrow NO_2 + NO_3$	$K26 = 2.5 \times 10^{13} \exp(-9650/T)$	118
R27	$N_2O_5 + h\nu \rightarrow 2NO_2 + O$	$J27 = 2 \times 10^{-5}$	118
R28	$N_2O_5 + H_2O \rightarrow 2HNO_3$	$K28 < 1 \times 10^{-21}$	119 and 184
R29	$NO + NO_2 + H_2O \rightarrow 2HNO_2$	$K29 = 1 \times 10^{-34}$	248
R30	$NO + OH \overset{M}{\rightarrow} HNO_2$	$K30 = 1.5 \times 10^{-12}$	238
R31	$H_2O_2 + NO \rightarrow HNO_2 + OH$	$K31 \leq 5 \times 10^{-20}$	83
R32	$HNO_2 + h\nu \rightarrow NO + OH$	$J32 = 0.1 \times J21$	117
R33	$OH + NO_2 \overset{M}{\rightarrow} HNO_3$	$K33 = 5 \times 10^{-12}$	228
R34	$HNO_3 + h\nu \rightarrow OH + NO_3$	See Table VII	119
R35	$HNO_3 + OH \rightarrow H_2O + NO_3$	$K35 = 1 \times 10^{-12} \exp(-550/T)$	Assume a pre-exponential of 1×10^{-12} and f[illegible] to value measur[illegible] by Hussain and Norrish (106)
R36	$HNO_3 \rightarrow$ rainout and particulate removal	$K36 = 2 \times 10^{-6}$	256
R37	$O + O_2 + M \rightarrow O_3 + M$	$K37 = 1 \times 10^{-34} \exp(500/T)$	103
R38	$O_3 + h\nu(\lambda < 3100\mathring{A}) \rightarrow O(^1D) + O_2(^1\Delta g)$	See Table VII	Calculated wit[illegible] absorption cros[illegible] section data compiled by Ackerman (1)
R39	$O_3 + h\nu(3000\mathring{A} < \lambda < 11800\mathring{A}) \rightarrow O_2(?) + O(^3P)$	See Table VII	

TABLE VI
(cont.)

	Reaction	Rate Coefficient	Reference
R40	$O_3 + h\nu(3100Å < \lambda < 4110Å) \rightarrow O(^1D) + O_2(^3\Sigma g)$	See Table VII	
R41	$O(^1D) + M \rightarrow O(^3P) + M$	$K41 = 5 \times 10^{-11}$	Recommended by Schiff (217)
R42	$O(^1D) + H_2O \rightarrow 2OH$	$K42 = 5 \times 10^{-10}$	190
R43a	$H_2C{=}O + h\nu \rightarrow HCO + H$	See Table VII	34
R43b	$H_2C{=}O + h\nu \rightarrow H_2 + CO$	See Table VII	34
R44	$HCO + O_2 \rightarrow HO_2 + CO$	$K44 = 1 \times 10^{-13}$	47
R45	$H + O_2 + M \rightarrow HO_2 + M$	$K45 = 3 \times 10^{-32}$	135
R46	$OH + CO \rightarrow CO_2 + H$	$K46 = 2.1 \times 10^{-13}\exp(-115/T)$	85
R47	$OH + O_3 \rightarrow HO_2 + O_2$	$K47 = 5 \times 10^{-14}$	See Section IV.D.2
R48	$OH + H_2C{=}O \rightarrow H_2O + HCO$	$K48 = 1.5 \times 10^{-11}$	177
R49	$OH + H_2 \rightarrow H_2O + H$	$K49 = 6.8 \times 10^{-12}\exp(-2000/T)$	85
R50	$OH + H_2O_2 \rightarrow HO_2 + H_2O$	$K50 = 4 \times 10^{-13}T^{1/2}\exp(-600/T)$	84
R51	$OH + CH_4 \rightarrow CH_3 + H_2O$	$K51 = 5.5 \times 10^{-12}\exp(-1900/T)$	86
R52	$CH_3 + O_2 + M \rightarrow CH_3O_2 + M$	$K52 = 8 \times 10^{-32}$	93
R53	$CH_3O_2 + NO \rightarrow CH_3O + NO_2$	K53 = K20	Adopted
R54	$CH_3O + O_2 \rightarrow HO_2 + H_2C{=}O$	$K54 = 1.7 \times 10^{-13}\exp(-3250/T)$	93
R55a	$CH_3O_2 + CH_3O_2 \rightarrow 2CH_3O + O_2$	$3 \times 10^{-11} > K55 > 3 \times 10^{-14}$ (R55a, R55b)	See Section IV.D.2
R55b	$CH_3O_2 + CH_3O_2 \rightarrow CH_3OH + H_2C{=}O$		
R56	$CH_3O_2 + HO_2 \rightarrow CH_3O_2H + O_2$	K56 = K63	Adopted

TABLE VI
(cont.)

Reaction	Rate Coefficient	Reference
R57 $CH_3O_2H + h\nu \rightarrow CH_3O + OH$	J57 = J71	Adopted
R58a $CH_3O_2 + NO \rightarrow CH_3O_2NO$	Not used in these calculations	
R58b $CH_3O_2 + NO \rightarrow H_2C{=}O + HNO_2$	Not used in these calculations	
R59a $CH_3O_2 + NO_2 \rightarrow CH_3O_2NO_2$	Not used in these calculations	
R59b $CH_3O_2 + NO_2 \rightarrow H_2C{=}O + HNO_3$	Not used in these calculations	
R60 $2CH_3O_2NO + O_2 \rightarrow 2CH_3ONO_2 + O_2$	Not used in these calculations	
R61 $CH_3O_2NO_2 + NO_2 \rightarrow CH_3ONO_2 + NO_3$	Not used in these calculations	
R62 $CH_3ONO_2 + h\nu \rightarrow CH_3O + NO_2$	Not known	
R63 $HO_2 + HO_2 \rightarrow H_2O_2 + O_2$	$K63 = 3.3 \times 10^{-12}$	157
R64 $OH + CH_3O_2H \rightarrow CH_3O_2 + H_2O$	K64 = K50	Adopted
R65 $HO_2 + OH \rightarrow H_2O + O_2$	$K65 = 2 \times 10^{-10}$	100
R66 $OH + OH \rightarrow H_2O + O$	$K66 = 2.5 \times 10^{-12}$	135
R67 $OH + OH + M \rightarrow H_2O_2 + M$	$K67 = 3.7 \times 10^{-30}$	33
R68 O_3 + α-pinene → ?	$K68 < 1.3 \times 10^{-16}$	211
R69 O_3 + β-pinene → ?	$K69 < 4 \times 10^{-17}$	211
R70 O_3 + isoprene → ?	$K70 \sim 10^{-17}$	79
R71 $H_2O_2 + h\nu \rightarrow 2OH$	See Table VII	222
R72 H_2O_2, CH_3O_2 → rainout and particulate removal	$K72 = 2 \times 10^{-6}$	256

TABLE VII
Noon Profiles for 30°N

	Summer							Winter						
Altitude (km)	$n(NO_2)/n(NO)$	$n[O(^1D)]$ (molecule cm^{-3})	J34 (sec^{-1})	J38 + J40 (sec^{-1})	J43a (sec^{-1})	J43b (sec^{-1})	J71 (sec^{-1})	$n(NO_2)/n(NO)$	$n[O(^1D)]$ (molecule cm^{-3})	J34 (sec^{-1})	J38 + J40 (sec^{-1})	J43a (sec^{-1})	J43b (sec^{-1})	J71 (sec^{-1})
0	2.2	1.4^{-2}[a]	7.7^{-7}	2.4^{-5}	4.0^{-5}	1.02^{-4}	3.2^{-6}	1.7	3.0^{-3}	3.3^{-7}	6.8^{-6}	2.4^{-5}	6.9^{-5}	1.9^{-6}
1			7.7^{-7}	2.5^{-5}	4.0^{-5}	1.02^{-4}	3.3^{-6}			3.4^{-7}	7.1^{-6}	2.5^{-5}	7.0^{-5}	1.9^{-6}
2	1.7	1.9^{-2}	7.8^{-7}	2.6^{-5}	4.0^{-5}	1.02^{-4}	3.3^{-6}	1.4	3.8^{-3}	3.5^{-7}	7.4^{-6}	2.6^{-5}	7.2^{-5}	2.0^{-6}
3			7.9^{-7}	2.6^{-5}	4.0^{-5}	1.02^{-4}	3.3^{-6}			3.6^{-7}	7.7^{-6}	2.6^{-5}	7.3^{-5}	2.0^{-6}
4	1.4	2.4^{-2}	8.0^{-7}	2.7^{-5}	4.0^{-5}	1.02^{-4}	3.3^{-6}	0.91	4.3^{-3}	3.7^{-7}	8.0^{-6}	2.7^{-5}	7.5^{-5}	2.1^{-6}
5			8.1^{-7}	2.7^{-5}	4.0^{-5}	1.03^{-4}	3.3^{-6}			3.9^{-7}	8.3^{-6}	2.8^{-5}	7.7^{-5}	2.1^{-6}
6	1.1	3.0^{-2}	8.2^{-7}	2.8^{-5}	4.1^{-5}	1.03^{-4}	3.3^{-6}	0.60	4.7^{-3}	3.9^{-7}	8.5^{-6}	2.8^{-5}	7.8^{-5}	2.1^{-6}
7			8.2^{-7}	2.8^{-5}	4.1^{-5}	1.03^{-4}	3.3^{-6}			4.1^{-7}	8.9^{-6}	2.9^{-5}	8.0^{-5}	2.2^{-6}
8	0.79	3.6^{-2}	8.3^{-7}	2.9^{-5}	4.1^{-5}	1.03^{-4}	3.4^{-6}	0.47	6.5^{-3}	4.2^{-7}	9.2^{-6}	2.9^{-5}	8.1^{-5}	2.2^{-6}
9			8.4^{-7}	2.9^{-5}	4.1^{-5}	1.03^{-4}	3.4^{-6}			4.3^{-7}	9.5^{-6}	3.0^{-5}	8.2^{-5}	2.3^{-6}
10	0.54	4.2^{-2}	8.5^{-7}	3.0^{-5}	4.1^{-5}	1.03^{-4}	3.4^{-6}	0.43	1.1^{-2}	4.4^{-7}	9.8^{-6}	3.0^{-5}	8.4^{-5}	2.3^{-6}
11			8.6^{-7}	3.0^{-5}	4.1^{-5}	1.03^{-4}	3.4^{-6}			4.5^{-7}	1.01^{-5}	3.1^{-5}	8.5^{-5}	2.4^{-6}
12	0.37	5.4^{-2}	8.6^{-7}	3.1^{-5}	4.1^{-5}	1.04^{-4}	3.4^{-6}	0.42	2.0^{-2}	4.6^{-7}	1.05^{-5}	3.2^{-5}	8.6^{-5}	2.4^{-6}
13			8.7^{-7}	3.1^{-5}	4.1^{-5}	1.04^{-4}	3.4^{-6}							
14			8.8^{-7}	3.2^{-5}	4.2^{-5}	1.04^{-4}	3.5^{-6}							
15	0.35	1.4^{-1}	8.9^{-7}	3.3^{-5}	4.2^{-5}	1.04^{-4}	3.5^{-6}							

[a] The superscript designates the power of 10 by which the number is to be multiplied.

A. Sulfur Chemistry

As was learned in Section II.D, not much is known quantitatively about either the global circulation of sulfur compounds or their atmospheric photochemistry. It is assumed that atmospheric sulfides and oxides are produced at the surface and then converted to sulfuric acid and sulfate salts, which are removed from the atmosphere by rainout and dry deposition.

1. Sulfur Dioxide (SO_2)

The photooxidation of SO_2 in air,

$$SO_2 + O_2 + h\nu \rightarrow SO_3 + O \qquad \text{(R1)}$$

was found to be very slow [Gerhard and Johnstone (78)]. Cadle and Allen (28) found the quantum yield of this reaction over the range 2800 to 4200 Å to be 2×10^{-3} and estimated an SO_2 residence time of at least 40 days, which is much longer than that estimated from regional data (see Section II.D).

It was found that magnesium salts would catalyze the oxidation of SO_2 in water droplets [Junge and Ryan (132)], and Johnstone and Coughanowr (122) determined an SO_2 residence time of a few hours when they used $MgSO_4$ as the catalyst. Under conditions more closely approaching those of a natural fog, Gall (69) found a three-day residence time.

Junge and Ryan suggested that dissolved NH_3 might be a more reasonable catalyst since the necessary metal catalysts had never been observed in normal fog.

Schuck and Doyle (221), Renzetti and Doyle (208), and Endow, Doyle, and Jones (57) showed that mixtures of olefin, NO_2, and SO_2 form considerable H_2SO_4 aerosol upon radiation, but Renzetti and Doyle (208) also pointed out that very little is formed if saturated hydrocarbons are substituted for the olefins.

Gartell, Thomas, and Carpenter (71) and Weber (249) both found very short lifetimes for SO_2 in the heavily polluted environment of a stack plume.

All the effective SO_2 oxidation reactions mentioned so far are applicable only in polluted environments and are most likely heterogeneous. While these reactions may be important for the global sulfur balance since combustion is a large source of atmospheric sulfur, they are not significant SO_2 oxidation reactions in the normal troposphere. There are, however, a number of homogeneous gas-phase reactions to consider.

The oxidation of SO_2 by O_3 does not go even in the presence of water vapor [Cadle and Allen (28)].

The slowness of the reaction with atomic oxygen

$$SO_2 + O + M \rightarrow SO_3 + M, \; K2 = 3.4 \times 10^{-32} \exp(-2240/RT) \tag{R2}$$

and the low densities of O in the troposphere make it an ineffective sink [Davis, Schiff, and Fischer (46c)].

Sidebottom, Badcock, Jackson, Calvert, Reinhardt, and Damon (227) have suggested that

$$SO_2 + h\nu \rightarrow {}^1SO_2^* \tag{R3}$$

followed by a quenching reaction

$${}^1SO_2^* + M \rightarrow {}^3SO_2^* + M \tag{R4}$$

would produce metastable ${}^3SO_2^*$, which would be much more reactive with O_2 and O_3. Assuming that all the ${}^3SO_2^*$ is oxidized, they have estimated that an SO_2 residence time as low as a few days is possible. This reaction scheme is, however, only hypothetical at this time.

Gall, Slater, and Calvert (70) suggested that

$$RO_2 + SO_2 \rightarrow RO + SO_3 \quad \text{(R5)}$$

might be important.

Payne, Stief and Davis (191a) report a rate coefficient for

$$HO_2 + SO_2 \rightarrow OH + SO_3 \quad \text{(R6)}$$

which has a range K6 ~ 1.5×10^{-15} to 8×10^{-16}. Using $\overline{n(HO_2)} = 3 \times 10^8$ [Levy (153)] as the daily average number density of the hydroperoxy radical and the upper limit for K6, we can estimate an SO_2 residence time as short as 24 days in the lower troposphere; this is not in harmony with previous estimates.

Payne, Stief, and Davis (191a) report a preliminary measurement of the rate coefficient for

$$SO_2 + OH + M \rightarrow HSO_3 + M \quad \text{(R6a)}$$

giving K6a ~ 1×10^{-31}. Using $\overline{n(OH)} = 3.7 \times 10^6$ [Levy (153)] we estimate a SO_2 residence time of 1 to 2 days in the lower troposphere if all HSO_3 is converted by further reactions to H_2SO_4 and $SO_4^=$ salts. This would be the dominant reaction path for removing SO_2 if R6a is as fast as proposed.

While SO_3 is thought to react rapidly with water to form sulfuric acid [Kellogg et al. (138) and Robinson and Robbins (214)],

$$SO_3 + H_2O \rightarrow H_2SO_4 \quad \text{(R7)}$$

there is some question as to whether such a reaction can take place without a surface. In the presence of NH_3, H_2SO_4 will rapidly form ammonium sulfate and bisulfate [Cadle and Robbins (32)]:

$$H_2SO_4 + 2NH_3 \rightarrow (NH_4)_2SO_4 \tag{R8}$$

or

$$H_2SO_4 + NH_3 \rightarrow (NH_4)\ HSO_4 \tag{R9}$$

2. Hydrogen Sulfide (H_2S)

Cadle and Ledford (30) measured the rate for the heterogeneous reaction

$$H_2S + O_3 \overset{\text{surface}}{\rightarrow\rightarrow\rightarrow\rightarrow} H_2O + SO_2 \tag{R10}$$

and found it to be very slow when extrapolated to a homogeneous gas-phase reaction.

Using the surface dependence of the reaction previously determined by Cadle and Ledford and Junge's (128) estimate of particulate densities in the troposphere, Robinson and Robbins (214) calculated that H_2S would have a residence time of 2 hr in the continental surface atmosphere and 28 hr in the troposphere as a whole. This calculation assumes that K10 depends only on the square root of the surface area and is independent of the nature of the surface.

The reaction of H_2S with atomic oxygen

$$H_2S + O \rightarrow OH + HS \tag{R11}$$

is not important, because n(O) is too small.

Niki, Brietenbrach, and Morris (184a) have estimated from preliminary data that the rate coefficient for

$$H_2S + OH \rightarrow "H_2O + HS" \qquad (R12)$$

lies in the range 5 to 10 x 10^{-12}. Since this measurement does not monitor products, it can only be assumed that HS is a product, but the extraction of H to form H_2O is well known for reactions of OH.

It is expected [Cadle and Allen (28)] that HS would be rapidly oxidized in the atmosphere to form a number of products such as SO_2, SO_3, and H_2SO_4. The mechanisms for such reactions have been studied at high temperatures in shock waves by Bradley and Dobson (25).

Using daily average number densities of the hydroxyl radical, $\overline{n(OH)} = 3.7 \times 10^6$, for the lower troposphere [Levy (153)], we find that the residence time for H_2S due to R12 ranges from 7 to 15 hr and is in harmony with previous estimates.

3. Dimethyl Sulfide

Niki, Brietenbrach, and Morris (184a) have estimated from limited preliminary data that the reaction of Me_2S with OH proceeds at a rate similar to that observed for H_2S. Thus it is possible that Me_2S also has a short residence time. The products of such a reaction in the troposphere are not known, though one can reasonably speculate that oxidation products similar to those expected for HS would result.

4. Sulfuric Acid and Sulfate Aerosol (H_2SO_4 and $SO_4^=$)

It is not believed that sulfuric acid and sulfate aerosols enter into normal tropospheric photochemical reactions. The general consensus (see Section II.D) is that they are removed directly by either dry deposition or rainout and serve as the ultimate sink for atmospheric sulfur.

5. Photochemical Model

From what is now known, we can treat the sulfur photochemistry independently of the bulk of the tropospheric photochemistry. The number densities of H_2S and SO_2 are too small to have any significant effect on the radical photochemistry, though the reactions of SO_2 and H_2S with HO_2 and OH, respectively, are important to the sulfur cycle. Although the heterogeneous oxidation reactions of SO_2 and H_2S may be important in polluted environments, the gas-phase oxidation reactions R6, R7, R8, and R12 should dominate in the troposphere as a whole. The sulfur photochemistry is summarized in Figure 1. The possible importance of Me_2S to the atmospheric sulfur cycle should be considered, and more experimental data are needed.

B. Nitrogen Chemistry

While considerably more data are available regarding the atmospheric reactions of nitrogen compounds than there are for sulfur, the data concerning global sinks, sources, and number densities are not extensive (see Section II.E). From what is known, the principal sources are natural production of NH_3, N_2O, and possibly NO and the combustion source of NO and NO_2. The principal sink is the rainout and dry deposition of NH_4^+ and NO_3^- salts and HNO_3. The only quantitative numbers available are Eriksson's (58) measurements of NH_4^+ and NO_3^- rainout and Robinson and Robbins' (216) estimate of the combustion source for NO and NO_2. A number of measurements of NO_2 and a few of NO and NH_3 exist in unpolluted areas, but these are not enough

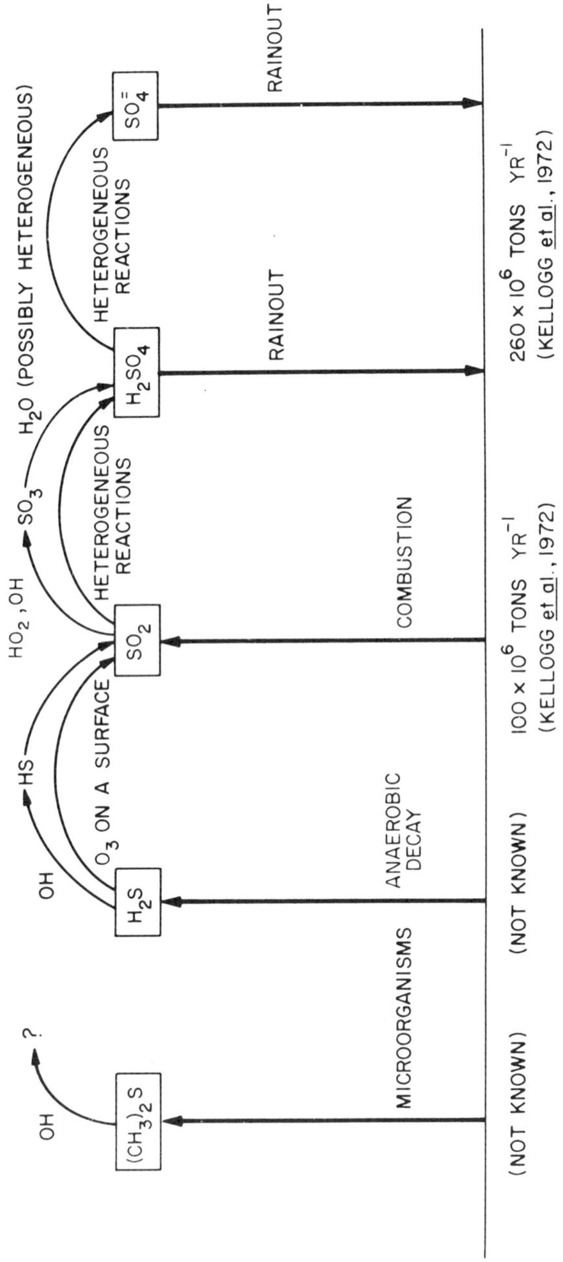

FIGURE 1. Tropospheric sulfur chemistry.

to provide a quantitative picture of the global distribution of nitrogen compounds. In particular, there are few, if any, measurements of the number densities of nitrogen compounds in the middle and upper troposphere.

1. Nitrous Oxide (N_2O)

The major source of N_2O is believed to be bacterial action in the soil (see Section II.E.1), but a reaction suggested by Bates and Witherspoon (12)

$$N_2 + O_3 \rightarrow N_2O + O_2 \tag{R13}$$

may be significant in the lower troposphere. Goody and Walshaw (82) found K13 $\leq 5 \times 10^{-28}$. Considering the large activation energy R13 should have, it might be an important source of N_2O in the lower troposphere, but it would not be sufficient to sustain the observed N_2O mixing ratio throughout the troposphere. McElroy (172) has suggested a reaction involving hot ground-state atomic oxygen,

$$O^*(^3P) + N_2 \rightarrow N_2O \tag{R14}$$

but no rate is known for either the reaction with N_2 or the initial formation of hot $O(^3P)$.

Currently the difficulties with N_2O photochemistry are either the source is too large or the sinks are too small. The following loss reactions exist for N_2O: the photodissociation reactions [Bates and Witherspoon (12) and Bates and Hays (11)]

$$N_2O + h\nu \rightarrow N_2 + O(^1D)\,\lambda < 3370\ \text{Å} \tag{R15a}$$

$$\rightarrow N_2 + O(^1S)\,\lambda < 2100\ \text{Å} \tag{R15b}$$

$$\rightarrow NO + N(^4S)\,\lambda < 2500\ \text{Å} \tag{R15c}$$

and the reactions with metastable atomic oxygen [Nicolet (181) and McElroy and McConnell (173)]

$$N_2O + O(^1D) \rightarrow NO + NO \qquad \text{(R16a)}$$

$$\rightarrow N_2 + O_2 \qquad \text{(R16b)}$$

Only R15a is effective in the troposphere, with an average photodissociation rate in the range J14a ≃ 3 to 4×10^{-10} [Bates and Hays (11)]. The solar flux is not available in the troposphere at the wavelengths required by R15b,c, and with $n(O^1D) < 0.2$ in the troposphere [Levy (154)], R16a,b are also unimportant.

Bates and Hays calculated a residence time of 70 yr for N_2O in the troposphere and found that its major sink was transport into the stratosphere followed by photodissociation. The results were confirmed by calculations of McElroy and McConnell who also found that R16a,b were unimportant loss reactions.

Schutz et al. (223) concluded from a study of the seasonal and latitudinal variations of $n(N_2O)$, as well as from its short-term fluctuations, that its lifetime may be 10 yr or less. This would require a much more efficient sink than is currently known.

2. Ammonia (NH_3)

No atmospheric sources of NH_3 are known (see Section II.D.4). Its reaction with nitric acid,

$$NH_3 + HNO_3 \rightarrow NH_4NO_3 \qquad \text{(R17)}$$

has been observed to be very fast, but it is believed to be a surface reaction [Niki (184)].

Based on measurements of NO_2 and NH_3 in Central America, Lodge and Pate (160) suggested that the oxidation of NH_3 to an oxide of nitrogen may be an important sink.

McConnell (170), Crutzen (43), and Stuhl and Niki (238) have proposed the reaction

$$NH_3 + OH \rightarrow NH_2 + H_2O \quad \text{(R18)}$$

followed by the rapid oxidation of NH_2 to either NO or NO_2. A preliminary determination of the rate for R18, $K18 = 1.5 \times 10^{-13}$, by Stuhl (236) would suggest that oxidation by OH may be very important, both as a sink for NH_3 and as an atmospheric source of NO or NO_2.

Using daily average values of n(OH), $\overline{n(OH)} = 3.7 \times 10^6$ [Levy (153)], we find a residence time of one month, comparable to that estimated for rainout and dry deposition (see Section II.D.4). The OH oxidation, along with rainout and dry deposition, would then provide an NH_3 residence time of two weeks, in harmony with the expectations of Robinson and Robbins (216) and short enough to allow the observed seasonal and latitudinal fluctuations.

3. Nitric Oxide and Nitrogen Dioxide (NO and NO_2)

While two N_2O dissociation reactions, R15c and R16a, are significant sources of NO in the stratosphere [Bates and Hays (11), Nicolet (181), and McElroy and McConnell (173)], they provide a tropospheric column production rate of only 10^7 molecules cm^{-2} sec^{-1} [McElroy and McConnell (173)], much less than even the average column production rate for combustion, 4×10^9 molecules cm^{-2} sec^{-1}.

The oxidation of NH_3 by OH, R18, may be an important atmospheric source if all the NH_2 is converted to NO and NO_2. Assuming that NH_3 has a mixing ratio of 6 ppb [Robinson and Robbins (216)] in the lower troposphere, we can use the preliminary value [Stuhl (236)] for K18 and the average daily profile for n(OH) [Levy (154)] to estimate a column production rate ranging from 8 to 40 x 10^9 molecules cm^{-2} sec^{-1} for a lower troposphere ranging from 1 to 5 km thick. Recent measurements by Murcray and Murcray (178), which failed to detect NH_3 near the tropopause, would suggest that NH_3 is not well mixed throughout the troposphere. If the resulting NH_2 is rapidly converted to NO and NO_2, R18 may provide 10 times as much tropospheric NO and NO_2 as combustion does. Recent calculations of McConnell (170) and Stuhl (236) support this result. The biological source suggested by Robinson and Robbins has never been measured and was used only to balance their nitrogen cycle. Kurylo has recently reported a measurement of the rate coefficient for R18 which is ~1/4 the value reported by Stuhl (236). While this would reduce the production of NO_x, the oxidation of NH_3 would still be a major source of atmospheric NO_x.

The two major reactive oxides of nitrogen rapidly reach a photostationary state via the following reactions:

$$NO + O_3 \rightarrow NO_2 + O_2 \qquad \text{(R19)}$$

$$NO + HO_2 \rightarrow NO_2 + OH \qquad \text{(R20)}$$

$$NO_2 + h\nu \rightarrow NO + O, \quad \lambda < 4047\ \text{Å} \qquad \text{(R21)}$$

The measurements of K19 have recently been summarized by Hampson et al. (91), who recommended K19 = 9 x 10^{-13} exp(-1200/T); Johnston (118) proposed an earlier measurement of Johnston and Crosby (120), K19 = 1.33 x 10^{-12} exp(-1250/T), which was made over the full range of tropospheric temperatures and agrees well with the more recent room-temperature measurements.

Payne, Stief, and Davis (191a) have determined a range for the value of K20, 9 to 1×10^{-13}, while Niki (184) and Westberg and Cohen (253) required K20 $\geq 2 \times 10^{-13}$ and 5×10^{-13}, respectively, to explain room-temperature smog-chamber experiments.

The calculations of J21 by Leighton (149), which assumed a decreasing quantum yield for wavelengths greater than 3660 Å, are well fitted by $J21 = 1 \times 10^{-2} \exp [-0.38 \sec (\chi)]$. This assumption is supported by the data of Pitts, Sharp, and Chan (194).

Using R19, R20, and R21, we derive an expression for the photostationary relationship between n(NO) and $n(NO_2)$,

$$\frac{n(NO_2)}{n(NO)} = \frac{n(O_3)\ K19 + n(HO_2)\ K20}{J21} \qquad (3)$$

which will increase with increasing solar zenith angle, resulting in the rapid conversion of NO to NO_2 at night, and will decrease with increasing altitude. In the troposphere, (3) adjusts in a few minutes to variations in incident solar radiation and possible fluctuations in $n(O_3)$ and $n(HO_2)$.

Altitude profiles for eq. (3) are shown in Table VII. Owing to the large activation energy for K19, the oxidation of NO slows with increasing altitude, resulting in the daytime dominance of NO over NO_2 in the middle and upper troposphere.

The loss reactions for NO and NO_2 involve the formation of acids and higher oxides and will be discussed next.

4. Higher Oxides of Nitrogen (NO_3 and N_2O_5)

a. Nitrogen Trioxide (NO_3). NO_3 is produced by the reaction between oxone and nitrogen dioxide,

$$NO_2 + O_3 \rightarrow NO_3 + O_2 \qquad (R22)$$

which is quite slow even at 294°K, K22 = 3 to 7 x 10^{-17} [Ford, Doyle, and Endow (65), Johnston and Yost (121)]. Measurements by Niki (184) fall in between, while some new measurements by Johnston (119) suggest it may be slower. A recent measurement by Davis, Prusacyzk, Dwyer, and Kim (46a) gives 3.5 x 10^{-17} while Ghormley, Ellsworth, and Hochanadel (78a) report K22 = 7.8 x 10^{-17}. Using limited data taken over a short temperature range, 286 to 320°K, Johnston and Yost determined an activation energy of 7 kcal, giving K22 = 1 x 10^{-11} exp(-3500/T). While R22 is highly temperature dependent, the data are too limited to make an accurate determination of the activation energy, and 7 kcal should be considered an estimate. Measurements over the complete range of tropospheric temperatures are in progress [Johnston (119)].

The important loss reactions,

$$NO_3 + h\nu \rightarrow NO + O_2 \qquad (R23a)$$

$$\rightarrow NO_2 + O \qquad (R23b)$$

$$NO_3 + NO \rightarrow NO_2 + NO_2 \qquad (R24)$$

$$NO_3 + NO_2 \rightarrow N_2O_5 \qquad (R25)$$

are all fast.

By means of absorption coefficients measured at a few wavelengths at high temperatures [Schott and Davidson (220)] and an old spectrogram [Jones and Wulf (123)], Crutzen (42) estimated J23a and J23b = 5 x 10^{-2}. Observing a sharp increase in NO_3 absorption at the threshold for R23a and the presence of NO as a primary product, Johnston (119) believes that R23a is the dominant path for photodissociation and estimates J23a = 5 to 1 x 10^{-2}.

Values for K24 range from 1 x 10^{-11} [Johnston (117)] to 2 x 10^{-10} [Hussain and Norrish (106)] at room temperature with a possible activation energy of 1 kcal [Ashmore and Burnett (7) and Schott and Davidson (220)]. Harker and Johnston (91a) report K24 - 8.7 x 10^{-12}.

Johnston (117) estimated K25 = 3 x 10^{-12}, while Schott and Davidson found K25 = 6 x 10^{-13} for what is an effective two-body reaction in the troposphere.

b. Dinitrogen Pentoxide (N_2O_5). The source of N_2O_5 is the NO_3 loss reaction, R25, and the loss reactions are

$$N_2O_5 \rightarrow NO_2 + NO_3 \qquad (R26)$$

$$N_2O_5 + h\nu \rightarrow 2NO_2 + O \qquad (R27)$$

$$N_2O_5 + H_2O \rightarrow 2HNO_3 \qquad (R28)$$

Johnston (118) reported K26 = 0.3 at 300°K with a large activation energy of 19.3 kcal and estimated, on the basis of absorption spectra measurements by Jones and Wulf (123), an average daytime value for J27 of 2 x 10^{-5} in the lower stratosphere. This value of J27 will serve as an upper limit for the troposphere and will be important only in the upper troposphere, where R26 and R28 are slow as a result of their large activation energies. Jaffe and Ford (114) reported K28 = 2 x 10^{-18} at 300°K, but it is now believed that their experiment was surface catalyzed and that the true gas-phase reaction is much slower, K28 $<$ 1 x 10^{-21} [Johnston (119) and Niki (184)]. No activation energy has been measured, but Johnston (118) estimates at least 8 kcal.

5. Acids (HNO_2, HNO_3)

a. Nitrous Acid (HNO_2). The two important production mechanisms are

$$NO + NO_2 + H_2O \rightarrow 2HNO_2 \quad (R29)$$

and

$$NO + OH \xrightarrow{M} HNO_2 \quad (R30)$$

Wayne and Yost (248) measured K29 = 1×10^{-34} at room temperature; Johnston (119) found two-thirds that value. No activation energy has been measured, but this reaction, with its dependence on $n(H_2O)$, will not be important in the middle and upper troposphere, where the dependence would be significant. Stuhl and Niki (238) found that while R30 is a three-body reaction at low pressures, it is an effective two-body reaction at tropospheric pressures and they recommend K30 = 1.5×10^{-12}. Their low-pressure three-body values are in reasonable agreement with preliminary results of Anderson and Kaufman (5). Morley and Smith (176a) and Westenberg and deHaas (253a) have also measured K30. A possible source of HNO_2 suggested by Nicolet (180)

$$H_2O_2 + NO \rightarrow HNO_2 + OH \quad (R31)$$

is much too slow, K31 $\leq 5 \times 10^{-20}$ [Gray, Lissi, and Heicklen (83)].

The dominant loss reaction is

$$HNO_2 + h\nu \rightarrow NO + OH \quad (R32)$$

Johnston (117) recommended J32 = 0.1 x J21 on the basis of nitrite-absorption data, but quantitative measurements of the HNO_2 absorption cross section are still in progress.

b. Nitric Acid (HNO_3). The major tropospheric source of HNO_3 is

$$OH + NO_2 \overset{M}{\rightarrow} HNO_3 \qquad (R33)$$

which approaches a two-body reaction in the troposphere, where Simonaitis and Heicklen (228) recently determined $K33 = 5 \times 10^{-12}$. An analysis of their measurements by Crutzen (43) finds that K33 decreases slightly to 4×10^{-12} at the tropopause. Measurements by Anderson and Kaufman (5), Stuhl and Niki (238), Morley and Smith (176a), Mulcahy and Smith (177a), and Westenberg and deHaas (253a) are in reasonable agreement with those of Simonaitis and Heicklen in the low-pressure three-body limit.

The N_2O_5 loss reaction, R28, previously considered to be a major source of HNO_3 in the troposphere [Levy (153)], is now believed to be very slow in the gas phase [Johnston (119); Niki (pr. c.)], and while it may be important under smog conditions, it was not included in these calculations.

Throughout the troposphere, the two significant gas-phase loss reactions are

$$HNO_3 + h\nu \rightarrow OH + NO_3 \qquad (R34)$$

and

$$HNO_3 + OH \rightarrow H_2O + NO_3 \qquad (R35)$$

Johnston (119) recently measured the absorption spectrum for HNO_3 in the ultraviolet and found that R34 proceeds with unit quantum efficiency. His absorption measurements are significantly less than the earlier measurements of Dalmon (44) for $\lambda > 2900$ Å, with the result that J34 is reduced considerably from previous estimates [Leighton (149)]. The altitude profile for J34 is given in Table VII, and the average daily values range from 2 to 3 x 10^{-7} in the troposphere. Hussain and Norrish (106) have measured K35 = 1.6 x 10^{-13} at 298°K, but no activation energy has been determined. It should be noted that R35 serves as an indirect means of converting NO_2 to NO_3.

. Using a daily average value of n(OH) in the lower troposphere, $\overline{n(OH)}$ = 3.7 x 10^6 [Levy (154)], we find that R34 and R35 provide a photochemical lifetime of 14 days, considerably longer than the current estimates of rainout and particulate lifetimes in the lower troposphere. Levy (153) has determined the upper limit for the rate of removal of HNO_3 by collisions with particulates to be 3 x 10^{-4}, but this assumes a surface efficiency similar to that found for atoms and it is likely that HNO_3 is much less efficiently removed. This particulate loss rate drops off rapidly with altitude, if we assume a 1-km scale height for particulate number density in the lower troposphere [Elterman (56)]. Wofsy et al. (256) have adopted a rainout rate for H_2O_2 and $H_2C{=}O$ of 2 x 10^{-6} in the first 4 km with a decrease to 0 at 8 km, and a similar mechanism is possible for HNO_3, which is also quite soluble.

As will be shown in later sections, the adoption of particulate removal or rainout as the dominant loss mechanism in the lower troposphere leads to a rainout of

NO_3^-, which greatly exceeds the observations of Eriksson (58). We will, however, include a general removal mechanism

$$HNO_3 \rightarrow \text{rainout and particulate removal} \qquad \text{(R36)}$$

in the model and consider its effects. The value K36 = 2×10^{-6} is reasonable for $Z \leq 5$ km since it leads to a six-day lifetime in the lower troposphere, consistent with measured lifetimes for atmospheric particulates. For higher altitudes, K36 = $2 \times 10^{-6} \exp[(Z - 5)12]$ will be used to provide a smooth drop off in the loss rate with increasing altitude.

6. Equations

We can write the following time-dependent equations for the number densities of NO_3, N_2O_5, HNO_2, and HNO_3:

$$\frac{d\ n(NO_3)}{dt} = n(NO_2)\ n(O_3)K22 + n(HNO_3)n(OH)K35 + n(N_2O_5)K26 - n(NO_3)[J23 + n(NO)K24 + n(NO_2)K25] \qquad (4)$$

$$\frac{d\ n(N_2O_5)}{dt} = n(NO_3)n(NO_2K25 - n(N_2O_5)[K26 + J27 + n(H_2O)K28] \qquad (5)$$

$$\frac{d\ n(HNO_2)}{dt} = 2n(NO)n(NO_2)n(H_2O)K29 + n(OH)n(NO)K30 - J32\ n(HNO_2) \qquad (6)$$

$$\frac{d\ n(HNO_3)}{dt} = n(OH)n(NO_2K33 - n(HNO_3)[J34 + n(OH)K35 + K36] \quad (7)$$

In theory, it is also possible to write a time-dependent equation for n(NO) + n(NO_2), but it is not possible at this time to quantify the major source, NH_3 oxidation by OH. Therefore, calculations in this chapter will fix the sum of n(NO) and n(NO_2), and their ratio [eq. (3)] will be allowed to vary. This will be discussed in more detail in Section V.

7. Atmospheric Cycle

The general photochemical cycle for nitrogen compounds in the troposphere is shown in Figure 2, along with the nighttime cycle. While the large number of important reactions—in particular, those involving OH—make a quantitative treatment impossible without the solution of a set of coupled nonlinear equations, several significant qualitative features can be extracted from Figure 2 with the help of reaction-rate data. The main accomplishment of the atmospheric nitrogen cycle is the conversion of nitrogen oxides, the input form of atmospheric nitrogen, to nitric acid, the major loss form. The important path for this is R33, and as we will see in Section VI, HNO_3 is the major reactive nitrogen species in the troposphere. Its actual number density will depend on n(H_2O), the time of year, the total column density of O_3, and the importance of the rainout loss mechanism; the last can only be estimated at this time. NO_3, N_2O_5, and HNO_2, all of which have fast daytime loss processes, rapidly achieve stationary-state number densities relative to the dominant oxides n(NO) and n(NO_2). The large activation energy of R22 causes the production of NO_3 and

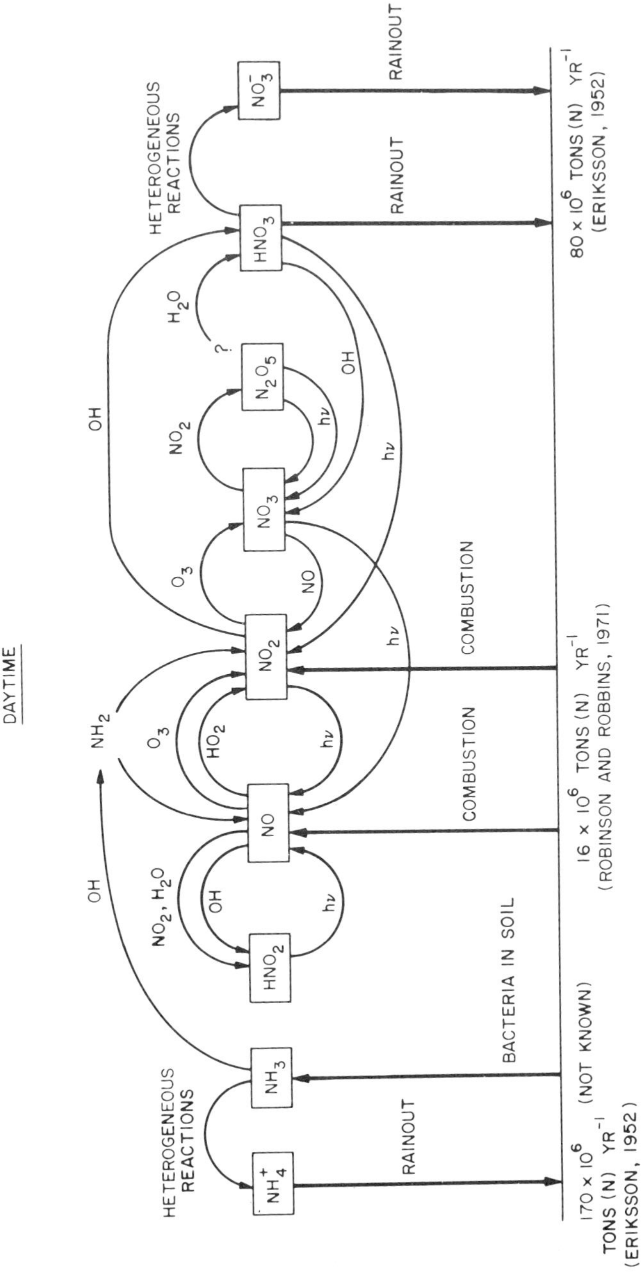

FIGURE 2a. Daytime tropospheric nitrogen chemistry.

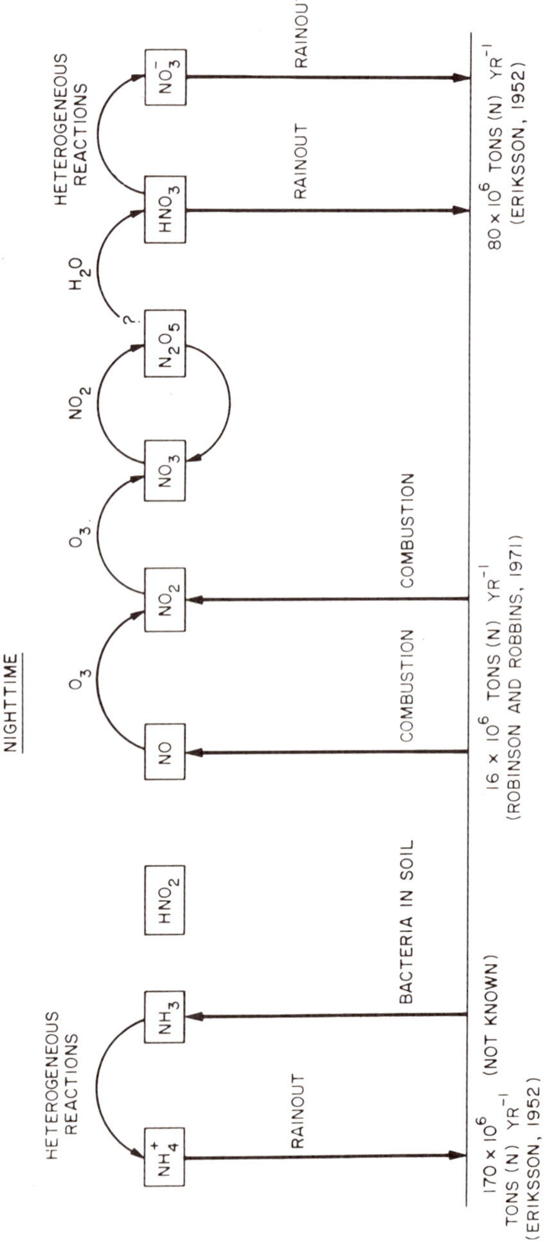

FIGURE 2b. Nighttime tropospheric nitrogen chemistry.

N_2O_5 to drop off rapidly with increasing altitude. While the loss paths for NO_3 remain fast, the major loss path for N_2O_5, R26, has a large activation energy and also decreases rapidly. The net result is that the daytime number density of NO_3 decreases sharply with altitude, while that of N_2O_5 does not change greatly. Since J32 is not much affected by altitude, n(HNO_2) behaves much as n(OH) does.

The nighttime chemistry is greatly simplified, although there are some interesting aspects. The production and removal of HNO_2 ceases since NO is rapidly converted to NO_2 once R21 is cut off. The production of HNO_3 ceases except for the small possible production via N_2O_5. Since the rainout rate is slow, there is no significant change in n(HNO_3) overnight. At low altitudes, K22 is fast enough to convert considerable NO_2 to NO_3, so that by midnight NO_3 and N_2O_5 are important. K22 slows with increasing altitude to such an extent that NO_2 regains dominance. At the same time, the dissociation of N_2O_5 back to NO_3 is also slowing, with the result that in the upper troposphere N_2O_5 becomes the dominant higher oxide, although it is considerably less numerous than NO_2. While NO_3 and N_2O_5 are major nighttime oxides at low altitudes, they are rapidly converted back to NO and NO_2 via R23a and R24 in the early morning, and the daytime chemistry is not affected. The conversion of N_2O_5 to NO_2 is considerably slower in the upper troposphere, but the NO_2 number density is not significantly depleted at night, because of the slowness of K22.

C. Oxygen Chemistry

1. Ozone

It is generally believed that the source of tropospheric ozone is injection from the stratosphere and the sink is loss at the earth's surface, both processes having a flux in the range 5 x 10^{10} to 1 x 10^{11} molecules cm^{-2} sec^{-1}; Crutzen (43), however, has recently suggested possible tropospheric sinks and sources for ozone that may be quite large. The reaction mechanisms will be presented in this section, and detailed calculations, later.

a. Tropospheric Source. The mechanism of atmospheric O_3 formation is similar to that reported for a photochemical smog system [see, for example, Leighton (149)]. It requires the oxidation of NO to NO_2 without the use of O_3; the resulting photodissociation of NO_2 will produce an extra O_3. The following sequence accomplishes just that:

$$HO_2 + NO \rightarrow NO_2 + OH \quad \text{(R20)}$$

followed by

$$NO_2 + h\nu \rightarrow NO + O \quad \text{(R21)}$$

and

$$O + O_2 + M \rightarrow O_3 + M \quad \text{(R37)}$$

K20 and J21 have already been discussed. Recent measurements by Stuhl and Niki (237) at 300°K, K37 = 5.6 x 10^{-34}, are in excellent agreement with those by Huie, Herron, and Davis (103) over the temperature range 200 to 350°K, K37 = 1 x 10^{-34} exp(500/T). This reaction is very fast in the troposphere. Using an average daily $n(HO_2)$ profile previously calculated [Levy (154)] and assuming a measured O_3 profile (see Section II.C), a tropospheric O_3 production as high as 9 x 10^{11} molecules cm^{-2} sec^{-1} can be estimated; this greatly exceeds the transport flux from the stratosphere.

Another possible source involves the oxidation of NO by CH_3O_2, but this reaction is still in question and HO_2 is the major peroxy radical in the troposphere [Levy (153,154)].

b. Tropospheric Sink. As Crutzen (43) suggested, the series of oxidations of NO and NO_2 by O_3 that results in NO_3 can serve as a sink for O_3 since one of the loss paths—photodissociation—results in no reformation of O or O_3. The sequence is as follows:

$$NO + O_3 \rightarrow NO_2 + O_2 \qquad \text{(R19)}$$

followed by

$$NO_2 + O_3 \rightarrow NO_3 + O_2 \qquad \text{(R22)}$$

and

$$NO_3 + h\nu \rightarrow NO + O_2 \qquad \text{(R23a)}$$

These three reactions result in the loss of $2O_3$. By using a calculated $n(NO_3)$ profile [Levy (154)] and values of J23a ranging from 5 to 1×10^{-2}, a daytime loss rate ranging from 5 to 1×10^{11} molecules cm^{-2} sec^{-1} can be estimated. This is a lower limit since the loss due to early morning photodissociation of NO_3 formed during the night has not been considered. Again this loss rate greatly exceeds the O_3 fluxes measured at the ground.

c. Atmospheric O_3 Cycle. The photochemical sources and sinks are pictured in Figure 3; more detailed calculations will be presented in Section VI. It should be noted that while most of the O_3 photodissociations result in $O(^3P)$ formation and no net loss of O_3, a small amount is lost in the production of OH.

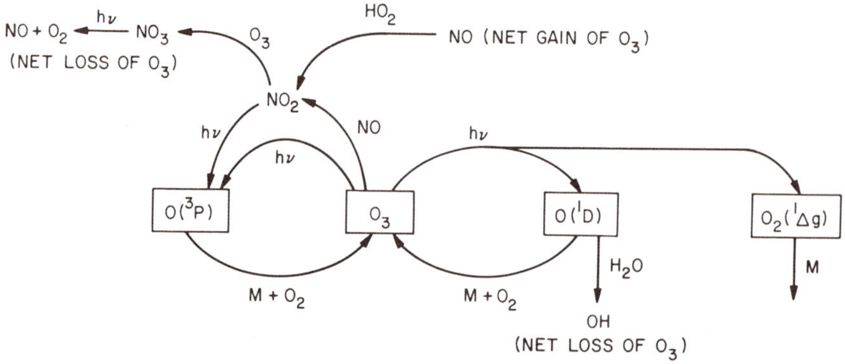

FIGURE 3. Tropospheric oxygen chemistry.

2. Metastable Molecular Oxygen [$O_2(^1\Delta g)$]

The possible reactions of $O_2(^1\Delta g)$ in photochemical smog have been of interest for a number of years [see Pitts (193)]. Calculations by Kummler and Bortner (145) and Kummler, Bortner, and Baurer (146) set an upper limit of $n[O_2(^1\Delta g)] = 6 \times 10^{-6}\ n(O_3)$ for production by Hartley photolysis of O_3:

$$O_3 + h\nu\ (\lambda < 3100\ \text{Å}) \rightarrow O(^1D) + O_2(^1\Delta g) \qquad \text{(R38)}$$

Frankiewicz and Berry (67) have recently calculated the rate of production of $O(^1\Delta g)$ from the quenching of photoexcited NO_2 by ground-state O_2; they found $n[O_2(^1\Delta g)]$ $4 \times 10^{-6}\ n(NO_2)$. Using normal tropospheric values of $n(O_3)$ and $n(NO_2)$, we find that $n[O_2(^1\Delta g)] < 10^7$ throughout the troposphere. Herron and Huie (99) determined that $O_2(^1\Delta g)$ was not very reactive and had at best a relative reactivity of 10^{-3} when compared to $O(^3P)$. Thus $O_2(^1\Delta g)$ does not play a part in the photochemistry of the troposphere.

3. Ground-State Atomic Oxygen [$O(^3P)$]

The major sources of $O(^3P)$ are ozone photolysis in the Hartley continuum and Chappuis bands,

$$O_3 + h\nu \; (3000 \text{ Å} < \lambda < 11800 \text{ Å}) \rightarrow O_2(?) + O(^3P) \qquad \text{(R39)}$$

and NO_2 photodissociation, R21. J39 never exceeds 5×10^{-4} sec^{-1} in the troposphere [Levy (154)]. The major sink is recombination with O_2 to form O_3, R37. The steady-state number density of $n[O(^3P)]$ is given by

$$n[O(^3P)] = \frac{n(O_3)\; J39 + n(NO_2)\; J21}{n(M)\; n(O_2)\; K37} \qquad (8)$$

which, while increasing with altitude, never exceeds 1×10^5 molecules cm^{-3} in the troposphere. Therefore $O(^3P)$ also plays no role in tropospheric photochemistry.

4. Metastable Atomic Oxygen [$O(^1D)$]

Even though metastable atomic oxygen has a very small number density in the troposphere, it is, as we will see, the driving force behind the photochemistry.

The source of $O(^1D)$ is O_3 photolysis in the Hartley continuum, and possibly in the Huggins bands,

$$O_3 + h\nu \; (\lambda < 3100 \text{ Å}) \rightarrow O(^1D) + O_2(^1\Delta g) \qquad \text{(R38)}$$

and

$$O_3 + h\nu \; (3100 \text{ Å} < \lambda < 4110 \text{ Å}) \rightarrow O(^1D) + O_3(^3\Sigma g) \qquad \text{(R40)}$$

The absorption cross section has been well measured for the wavelength region of tropospheric interest [Watanabe et al. (247), Inn and Tanaka (109), and Vigroux (246)], and these data have been compiled by Ackerman (1).

The important and yet incompletely resolved question is that of the quantum yield for $O(^1D)$ production by R38 and R40, $\phi[O(^1D)]$. In a recent review, Schiff (217) examined the available data and found that while for $\lambda < 3000$ Å, $[O(^1D)]$ appears to be 1, the results for $\lambda < 3000$ Å are not conclusive. Demore and Raper (48) determined that $\phi[O(^1D)]$ was 1 for $\lambda < 3000$ Å and then dropped to 0.4 to 0.5 for 3000 Å $\leq \lambda \leq$ 3130 Å. Schiff interpreted their data to give $\phi[O(^1D)] = 0.2$ at 3130 Å. Demore and Raper got $\phi[O(^1D)] > 0.1$ as far out as 3340 Å. on the other hand, Jones and Wayne (124) found $\phi[O(^1D)] =$ 1 for $\lambda < 3000$ Å, dropping to 0.1 at 3130 Å and 0.02 or 0.0 at 3340 Å. To add further complications, Castellano and Schumacher (35) obtained $\phi[O(^1D)] = 1.0$ at 3130 Å. None of these experiments is yet definitive, but it would appear that some $O(^1D)$ is produced by the spin-forbidden photodissociation, R40. For all calculations in this chapter, we will use $\phi[O(^1D)] = 1.0$ for $\lambda < 3000$ Å and 0.5 to 0.1 for 3000 Å $\leq \lambda \leq$ 3150 Å, with a cutoff at 3150 Å. These would appear to be conservative choices. The effective values for J38 and J40, which already include $\phi[O(^1D)]$ in their calculation, are given in Table VII.

The two important tropospheric reactions of $O(^1D)$ are

$$O(^1D) + M \rightarrow O(^3P) + M \qquad \text{(R41)}$$

and

$$O(^1D) + H_2O \rightarrow 2OH \qquad \text{(R42)}$$

The current data for these rates have been reviewed by Schiff (217), who recommended K41 = 5 ± 4 x 10^{-11}. Measurements for M = N_2 have been made by Noxon (185); Young, Black, and Slanger (257), and Snelling and Bair (232). Many groups have measured the relative quenching efficiency of N_2 and O_2 and have found O_2 to be slightly more effective.

A number of measurements have been made of the rate for R42. Paraskevopoulos and Cvetanovic (190) found K42 = 5 x 10^{-10}; Scott and Cvetanovic (224) obtained K42 = 2 x 10^{-10}; and Lissi and Heicklen (156) reported a value in between. The measurement by Biedenkapp, Hartshorn, and Bair (20) has recently been increased to bring it into agreement with others [Bair (9)].

The loss reactions for $O(^1D)$ are very fast, thus allowing us to treat its number density in the stationary-state approximation:

$$n[O(^1D)] = \frac{n(O_3)\ (J38 + J40)}{K41\ n(M) + K42\ n(H_2O)} \qquad (9)$$

The value of $n[O(^1D)]$ is very dependent on the total column of O_3 above it, as well as on the local value of $n(O_3)$, both of which vary greatly with season and latitude (see Table IV). Values of $n[O(^1D)]$ are given for a number of latitudes and seasons in Table VII. Owing to its rapid quenching, $n[O(^1D)]$ is always less than 1.0 in the troposphere [Levy (154)], but the large value of $n(H_2O)$, particularly in the lower troposphere, makes $O(^1D)$ the major source of radicals in the troposphere.

5. Atmospheric Cycle

The atmospheric cycle for the oxygen species is shown in Figure 3. At night there is no photolysis, and the only reaction that takes place is the oxidation of

NO_2 to NO_3 by O_3. This does not appreciably change $n(O_3)$, since it is much larger than $n(NO_2)$. The values of $n[O(^3P)]$, $n[O(^1D)]$, and $n[O_2(^1\Delta g)]$ all go to zero.

D. Radicals (OH, HO_2, CH_3O, and CH_3O_2)

While the radical number densities in the troposphere are quite small, these species are highly reactive and very important to the photochemistry. A large body of atmospheric data concerning sinks, sources, and concentrations of CH_4, $H_2C{=}O$, CO, H_2, HNO_3, and others are explained by their presence, as has been shown in Sections IV.A and IV.B and will be discussed in Sections IV.E, IV.F, and IV.G.

1. Production

The major source of radicals is the reaction of $O(^1D)$ with H_2O,

$$O(^1D) + H_2O \rightarrow 2OH \qquad \text{(R42)}$$

which was discussed in Section IV.C.4. While $n[O(^1D)]$ increases with altitude in the lower troposphere as a result of increased $n(O_3)$ and greater penetration of ultraviolet radiation with $\lambda < 3130$ Å (see Table VII), $n(H_2O)$ sharply decreases (see Table III). The net result is that the production of OH decreases with increasing altitude, roughly proportional to $n(H_2O)$.

Another significant source is the photolysis of formaldehyde [Calvert, Kerr, Demerjian, and McQuigg (34) and Levy (153)], a byproduct of the OH oxidation of CH_4 [Levy (152)]. The sequence of reactions is

$$H_2CO + h\nu \rightarrow HCO + H \qquad \text{(R43a)}$$

followed by

$$HCO + O_2 \rightarrow HO_2 + CO \qquad \text{(R44)}$$

and

$$H + O_2 + M \rightarrow HO_2 + M \qquad \text{(R45)}$$

We will discuss J43a,b and K44 in Section IV.E. For K45, Schofield (218) recommended $3 \times 10^{-32}(273/T)^{1.3}$; Kaufman (135) 3×10^{-32} at 300°K; and Baulch et al. (15) 4×10^{-33} $\exp(500/t)$. All these values were for M = Ar, and it is thought that K45 may be somewhat faster for M = air. Wong and Davis (256a) report K45 = $1.9 \times 10^{-32}\exp(685/RT)$. However, R45 is very fast in the troposphere, and we will assume that all hydrogen produced immediately forms HO_2. The rate of radical production via the $H_2C{=}O$ path also drops sharply with altitude since $n(H_2C{=}O)$ depends on the oxidation of CH_4 by OH, which has a large activation energy. This will be discussed in detail in Section IV.E.

The dark reaction

$$NO + NO_2 + H_2O \rightarrow 2HNO_2 \qquad \text{(R29)}$$

followed by photolysis,

$$HNO_2 + h\nu \rightarrow OH + NO \qquad \text{(R32)}$$

is a small and insignificant source of OH in the troposphere

The photolyses of HNO_3, H_2O_2, and CH_3O_2 also produce OH, but since radicals are required for the formation of these molecules, there is no net production. Although the sums of production and loss cycles of these three species serve as net radical sinks when averaged over the full day, they do serve as net sources at dawn and dusk. In this manner they buffer sharp variations in radical production near dawn and dusk; these variations are caused by the dependence of J43 and $n[O(^1D)]$ on ultraviolet flux, whose extinction is particularly sensitive to large values of the solar zenith angle.

The important reaction paths for radical production are given in Figure 4a. The magnitude of the total

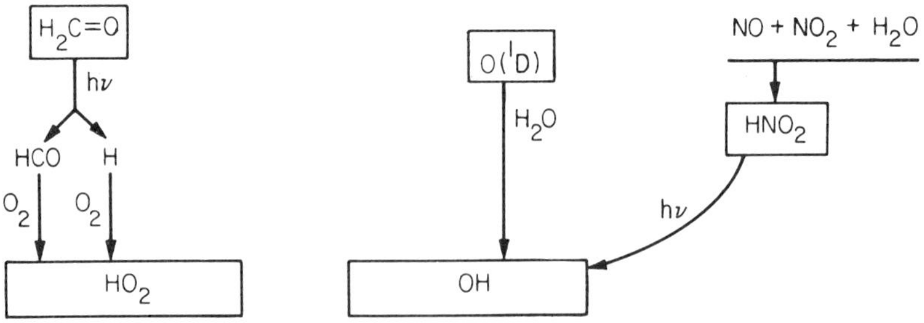

FIGURE 4a. Tropospheric radical production paths.

production rate, to a large extent, is controlled by $n(H_2O)$, with the indirect effect of temperature on $n(H_2C{=}O)$ being of some importance. The rate of production of radicals drops rapidly with altitude. There is also a smaller though significant effect on these parameters of season and latitude through the dependence of $n[O(^1D)]$ (see Table VII). The dominating influence, however, is the rapid decrease in $n(H_2O)$ with altitude.

2. Interconversion

A large number of atmospheric reactions that involve radicals as reactants form one or more different radical species as products. These reactions can be combined to form chain reactions, which then serve as the major source for $H_2C{=}O$, CO, and H_2 and the major sinks for CH_4, CO, and H_2. While these reactions result in no net radical loss, they do redistribute the individual radical number densities.

The dominant radical chain reaction in the troposphere is the oxidation of CO by OH in the presence of O_2,

$$OH + CO \rightarrow CO_2 + H \qquad (R46)$$

followed by

$$H + O_2 + M \rightarrow HO_2 + M \qquad (R45)$$

to give HO_2, which in turn oxidized NO,

$$HO_2 + NO \rightarrow OH + NO_2 \qquad (R20)$$

and reforms OH.

Baulch et al. (13) reviewed the data for R46 and recommended K46 = $9 \times 10^{13} \exp(-540/T)$, while Greiner (85) measured $2.1 \times 10^{-13} \exp(-115/T)$ over the temperature range 300 to 500°K. Westenberg and de Haas (253b) report 1.3×10^{-13} and Stuhl and Niki (238) report 1.4×10^{-13} at room temperature. Davis, Schiff, and Fischer (46c) have recently measured K46 = $2.1 \times 10^{-13} \exp(-180/RT)$. Since Baulch et al.'s temperature data are taken mainly from combustion work, the activation energy estimated by Greiner and confirmed by Davis et al. (46c) will be used in the troposphere.

Recently Davis, Payne, and Stief (46b) measured the rate for R20 at room temperature; they recommended a range of values K20 = 9 to 1×10^{-13}. To explain smog-chamber data, Niki (184) required K20 $\leq 2 \times 10^{-13}$ and Westberg and Cohen (253) K20 = 5×10^{-13}. An earlier preliminary measurement of Zafonte and Johnston (258) giving

$K20 \leq 2 \times 10^{-15}$ now appears to be in error. Simonaitis and Heicklen (226a) report $K20 > 1.5 \times 10^{-13}$.

Because of the small activation energy [Greiner (85)] and the constant mixing ratio for CO, the altitude variation in the rate of oxidation of CO by OH is only proportional to the variation in n(M). While R20 is likely to have a larger activation energy, the increase in the NO mixing ratio with altitude will compensate, thus leading to a similar altitude variation in the rate of oxidation of NO.

Other reactions that convert OH to HO_2 are

$$OH + H_2C{=}O \rightarrow HCO + H_2O \qquad (R48)$$

followed by

$$HCO + O_2 \rightarrow HO_2 + CO \qquad (R44)$$

$$OH + H_2 \rightarrow H + H_2O \qquad (R49)$$

followed by

$$H + O_2 + M \rightarrow HO_2 + M \qquad (R45)$$

and

$$OH + H_2O_2 \rightarrow HO_2 + H_2O \qquad (R50)$$

Morris and Niki (177) have found the rate for R48 at room temperature to be $K48 = 1.5 \times 10^{-11}$. They also discussed earlier measurements that they believe to be in error. No activation energy is known, but for such a fast reaction, it should be quite small. Greiner (85) measured the rate of reaction for R49 over the temperature range 300 to 500°K and found $K49 = 6.8 \times 10^{-12} \exp(-2000/T)$

earlier work reviewed by Kaufman (135), with K49 = 7×10^{-15} at 300°K, is in reasonable agreement. Recent measurements by Westenberg and de Haas (253b), 8×10^{-15}, and Stuhl and Niki (238), 7×10^{-15}, at room temperature are also in good agreement. The rate for R50 was measured over the same temperature range by Greiner (84), who found K50 = $4 \times 10^{-13}\ T^{1/2}\exp(-600/T)$, and by Baulch et al. (15), who recommended $1.7 \times 10^{-11}\exp(-900/T)$. These two values are in reasonable agreement, particularly at 300°K.

The reaction

$$OH + O_3 \rightarrow HO_2 + O_2 \qquad (R47)$$

was not included as a conversion path since Langley and McGrath (148) had reported that it was very slow (K47 < 1×10^{-16}). Recent measurements by Anderson and Kaufman (5a) (K47 = $1.3 \times 10^{-12}\exp(-1900/RT)$ and Demore (48a) (K47 = 8×10^{-14} at 298°K) show it to be much faster, but it still is not a significant conversion path and will not affect the radical number densities that are calculated later.

The photodissociation of H_2O_2, formed in an HO_2 + HO_2 loss reaction (see Figure 4d), serves as a minor path for converting HO_2 to OH.

It has been suggested by Westenberg (253e) that

$$HO_2 + CO \rightarrow CO_2 + OH \qquad (R50a)$$

might be an important conversion path since Westenberg and de Haas (253d) measured K50a = 1×10^{-12}. Recent measurements by Davis, Payne, and Stief (46b) (K50a < 10^{-19}) and Simonaitis and Heicklen (226a) (K50a < 5×10^{-18}) would suggest that this reaction is very slow. For that reason it was not included in the model.

Another path that was not included is

$$HO_2 + O_3 \rightarrow OH + 2O_2 \qquad (R50b)$$

Recent measurements by Demore (48a) ($K50b = 3 \times 10^{-15}$), Anderson and Kaufman (5a) ($K50b < 1 \times K50a$), and Simonaitis and Heicklen (226b) ($K50b \sim 0.1 \times 10^{-15}$) suggest that it may be a minor conversion path, but its exclusion will not significantly affect later calculations of radical number densities. Both R47 and R50b may make a significant contribution to the destruction of O_3.

These reactions are important in the lower troposphere, but not at higher altitudes because of the sharp drop in $n(H_2C{=}O)$ and $n(H_2O_2)$ [Levy (154)] and the large activation energy for R49 [Greiner (85)]. These reaction paths, along with the CO and NO paths, are shown in Figure 4b.

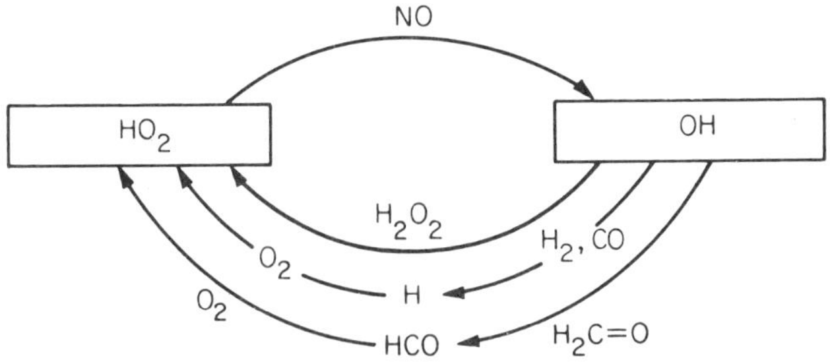

FIGURE 4b. Tropospheric radical interconversion cycles.

The final important radical-radical conversion path is the oxidation of CH_4. The first step is attack by OH,

$$OH + CH_4 \rightarrow CH_3 + H_2O \qquad (R51)$$

followed by

$$CH_3 + O_2 + M \rightarrow CH_3O_2 + M \qquad (R52)$$

Greiner measured the rate for R51 from 300 to 500°K and found K51 = 5.5 x 10^{-12}exp(-1900/T). An earlier room-temperature measurement of Wilson and Westenberg (255) gave K51 = 1 x 10^{-14}. Wilson (254) reports an activation energy of 5.5 kcal for R51 while Davis, Schiff, and Fischer (46c) have recently measured K51 = 2.5 x 10^{-12} exp(-3320/RT).

The large value of the activation energy for R51 causes K51 to decrease sharply with increasing altitude. Therefore, the chain mechanism for destroying CH_4 is effective only in the lower troposphere. The addition of O_2 to the methyl radical is very fast and can be considered instantaneous in the troposphere. In his review, Heicklen (93) reported K52 = 8 x 10^{-32}; measurements by Van den Bergh and Callear (245) and Basco, James, and James (10) gave limiting values of 2 x 10^{-12} and 5 x 10^{-13}, respectively, at high pressures.

The second step is not clear. By analogy with R47, Levy (152) proposed

$$CH_3O_2 + NO \rightarrow CH_3O + NO_2 \qquad (R53)$$

followed by

$$CH_3O + O_2 \rightarrow HO_2 + H_2C{=}O \qquad (R54)$$

It was assumed that K53 = K20. Spicer, Villa, Wiebe, and Heicklen (233) recently tried to study this reaction and found no evidence that it occurred. Their experiment was quite complicated, however, and open to other interpretations. Heicklen reported K54 = 2.6 x 10^{-18} at room temperature and recommended a more general expression, $K54 = 1.7 \times 10^{-13} \exp(-3250/T)$. While R54 is very slow, the large excess of O_2 over NO and NO_2 makes this path dominant in the troposphere.

An alternative path from CH_3O_2 to CH_3O is

$$CH_3O_2 + CH_3O_2 \rightarrow 2CH_3O + O_2 \qquad \text{(R55a)}$$

although

$$CH_3O_2 + CH_3O_2 \rightarrow CH_3OH + H_2C{=}O + O_2 \qquad \text{(R55b)}$$

is also possible. The Russell mechanism, R55b, dominates at room temperature in solution [Howard (101)], but Thomas and Calvert (241) and Heicklen and Johnston (94) have suggested the presence of R55a at room temperature in the gas phase. Based on the work of Mill, Mayo, Richardson, Irwin, and Allard (176) with n-butane, Hendry (95) believes that R55b is the preferred path at room temperature. Heicklen estimated an upper limit for K55a of 3×10^{-11}, but he considers that value to be too high. Howard and Ingold (102) found $K55 = 3.3 \times 10^{-14}$ when CH_3 is replaced by n-butyl and a much faster value, $K55 = 3.3$ to 6.6×10^{-13}, for primary peroxy radicals that contain π systems.

The rate of reaction for R55b, since it is dependent on the ease of transfer of the α-H of one of the peroxy compounds to the other, makes estimating K55b for CH_3O_2 from available rate data quite difficult, and it is not yet clear which path is dominant in the gas phase. From current data K55 would appear to fall between 3×10^{-11} and 3×10^{-14}, a spread of 10^3.

A third possibility is

$$CH_3O_2 + HO_2 \rightarrow CH_3O_2H + O_2 \qquad \text{(R56)}$$

followed by

$$CH_3O_2H + h\nu \rightarrow CH_3O + OH \quad \text{(R57)}$$

No rate has been measured for R56, nor has an absorption coefficient been determined for the photodissociation of methylperoxide, CH_3O_2H. The known rates for similar reactions involving HO_2 and H_2O_2 were used [Levy (153, 154)].

These various paths for the oxidation of CH_4, and the resultant production of $H_2C{=}O$ and conversion of OH to HO_2, are presented in Figure 4c.

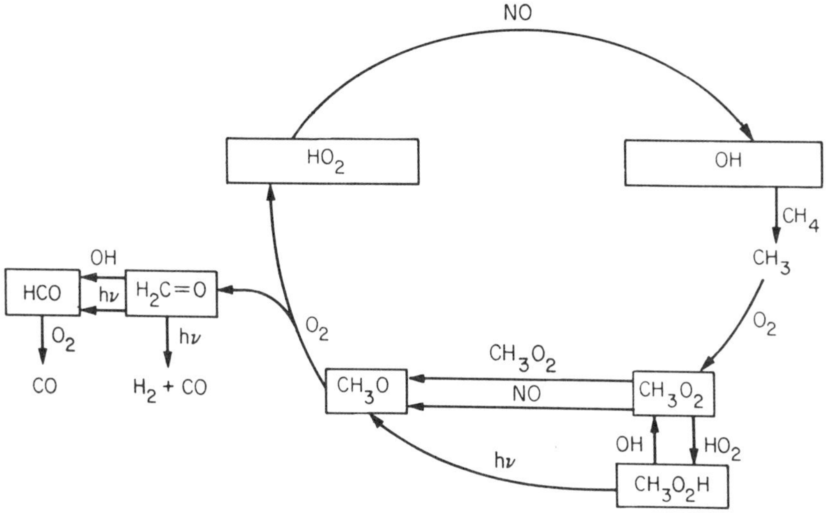

FIGURE 4c. Tropospheric methane oxidation cycle.

Crutzen (43) and Nicolet and Peetermans (183) have suggested a fourth possibility, based on the recent work of Spicer et al. (233), who reported

$$CH_3O_2 + NO \begin{cases} \rightarrow CH_3O_2NO & \text{(R58a)} \\ \rightarrow H_2C{=}O + HNO_2 & \text{(R58b)} \end{cases}$$

and

$$CH_3O_2 + NO_2 \begin{cases} \rightarrow CH_3O_2NO_2 & \text{(R59a)} \\ \rightarrow H_2C{=}O + HNO_3 & \text{(R59b)} \end{cases}$$

with K58 = 5 x 10^{-14}, 60% R58a and 40% R58b, and K59 = 5 x 10^{-14}, 80% R59a and 20% R59b. In Spicer et al.'s experiment, the only addition product observed was CH_3ONO_2, which they explained by assuming that the following multicenter rearrangements were rapid:

$$2CH_3O_2NO + O_2 \rightarrow 2CH_3ONO_2 + O_2 \qquad \text{(R60)}$$

and

$$CH_3O_2NO_2 + NO_2 \rightarrow CH_3ONO_2 + NO_3 \qquad \text{(R61)}$$

If we assume that photolysis of CH_3ONO gives

$$CH_3ONO_2 + h\nu \rightarrow CH_3O + NO_2 \qquad \text{(R62)}$$

This involved mechanism would have the same result as R53. It should be remembered that Spicer et al.'s experiment was extremely complicated and there is as yet no satisfactory explanation of their results. The sequence of reactions R58 to R62 was not included in any of the calculations discussed in this chapter.

To a first approximation, we can write

$$\frac{n(HO_2)}{n(OH)} = \frac{n(CO)K46 + n(CH_4)K51 + n(H_2)K49 + n(H_2C{=}O)K48 + n(H_2O_2)}{n(NO)K20} \qquad (10)$$

Because R51, R49, R48, and R50 all decrease rapidly with increasing altitude until they reach the middle troposphere, where they are no longer effective at all, the ratio $n(HO_2)/n(OH)$ will decrease in the lower troposphere until it reaches its limiting value of

$$\frac{n(HO_2)}{n(OH)} = \frac{n(CO)\ K46}{n(NO)\ K20} \tag{11}$$

which remains essentially constant up to the tropopause.

As a result of this altitude dependence, the total radical number density will decrease with altitude in the lower troposphere, and $n(OH)$ will remain relatively constant until $n(HO_2)/n(OH)$ reaches its limiting value [eq. (11)].

3. Loss

The significant radical loss paths are shown in Figure 4d. The important radical-radical elimination reactions are

$$HO_2 + HO_2 \rightarrow H_2O_2 + O_2 \tag{R63}$$

followed by R50;

$$HO_2 + CH_3O_2 \rightarrow CH_3O_2H + O_2 \tag{R56}$$

followed by

$$CH_3O_2H + OH \rightarrow H_2O + CH_3O_2 \tag{R64}$$

and

$$HO_2 + OH \rightarrow H_2O + O_2 \tag{R65}$$

A number of room-temperature measurements of the rate coefficient for R63 have been made: K63 = 3.6 x 10^{-12} [Paukert and Johnston (191)]; K63 = 3 x 10^{-12} [Foner and Hudson (63)]; K63 = 1.1 x 10^{-10} [Burgess and Robb (26)]; and K63 = 9.5 x 10^{-12} [Hochanadel, Ghormley, and Ogren (100)].

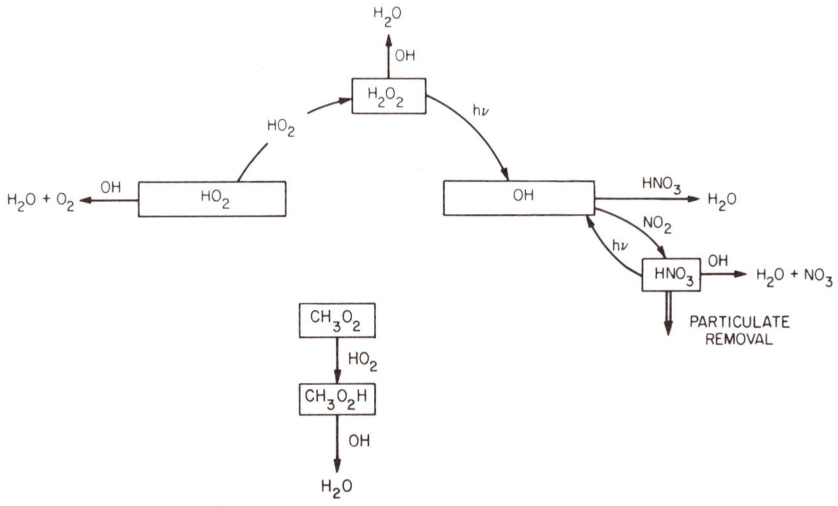

FIGURE 4d. Tropospheric radical loss paths.

Hampson et a.l (90, 91) and Lloyd (157) both recommended K63 = 3.3 x 10^{-12}, believing that the activation energy is either zero or small.

Neither K56 nor K64 has been measured, and it has been assumed that the rate coefficients are the same as those determined for R63 and R50.

Hochanadel et al. found K65 = 2 x 10^{-10} as the reaction rate for R65. Kaufman (135) estimated a lower limit, K65 > 1 x 10^{-11}, but does not believe that it is so fast as 2 x 10^{-10}. Considering the fact that Hochanadel et al. determined both K63 and K65 in the same experiment, their value for K65 may also be high.

The OH reactions

$$OH + OH \rightarrow H_2O + O \qquad (R66)$$

and

$$OH + OH + M \rightarrow H_2O_2 + M \qquad (R67)$$

are not important because of the large value for $n(HO_2)/n(OH)$.

Reactions R63, R56, and R65 are all important radical loss paths in the lower troposphere, but the decrease of $n(HO_2)/n(OH)$ with higher altitudes results in the increasing importance of the reaction involving OH, R65, which ultimately dominates in the upper troposphere.

The other two major radical loss reactions are

$$OH + HNO_3 \rightarrow H_2O + NO_3 \qquad (R35)$$

and

$$OH + NO_2 \xrightarrow{M} HNO_3 \qquad (R33)$$

followed by R35 or rainout. Their relative importance depends on the significance of particulate removal and rainout, R36, to HNO_3 loss [see Levy (154)].

The more effective R36 is, the smaller the percentage of HNO_3 lost by photolysis is and the more effective R33 as an OH loss mechanism is. At the same time decreasing $n(HNO_3)$ will lessen the importance of the other loss path, R35. A decrease in the effectiveness of rainout results in just the opposite.

This feedback mechanism serves to make the radical number densities in the lower troposphere insensitive to the inclusion or exclusion of HNO_3 rainout and to the resulting variations in $n(HNO_3)$ [Levy (154)]. With increasing altitude, this all breaks down as R33 becomes a more important path than R35, owing to the difference in activation energies.

At night, all the OH is immediately converted to HO_2 and CH_3O_2, and these two radicals decay via R63, R56, and R55 [Levy (153)].

4. Equations

The time-dependent behavior of each of the radical number densities $n(HO_2)$, $n(CH_3O_2)$, $n(OH)$, and $n(CH_3O)$ is represented by the following equations:

$$\begin{aligned}\frac{d\,n(HO_2)}{dt} &= 2n(H_2C{=}O)\ J43a + n(OH)\ [n(CO)\ K46 \\ &\quad + n(H_2O_2)\ K50 + n(H_2)\ K49 + n(H_2C{=}O)\ K48] \\ &\quad + n(CH_3O)\ n(O_2)\ K54 - n(HO_2)\ [2n(HO_2)\ K63 \\ &\quad + n(OH)\ K65 + n(NO)\ K20 + n(CH_3O_2)\ K56]\end{aligned} \qquad (12)$$

$$\frac{d\ n(CH_3O_2)}{dt} = n(OH)\ [n(CH_4)\ K51 + n(CH_3O_2H)\ K64] - [n(NO)\ K53 + n(HO_2)\ K56 + 2n(CH_3O_2)\ K55]\ n(CH_3O_2) \quad (13)$$

$$\frac{d\ n(OH)}{dt} = 2n(O^1D)\ n(H_2O)\ K42 + n(HO_2)\ n(NO)\ K20 + n(HNO_3)\ J34 + n(HNO_2)\ J32 + 2n(H_2O_2)\ J71 + n(CH_3O_2H)\ J\ 57 - n(OH)\ [n(NO_2)\ K33 + n(HNO_3)\ K35 + n(NO)\ K30 + n(CO)\ K46 + n(H_2O_2)\ K50 + n(H_2)\ K49 + n(CH_4)\ K51 + n(H_2C{=}O)\ K48 + n(CH_3O_2H)\ K64 + n(HO_2)\ K65] \quad (14)$$

$$\frac{d\ n(CH_3O)}{dt} = n(NO)\ n(CH_3O_2)\ K53 + 2n(CH_3O_2)^2\ K55a + n(CH_3O_2H)\ J57 - n(O_2)\ K54 \quad (15)$$

At night there is no OH production, and n(OH), along with all photodissociation rate coefficients, goes to zero, resulting in

$$\frac{d\ n(HO_2)}{dt} = n(CH_3O)\ n(O_2)\ K54 - n(HO_2)[2n(HO_2)\ K63 + n(CH_3O_2)\ K56] \quad (16)$$

$$\frac{d\ n(CH_3O_2)}{dt} = -n(CH_3O_2)\ [n(HO_2)\ K56 + 2n(CH_3O_2)\ K55] \quad (17)$$

$$\frac{d\ n(CH_3O)}{dt} = 2n(CH_3O_2)^2\ K55 - n(CH_3O)\ n(O_2)\ K54 \quad (18)$$

E. Carbon Chemistry

We considered above the role that carbon compounds such as CH_4, CO, and $H_2C{=}O$ play in the conversion of one radical to another, and we discovered that a number of radical chain reactions existed for the production and destruction of such compounds. Here we will develop a unified picture of atmospheric carbon chemistry, in which radicals play a major role.

1. Carbon Dioxide (CO_2)

While CO_2 is the major atmospheric carbon species, it has no significant atmospheric reactions in the troposphere. The general mechanisms for keeping it in balance in the lower atmosphere have already been discussed (see Section II.F.1). CO_2 photochemistry in the stratosphere and mesosphere has been reported in detail [Bates and Witherspoon (12), Hays and Olivero (92), and Wofsy et al. (256)].

2. Methane (CH_4)

The major sources of atmospheric CH_4—anaerobic decay of organic matter with smaller contributions from natural-gas emission and combustion—lead to an average column production rate of 1.5 to 4×10^{11} molecules cm^{-2} sec^{-1} [see Ehhalt (52)]. The major sink [Levy (152)] appears to be a chain reaction initiated by

$$CH_4 + OH \rightarrow CH_3 + H_2O \quad (R51)$$

which immediately forms the peroxy radical

$$CH_3 + O_2 + M \rightarrow CH_3O_2 + M \quad (R52)$$

The exact path by which the chain reaction is completed is not yet certain. As already discussed (Section IV.D.2), Levy (152-154) has suggested

$$CH_3O_2 + NO \rightarrow CH_3O + NO_2 \quad (R53)$$

followed by

$$CH_3O + O_2 \rightarrow HO_2 + H_2C{=}O \quad (R54)$$

with alternative paths for converting CH_3O_2 to CH_3O:

$$CH_3O_2 + CH_3O_2 \rightarrow 2CH_3O + O_2 \quad (R55a)$$

and

$$CH_3O_2 + HO_2 \rightarrow CH_3O_2H + O_2 \quad (R56)$$

followed by

$$CH_3O_2H + h\nu \rightarrow CH_3O + OH \quad (R57)$$

Spicer et al. (233) found that the photolysis of $CH_3N_2CH_3$ in an O_2, NO, NO_2 environment produces CH_3ONO_2, along with some $H_2C{=}O$ and HNO_2 or HNO_3 as the primary products of reactions between CH_3O_2 and NO or NO_2. A unique interpretation of their results is not possible at this time, but they concluded that R53 is unimportant.

If we assume the photolysis of CH_3ONO_2,

$$CH_3ONO_2 + h\nu \rightarrow CH_3O + NO_2 \quad \text{(R62)}$$

the findings of Spicer et al. will still result in either the conversion of CH_3O_2 to CH_3O or the direct production of $H_2C{=}O$ together with a photolytic source of OH.

In all cases, the oxidation of CH_4 will result in the production of $H_2C{=}O$ and another radical, either HO_2 or OH, to continue the chain reaction.

Therefore, the chain mechanism portrayed in Figure 4c is both the loss path for CH_4 and the source for atmospheric formaldehyde.

3. Formaldehyde ($H_2C{=}O$)

$H_2C{=}O$ is produced either by the oxidation of CH_3O, a product of the OH destruction of CH_4, or by CH_3O_2 [Spicer et al. (233)], the first product of CH_4 destructio[n]. For this chapter we will consider

$$CH_3O + O_2 \rightarrow H_2C{=}O + HO_2 \quad \text{(R54)}$$

to be the major source of tropospheric $H_2C{=}O$.

The important loss paths are photodissociation,

$$H_2C{=}O + h\nu \rightarrow HCO + H \quad \text{(R43a)}$$

and

$$H_2C{=}O + h\nu \rightarrow H_2 + CO \quad \text{(R43b)}$$

and reaction with OH,

$$H_2C{=}O + OH \rightarrow H_2O + HCO \quad \text{(R48)}$$

The absorption coefficients for the competing process, R43a and R43b, have been reported by Calvert et al (34), who gave average values of J43a and J43b for the surface atmospheres with the assumption that the quantum yield for R43a plus R43b is 1. Altitude profiles of J43a and J43b used in this calculation are given in Table VII. Noon values are J43a = 4×10^{-5} and J43b = 1×10^{-4}. The value of the rate coefficient for R48, K48 = 1.5×10^{-11} [Morris and Niki (177)], has already been discussed (see Section IV.D.2). In the lower troposphere, the two loss paths—photolysis and OH reaction—are comparable, while in the middle and upper troposphere, where n(OH) drops off, photolysis is dominant. At the same time, the destruction of CH_4 by OH, which has a large activation energy, slows greatly, resulting in a significant reduction in the production of $H_2C{=}O$ and therefore in $n(H_2C{=}O)$ itself.

4. Carbon Monoxide (CO)

All the loss paths for $H_2C{=}O$ lead to the formation of CO. The photolysis path R43b is a direct source, while the other photolysis path, R43a, and the OH path, R48, must be followed by

$$HCO + O_2 \rightarrow HO_2 + CO \quad \text{(R44)}$$

Demerjian, Kerr, and Calvert (47) reported K44 = 1×10^{-13}, which gives an instantaneous conversion of HCO to HO_2 and CO in the troposphere. The major sink for CO is

$$OH + CO \rightarrow CO_2 + H \quad \text{(R46)}$$

the rate coefficient for this reaction was discussed in Section IV.D.2.

5. Nonmethane Hydrocarbons

As discussed in Section II.F.4, there is very little quantitative knowledge concerning background levels of nonmethane hydrocarbons. Cavanagh et al. (37) reported fractional ppb levels, except for an anomalous 10 ppb for n-butanol. While local measurements that are quite high have been given for terpenes—particularly α-pinene, β-pinene, and isoprene—in areas of dense foliage, there are no worldwide estimates.

Ripperton et al. (211) measured the reaction rates for

$$O_3 + \alpha\text{-pinene} \rightarrow ? \qquad \text{(R68)}$$

$$O_3 + \beta\text{-pinene} \rightarrow ? \qquad \text{(R69)}$$

and found $K68 = 13 \times 10^{-17}$ and $K69 = 4 \times 10^{-17}$, respectively. However, it would appear that chain mechanisms were involved, so the true second-order rate should be slower. Stephens and Scott (235) reported that terpenes had high reactivities when compared with various hydrocarbons in polluted atmospheres, and Glasson and Tuesday (79) suggested that isoprenes had one-third the reactivity found for trans-2-butene yet twice that for ethylene.

Thus it would be reasonable to assume $K70 \geq 1 \times 10^{-17}$ for O_3 destruction of terpenes,

$$O_3 + \text{terpenes} \rightarrow ? \qquad \text{(R70)}$$

on the basis of current ozonolysis rates for reactive hydrocarbons [see Altshuller and Bufalini (4)].

Assuming the normal range for O_3 in the surface atmosphere, $n(O_3) = 5$ to 10×10^{11}, we would expect terpenes and other naturally emitted reactive hydrocarbons to have atmospheric lifetimes ranging from a few days to a few hours. An examination of the rate coefficients for OH with reactive hydrocarbons [see Wilson (254)] leads to the same conclusion. Thus it is likely that the reactive hydrocarbons—of which the terpenes seem to have the largest known natural source [see Rasmussen (199)]—would not be at all uniformly distributed around the globe, as CH_4 is, but would be found locally near their sources. While ozonolysis is not well understood and no reactions between OH and terpenes have been studied, it would be reasonable to assume that a considerable portion of the carbon in the reactive hydrocarbons ends up as CO, similar to the fate of CH_4.

Thus the reactive hydrocarbons with their very low steady-state concentration due to their high reactivity may be a source of CO comparable to CH_4 in regions where there is considerable vegetation. Estimates of the yearly production of terpenes run as high as 430×10^6 tons yr^{-1}, although these are tentative [see Rasmussen (199)], while estimates of CH_4 production range from 700 to 2000×10^6 tons yr^{-1} [see Ehhalt (52)].

6. Atmospheric Cycle

The tropospheric chemistry of carbon compounds is summarized in Figure 5. CH_4 is the dominant input of reactive carbon in the atmosphere, and the main action of atmospheric chemistry is to oxidize it to CO_2. Atmospheric $H_2C{=}O$ and CO are produced along the way. Combustion is only a small source of CO, and the total CH_4 contribution to atmospheric carbon is less than 1% of the atmospheric carbon exchanged by the CO_2 cycle. The contribution of reactive hydrocarbons such as terpenes is not clear, although they may provide a significant source of CO and

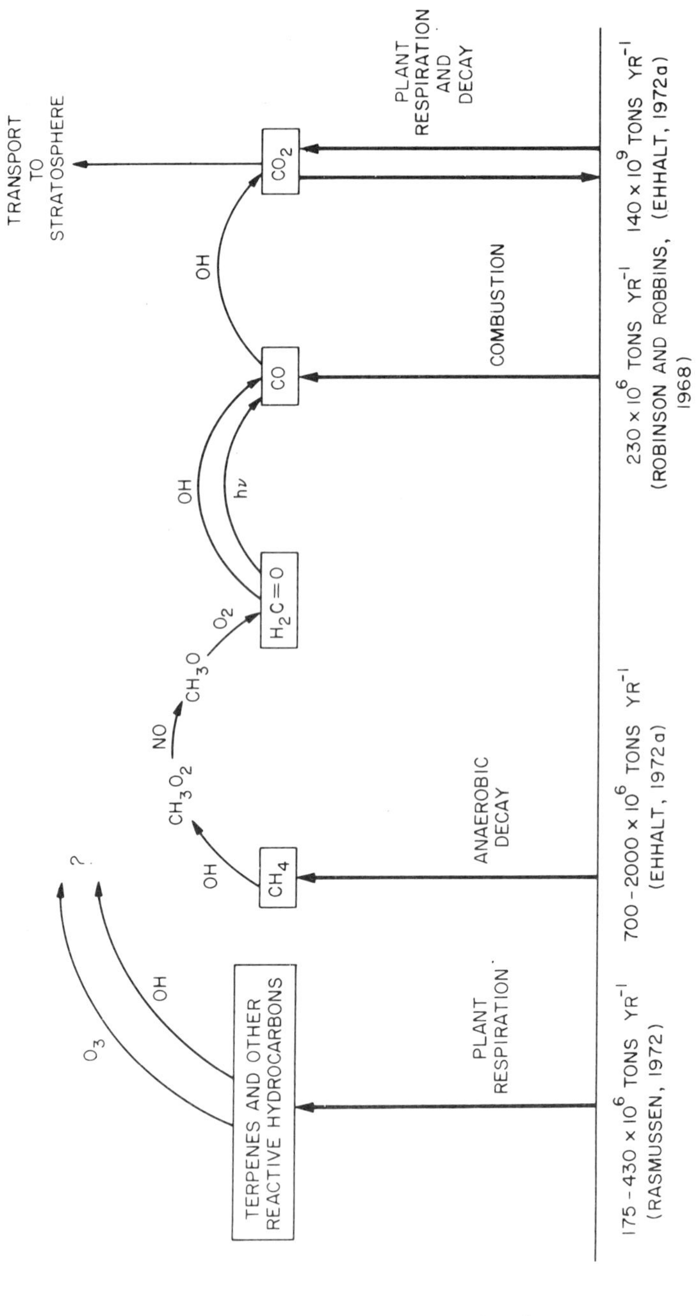

FIGURE 5. Tropospheric carbon chemistry.

may also be responsible for the surface removal of O_3 [Ripperton et al. (211)].

7. Equations

The time-dependent behavior of $n(H_2C{=}O)$ is given by

$$\frac{d\,n(H_2C{=}O)}{dt} = n(O_2)\ n(CH_3O)\ K54 - n(H_2C{=}O)\ [n(OH)\ K48 + J43a + J43b] \qquad (19)$$

There is essentially no production and loss at night. An upper limit for $n(H_2C{=}O)$ is given by [Levy (155)]

$$n(H_2C{=}O) < \frac{K51}{K48}\ n(CH_4) \qquad (20)$$

If we set

$$\frac{d\,n(CH_4)}{dt} = \frac{d\,n(CO)}{dt} = 0 \qquad (21)$$

we can write equations for the rates of production and loss of CH_4 and CO. Equation (21) is a reasonable assumption for CH_4, with a chemical lifetime of 2 to 7 yr in the troposphere, and is still applicable to CO, with a chemical lifetime of 0.1 to 0.3 yr. Another way of rationalizing this approach is to say that since photochemical production and loss rates are small relative to the ambient level, the fluctuations induced will not be significant.

We can write the following equations: the average daily loss rate for CH_4 is

$$\overline{L(CH_4)} = \overline{n(OH)}\ n(CH_4)\ K51 \qquad (22)$$

where $\overline{n(OH)}$ is the average daily OH number density. The average daily production rate for CO is

$$\overline{P(CO)} = \overline{L(CH_4)} \tag{23}$$

and its average daily loss rate is

$$\overline{L(CO)} = \overline{n(OH)}\ n(CO)\ K46 \tag{24}$$

At any given place in the troposphere, we can write $\overline{L(CH_4)} = \overline{L(CO)}$, which results in [Levy (155)]

$$n(CO) = \frac{K51}{K46}\ n(CH_4) \tag{25}$$

Therefore n(CO) is independent of n(OH), but it does depend on temperature because of the large activation energy of K51. At the surface, eq. (23) gives n(CO) = 0.1 ppm, in good agreement with measurements, but this value drops off rapidly with altitude. With its relatively long lifetime, CO is well mixed vertically in the troposphere and both $\overline{L(CH_4)}$ and $\overline{L(CO)}$ must be integrated over altitude. The result is that a vertically mixed CO based on CH_4 oxidation will give n(CO) = 0.07 ppm [Levy (155)] and another source—possibly terpenes—is needed.

F. Molecular Hydrogen (H_2)

The atmospheric source of H_2 is the photodissociation of $H_2C{=}O$,

$$H_2C{=}O + h\nu \rightarrow H_2 + CO \tag{R43b}$$

and the major loss path is reaction with OH,

$$OH + H_2 \rightarrow H_2O + H \qquad \text{(R49)}$$

H_2 is produced in the same CH_4 destruction cycle that is a source for $H_2C{=}O$ and CO. The natural source of H_2 is thought to be quite small [Koyama (142)].

The value of J43b has already been discussed (Section IV.E.3). Greiner (85) measured the rate for R49 over the temperature range 300 to 500°K and found K49 = $6.8 \times 10^{-12} \exp(-2000/T)$. Having a large activation energy, as does the similar reaction with CH_4, it is slow, even at room temperature, and H_2 has a long photochemical lifetime in the troposphere. Therefore we can set $d\,n(H_2)/dt = 0$ and write the following equations for average daily production and loss:

$$\overline{P(H_2)} = \overline{n(H_2C{=}O)\ J43b} \qquad (26)$$

and

$$L(H_2) = \overline{n(OH)}\ n(H_2)\ K49 \qquad (27)$$

For the average daily production, it is necessary to average the product of $n(H_2C{=}O)$ and J43b since both vary with time.

G. Peroxides (H_2O_2) and CH_3O_2H)

As shown in Figure 4d, hydrogen peroxide (H_2O_2) and methyl peroxide (CH_3O_2H) are products of major radical loss paths,

$$HO_2 + HO_2 \rightarrow H_2O_2 + O_2 \qquad \text{(R63)}$$

and

$$CH_3O_2 + HO_2 \rightarrow CH_3O_2H + O_2 \quad \text{(R56)}$$

The rate coefficient for R63 was discussed in Section IV.D.3, and we have assumed that K56 = K63 since no data for R56 are available.

The important loss reactions are photodissociation and the OH reactions:

$$H_2O_2 + h\nu \rightarrow 2OH \quad \text{(R71)}$$

$$CH_3O_2H + h\nu \rightarrow CH_3O + OH \quad \text{(R57)}$$

$$H_2O_2 + OH \rightarrow H_2O + HO_2 \quad \text{(R50)}$$

$$CH_3O_2H + OH \rightarrow H_2O + CH_3O_2 \quad \text{(R64)}$$

Since there are no gas-phase absorption data for H_2O_2 at wavelengths greater than 2900 Å, it was necessary to use the liquid-phase absorption measurements compiled by Schumb, Gatterfield, and Wentworth (222). Altitude profiles for J71 are given in Table VII. No absorption data exist for CH_3O_2H, so we assumed J57 = J71. The rate coefficient for R50 was discussed in Section IV.D.2. Again there are no data for CH_3O_2H, so K64 = K50 was assumed.

Wofsy et al. (256) have included a rainout-removal mechanism for the peroxides, but they have a photochemical lifetime of three days or less in the lower troposphere and rainout should not be important. However, we will include this removal mechanism for completeness.

$$\left.\begin{matrix} H_2O_2 \\ CH_3O_2H \end{matrix}\right. \rightarrow \text{rainout} \quad \text{(R72)}$$

and set K72 = K36 = 2×10^{-6}.

The time-dependent behaviors of $n(H_2O_2)$ and $n(CH_3O_2H)$ are described by

$$\frac{d\ n(H_2O_2)}{dt} = n(HO_2)^2\ K63 - n(H_2O_2)$$
$$[J71 + K72 + n(OH)\ K50] \tag{28}$$

and

$$\frac{d\ n(CH_3O_2H)}{dt} = n(HO_2)\ n(CH_3O_2)\ K56 - n(CH_3O_2H)$$
$$[J57 + K72 + n(OH)\ K64] \tag{29}$$

V. CALCULATIONS

As mentioned in the introduction, no detailed treatments of tropospheric photochemistry had been undertaken until recently. In this section we will briefly discuss past calculations and then outline those performed here.

A. Previous Calculations

In an extension of analytic techniques used by Leovy (150) and Nicolet (180) for stratospheric calculations, Levy (152) developed a simple stationary-state photochemical model of tropospheric radical chemistry. Assuming R42 to be the sole source of radicals and assuming radical-radical reactions to be the sole loss paths, with $R63 = R66 = \frac{1}{2}\ R65$, the following expression was derived:

$$n(OH_2) + n(OH) = \left[\frac{n[O(^1D)]\ n(H_2O)\ K42}{K63}\right]^{1/2} \tag{30}$$

If we consider only the interconversion paths involving CO, CH_4, O_3, and NO, two other relations can be derived:

$$\frac{n(HO_2)}{n(OH)} = \frac{n(CO)\ K46 + n(CH_4)\ K51 + n(O_3)\ K47}{n(NO)\ K20} \tag{31}$$

and

$$n(CH_3O_2) = 33\ n(OH) \tag{32}$$

While the value of K42 used in the calculations [Levy (152) was the incorrect one reported by Biedenkapp et al (20), the role of $H_2C{=}O$ as a radical source and in the conversion of OH to HO_2 was neglected, and the radical loss reactions involving HNO_3 were not included, these expressions did produce qualitatively correct results for the surface atmosphere. Radical number densities approaching 5 x 10^8, a CO lifetime of ~0.2 yr, and $n(H_2C{=}O)$ ~1 ppb were predicte

McConnell et al. (171) then included this steady-state radical photochemistry in a one-dimensional eddy-diffusion treatment of CH_4, which they then solved for the noon conditions of the Levy model. They calculated the total column loss rate for CH_4 and, on the assumption that it was all converted to CO, estimated a production of CO consistent with a CO residence time in the troposphere of 0.3 yr.

A more detailed photochemical model of the surface atmosphere developed by Levy (153) coupled the radical, carbon, and nitrogen chemistry together. As in the simpler model, the number densities of the major atmospheric species, N_2, O_2, H_2O, CH_4, CO, O_3, and NO + NO_2, were held constant at values appropriate to a summer latitude of 34°N, and hourly number densities were calculated for HO_2, OH, CH_3O_2, CH_3O, $H_2C{=}O$, NO, NO_2, NO_3, N_2O_5, HNO_2, and HNO_3, along with average daily values of

$H_2C{=}O$, CH_3O_2H, and HNO_3 when particulate removal was excluded. The number densities were all calculated in the steady-state approximation. With the inclusion of the faster value of K42, the total radical number density approached 1.7 x 10^9, while n($H_2C{=}O$), n(H_2O_2), and n(CH_3O_2H) were in the ppb range. The inclusion of the reaction

$$N_2O_5 + H_2O \rightarrow 2HNO_3 \qquad \text{(R28)}$$

now believed to be very slow, led to a large production of HNO_3 and a steady-state number density approaching 0.1 ppm. A particulate-removal mechanism was then introduced; this greatly reduced n(HNO_3) but, in turn, resulted in a removal rate for NO_3^- that greatly exceeded the measurements of Eriksson (58). Even with the exclusion of R28 as an HNO_3 source, this problem still exists in the tropospheric nitrogen chemistry. In this model,

$$CH_3O_2 + NO \rightarrow CH_3O + NO_2 \qquad \text{(R53)}$$

was the important path for converting CH_3O_2 to CH_3O.

Wofsy et al. (256) then performed a one-dimensional eddy-diffusion calculation of the atmospheric profiles for n(CH_4) and n(CO), with steady-state treatment of the radical photochemistry. They used previously calculated profiles of nitrogen compounds, along with assumed n(CH_4) and n(CO) profiles, to generate steady-state radical profiles. They then performed a one-dimensional eddy-diffusion calculation of n(CH_4), assuming the previously calculated radical profiles, followed by a one-dimensional eddy-diffusion calculation of n(CO), which used the n(CH_4) profile as a CO source. This three-step procedure was iterated until both the n(CH_4) and the n(CO) profiles converged.

The tropospheric photochemistry included in the model was essentially that of Levy (152,153), except that R53 was excluded. R55a and R56 followed by R57 were relied on to convert CH_3O_2 to CH_3O.

The resulting n(OH) profiles, calculated for summer and winter latitudes of 30°N with a zenith angle of 60°, were in harmony with the tropospheric calculations of Levy (154), where average daily values for n(OH) were determined explicitly by averaging steady-state values of n(OH) calculated at 1-hr intervals.

In the calculations for the full troposphere [Levy (154)], conditions similar to those in the previous surface calculation [Levy (153)] were assumed. The altitude profiles for the major species, H_2O, CH_4, H_2, CO, NO + NO_2, and O_3 were fixed at values appropriate to a summer latitude of 34°N (50% humidity for H_2O; single tropospheric mixing ratios for CH_4, H_2, CO, and NO + NO_2; and a measured profile for O_3). A set of simultaneous nonlinear equations was then solved hourly for the steady-state number densities of NO, NO_2, NO_3, N_2O_5, HNO_2, CH_3O, CH_3O_2, $H_2C{=}O$, OH, and HO_2, and average daily profiles for the long-lived species, HNO_3, H_2O_2, and CH_3O_2H, were calculated.

Within the steady-state approximation, the total column loss rate for CH_4 and the total column production and loss rates for CO [Wofsy et al. (256) and Levy (155)] and for H_2 [Levy (155)] have been calculated. The resulting residence times and tropospheric mixing ratios are in harmony with recent experimental results, with the exception of the n(CO) mixing ratio. This ratio is still too small, and it would appear that another large natural source of CO is required.

B. Current Model

In principle, the optimal photochemical model would treat all the major species together with the sulfur, nitrogen, carbon, and radical chemistry; however, that is not yet possible nor is it completely necessary.

1. Atmospheric Conditions

Experimental measurements and the recent calculations of Wofsy et al. (256) show that $n(CH_4)$ is well represented by a single mixing ratio and n(CO) is well mixed vertically, although its mixing ratio may be latitude dependent. Insufficient data concerning the altitude profiles and global number densities of NO, NO_2, and NH_3 exist to permit a detailed understanding of their tropospheric mixing. It would appear that NH_3 is not well mixed vertically, while the sum of NO + NO_2 may be. Considerable evidence points to their number densities being latitude dependent. A complete model of tropospheric nitrogen chemistry would require quantitative values for the natural sources of NH_3, NO, and NO_2, but these data are simply not available.

It is well known that $n(H_2O)$ and $n(O_3)$ are highly variable, but their inclusion as variables in the model would require that the models of global atmospheric motion be coupled to photochemistry, and this is not yet possible. Therefore at this time, it is necessary to fix $n(H_2O)$ and $n(O_3)$ profiles at representative values (see Tables III and IV) and to assume a constant mixing ratio for n(NO) + $n(NO_2)$ that is representative of available measurements (see Section II.E.2). $n(CH_4)$ and

$n(H_2)$ are well represented by a single mixing ratio, while n(CO) is well represented by a latitude-dependent one.

2. Numerical Techniques

The most general approach would be to solve the time-dependent, coupled, nonlinear differential equations for all remaining species; adequate numerical techniques are available [Gear (73,74)] and have been applied to photochemical smog problems [see, for example, Gelinas (75)]. Preliminary time-dependent calculations have been reported by Kummler and Baurer (144). For the calculations here, techniques similar to those employed previously [Levy (153,154)] will be used. Preliminary calculations were also performed for the full set of time-dependent equations using a program developed by Gear (73,74) and the results agreed well with the more approximate steady-state calculation. The steady-state calculations did exaggerate the decay of some species at sunset as well as their growth at sunrise but these effects were minor.

The species of interest are separated into two categories, according to their chemical lifetimes: those exceeding one day, and those less than one hour. The sole exception, $H_2C{=}O$, does not have a rapidly varying number density and can be included in the latter group. This separation enables a steady-state solution of the number densities to be made at hourly intervals and generates a reasonably accurate picture of the time-dependent behavior. Quantitative inaccuracies exist near dawn and dusk, where dissociation rates are changing rapidly and chemical residence times may exceed 1 hr, but the overall daytime picture is accurate. By using assumed values for the long-lived species, $n(HNO_3)$, $n(H_2O_2)$, and $n(CH_3O_2H)$, the simultaneous equations for the short-lived species can be solved at 1-hr intervals. Corrected values of n(HNO_3), n(H_2O_2), and n(CH_3O_2H) are then calculated from average daily values for n(OH) and the products n(OH) n(NO_2), n(HO_2) n(HO_2), and n(HO_2) n(CH_3O_2). The equations for the short-lived species are then solved with the corrected values for the long-lived species. This iterative procedure is continued until all number densities can be reproduced with an error $\leq \pm 1\%$.

3. Equations

The sulfur chemistry has not been included in the model, since it does not appear to have a significant effect on the radical chemistry. If, however, the preliminary estimates of K6a are correct, the oxidation of SO_2 by OH is an important loss path for both OH and SO_2 and the sulfur chemistry will then be coupled to the radical chemistry. It will not be the dominant radical loss path but it would be an important one. From what little is known about the gas-phase chemistry of sulfur, chemical lifetimes for H_2S and SO_2 will be estimated from n(OH) and n(HO_2) profiles calculated separately.

NH_3 is an important source of NO and NO_2, but it does not directly affect the radical chemistry, and with the fixing of the NO + NO_2 profile, it is not a necessary input to the nitrogen chemistry. In a detailed treatment of nitrogen chemistry that includes NO and NO_2 sources, NH_3 would be incorporated, but such a model is not quantitatively possible at this time.

Solving the time-dependent equations (4) to (7), (12) to (15), (19), (28), and (29) in the steady-state approximation (dn/dt = 0), we find the following sets of equations.

For the short-lived species

$$n(NO_3) = \frac{n(NO_2)\ n(O_3)\ K22 + n(HNO_3)\ n(OH)\ K35 + n(N_2O_5)\ K26}{J23 + n(NO)\ K24 + n(NO_2)\ K25} \quad (33)$$

$$n(N_2O_5) = \frac{n(NO_3)\ n(NO_2)\ K25}{K26 + J27 + n(H_2O)\ K28} \quad (34)$$

$$n(HNO_2) = \frac{2n(NO_2)\ n(NO)\ n(H_2O)\ K29 + n(OH)\ n(NO)\ K30}{J32} \quad (35)$$

$$n(CH_3O) = \frac{n(NO)\ n(CH_3O_2)\ K53 + 2n(CH_3O_2)^2\ K55a + n(CH_3O_2H)\ J57}{n(O_2)\ K54} \quad (36)$$

$$n(CH_3O_2) = \frac{-B - (B^2 - 4AC)^{1/2}}{2A} \quad (37)$$

where

$$A = -2K55$$

$$B = n(NO)\ K53 + n(HO_2)\ K56$$

and

$$C = n(OH)\ [n(CH_4)\ K51 + n(CH_3O_2H)\ K64]$$

$$n(H_2C{=}O) = \frac{n(O_2)\ n(CH_3O)\ K54}{n(OH)\ K48 + J43a + J43b} \quad (38)$$

$$n(OH) = \frac{\{2n[O(^1D)]\ n(H_2O)\ K42 + n(HO_2)\ n(NO)\ K20 + n(HNO_3)\ J34 + n(HNO_2)\ J32 + 2n(H_2O_2)\ J71 + n(CH_3O_2H)\ J57\}}{[n(NO_2)\ K33 + n(NO)\ K30 + n(HNO_3)\ K35 + n(CO)\ K46 + n(H_2O_2)\ K50 + n(H_2)\ K49 + n(CH_4)\ K51 + n(H_2C{=}O)\ K48 + n(CH_3O_2H)\ K64 + n(HO_2)\ K65]} \quad (39)$$

and

$$n(HO_2) = \frac{-B - (B^2 - 4AC)^{1/2}}{2A} \tag{40}$$

where

$A = -2K63$

$B = n(OH)\ K65 + n(NO)\ K20 + n(CH_3O_2)\ K56$

$C = 2n(H_2C{=}O)\ J43a + n(CH_3O)\ n(O_2)\ K54 + n(OH)$
$[n(CO)\ K46 + n(H_2O_2)\ K50 + n(H_2)\ K49 + n(H_2C{=}O)\ K48]$

For the long-lived species:

$$n(HNO_3) = \frac{n(OH)\ n(NO_2)\ K33}{J34 + n(OH)\ K35 + K36} \tag{41}$$

$$n(H_2O_2) = \frac{n(HO_2)^2\ K63}{J71 + K72 + n(OH)\ K50} \tag{42}$$

and

$$n(CH_3O_2H) = \frac{n(HO_2)\ n(CH_3O_2)\ K56}{J57 + K72 + n(OH)\ K64} \tag{43}$$

At night, with n(OH) and n(NO) vanishingly small, we find the following steady-state equations for NO_3 and N_2O_5:

$$n(NO_3) = \frac{n(NO_2)\ n(O_3)\ K22 + n(N_2O_5)\ K26}{n(NO_2)\ K25} \tag{44}$$

and

$$n(N_2O_5) = \frac{n(NO_3)\ n(NO_2)\ K25}{K26 + n(H_2O)\ K28} \tag{45}$$

The nighttime production and loss reactions for HNO_2, $H_2C{=}O$, HNO_3, H_2O_2, and CH_3O_2H are all slow, so their number densities are not significantly affected by nighttime chemistry.

The CH_3O radical, represented by

$$n(CH_3O) = \frac{2n(CH_3O_2)^2 \ K55a}{n(O_2) \ K54} \quad (46)$$

always has a small number density. The other two radicals, HO_2 and CH_3O_2, have only loss reactions at night, and their time-dependent equations (16) and (17) can be solved in an approximate manner [Levy (153)] to show a slow decay through the night

VI. RESULTS

By use of the techniques described in Section V.B.2, steady-state calculations were performed for both a summer and a winter latitude of 30°N with the set of equations in Section V.B.3. The values adopted for those constituents with fixed altitude profiles were consistent with the above conditions. The calculations were made with and without the inclusion of rainout as a loss path for long-lived species.

The dependence of tropospheric chemistry on season and altitude will be covered later in this section. The effects of latitude can be inferred from Tables I to IV. Latitude-dependent variations have their major impact on the production of radicals.

The increase in the total column of O_3 with increasing latitude (see Table IV) causes a decrease in solar radiation with $\lambda \leq 3150$ Å, which results in a decrease in the production of $O(^1D)$ [see eqs. (1) and (2), R38, and R40]. The local variation in $n(O_3)$ may at times offset this effect. The decrease in $n(H_2O)$ with increasing latitude (only for latitudes > 30° in the summer) (see

Table III) will result in a decreased production of OH via R42. The solar zenith angle, an important factor in the production of $O(^1D)$, varies much more with season than with latitude. The increase in total density (see Table II) with increasing latitude will have a small effect on $n[O(^1D)]$ [see eq. (9)].

The net results of these effects are as follows: in the summer, radical production and number density are relatively constant over the latitude range 0 to 30°N and slowly decrease for latitudes > 30°N. In the winter, radical production and number density decrease with increasing latitude over the complete range, with a sharp drop in the polar region. Except for the winter high latitudes, these variations are no greater than the seasonal variations at 30°N.

The final latitude-dependent variable—temperature—decreases slowly with increasing latitude and will result in a slight decrease in those reaction rates with large activation energies, the most important one being CH_4 oxidation. Again, in the winter polar region, the effects are much larger. In general the chemistry of the winter polar region is anomalous.

A. Radicals (OH, HO_2 and CH_3O_2)

Both the noontime and the average daily profiles of n(OH) are shown in Figure 6, and the noontime profiles of $n(HO_2)$ and $n(CH_3O_2)$, in Figure 7.

With the reaction of $O(^1D)$ and H_2O as the major source of radicals, the sharp drop in their number densities with increasing altitude is to be expected when the altitude profile for $n(H_2O)$ is considered (see Table III). The production of $n(CH_3O_2)$ due to R51, which has a 3.8-kcal activation energy, drops off the most rapidly and

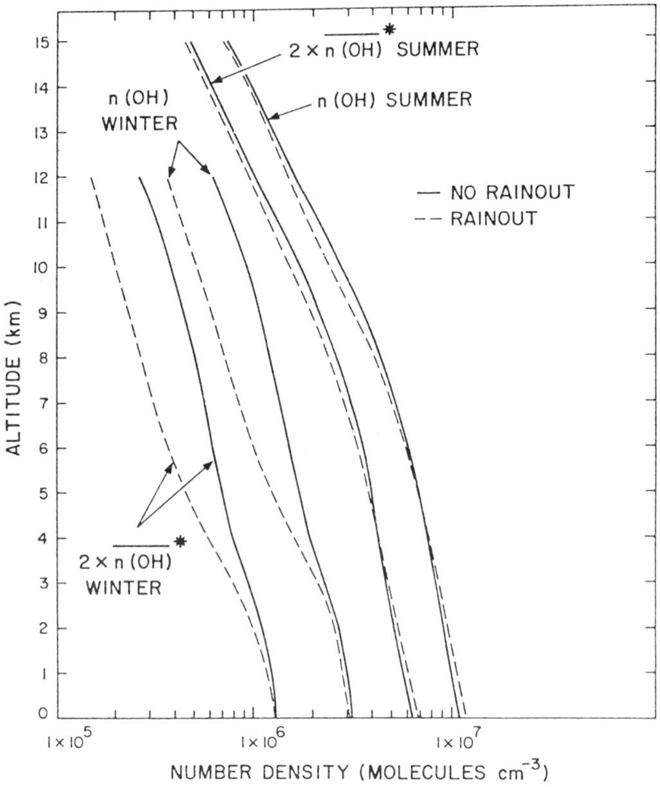

FIGURE 6. Noontime and average daily n(OH) altitude profiles.

becomes quite small in the cold upper troposphere. The decrease in n(OH) with increasing altitude is slower in the lower troposphere than that for the other radicals because its rate of conversion to HO_2 via R51 is decreasing (see Section IV.D.2). In a previous calculation, which assumed a constant relative humidity [Levy (154)], n(OH) changed very little in the first 5 km. In this calculation, the relative humidity decreases with increasing altitude, so there is also some decrease in n(OH). While $n[O(^1D)]$ increases with altitude, it

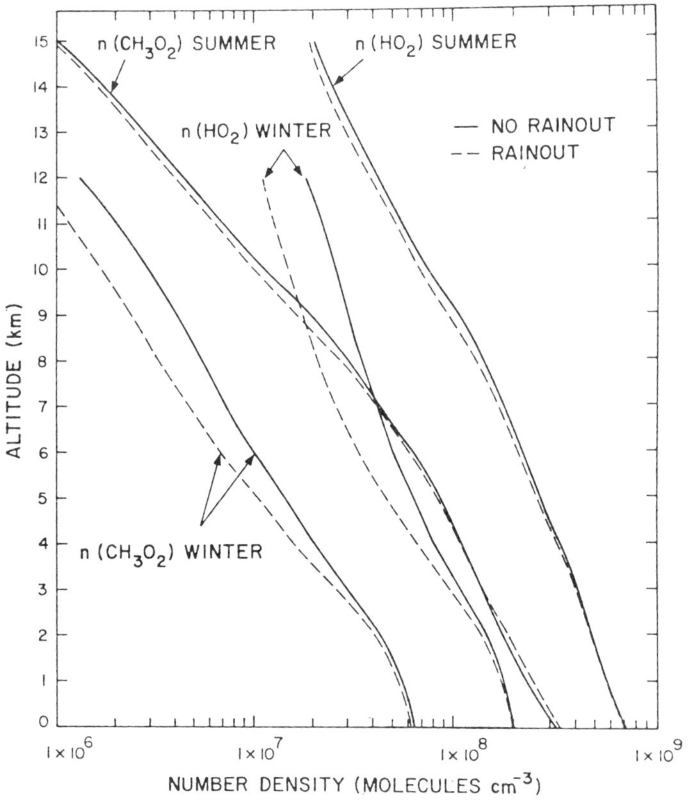

FIGURE 7. Noontime $n(HO_2)$ and $n(CH_3O_2)$ altitude profiles.

is not enough to balance the drop in $n(H_2O)$. At the tropopause and above in the lower stratosphere, radical number densities will again start to increase because $n(H_2O)$ does not change much and $n(O_3)$ increases sharply.

The winter noontime values of n(OH), $n(HO_2)$, and $n(CH_3O_2)$ decrease by a factor 3.5 from those calculated for the summer. The primary cause is an increase in the solar zenith angle, which decreases the intensity of ultraviolet radiation required to produce $O(^1D)$. Because of lower winter temperatures, $n(H_2O)$ also decreases but has only a small effect [see Wofsy et al. (256)]. The daily average value of n(OH), $\overline{n(OH)}$, shows an even greater decrease owing to the shorter winter day.

The inclusion of a rainout loss mechanism for long-lived species—in particular, HNO_3—has little effect on the summer radical profiles for the following reasons: in the lower and middle troposphere, the radical number densities are high enough to make radical-radical reactions the dominant loss path. With n(OH) large enough to make photodissociation unimportant as an HNO_3 loss reaction, the inclusion of rainout, while having some effect on $n(HNO_3)$, does not significantly enhance the effectiveness of R33 as an OH loss path. In the upper troposphere, where R33 and R35 become major radical loss reactions, rainout is too slow to strongly affect $n(HNO_3)$, whose major loss path is photodissociation.

In the winter, radical number densities are much smaller and OH reactions with NO_2 and HNO_3 become important loss paths, particularly in the middle and upper troposphere. With $\overline{n(OH)}$ being much smaller, photodissociation, which reforms an OH (see Figure 4d), is an important HNO_3 loss reaction. If rainout is included, the OH loss mechanism via R33 is greatly enhanced, with a resulting decrease in radical density. Since the balance between HNO_3 loss paths that destroy and those that conserve radicals is quite sensitive to the presence of rainout in the winter, even in the upper troposphere where it is small, the winter radical profiles show a similar sensitivity.

B. Peroxides (H_2O_2 and CH_3O_2H)

Altitude profiles for $n(H_2O_2)$ and $n(CH_3O_2H)$ are plotted in Figure 8. The peroxide sources are the radical-

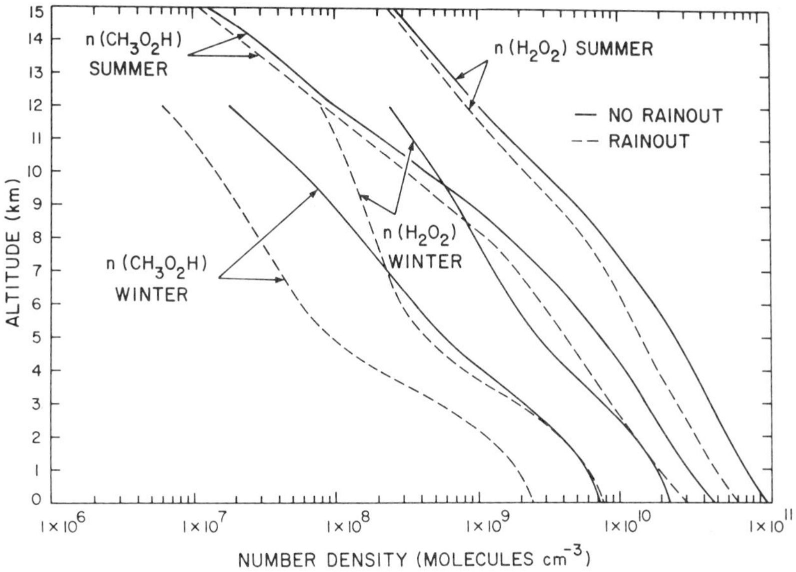

FIGURE 8. $n(H_2O_2)$ and $n(CH_3O_2H)$ altitude profiles.

radical loss reactions R56 and R63, and their production is proportional to $n(HO_2)^2$ and $n(HO_2) \cdot n(CH_3O_2)$. This strong dependence on the peroxy radical concentrations shows up both in the sharp drop in peroxide number density with increasing altitude and in the large decrease in the winter profiles relative to summer. Since the summer peroxy radical profiles are not affected by the inclusion of rainout, the difference between summer profiles represents the increase in peroxide loss rate due to rainout. This difference decreases with increasing altitude, as it should. In the winter, both the direct effect of rainout—increased loss rate—and the indirect effect—reduction of peroxy radical number density—are observed. While both peroxides are present in the ppb range in the lower troposphere, their number density drops off rapidly in the upper troposphere.

C. Sulfur Compounds (H_2S and SO_2)

There is not yet a great deal of atmospheric sulfur chemistry to consider. From daily average values of n(OH) and $n(HO_2)$, it is possible to estimate the lower troposphere residence times of H_2S and SO_2:

$$\tau_{H_2S} = 1/K12\overline{n(OH)} \quad (47)$$

and

$$\tau_{SO_2} = 1/K6\overline{n(HO_2)} \quad (48)$$

Since $\overline{n(OH)}$ and $\overline{n(HO_2)}$ are not greatly affected by the inclusion of particulate loss in the lower troposphere, we will use an average summer and winter value for each. The results are given in Table VIII. These values were calculated with $K6 = 9 \times 10^{-16}$.

TABLE VIII
Calculated H_2S and SO_2 Residence Times

Altitude (km)	Summer		Winter	
	τ_{H_2S}	τ_{SO_2}	τ_{H_2S}	τ_{SO_2}
0	8-17 hr	$\geq$ 45 days	2-4 days	$\geq$ 200 days
2	10-21 hr	$\geq$ 70 days	2-4 days	$\geq$ 280 days
4	13-26 hr	$\geq$ 97 days	3-6 days	$\geq$ 580 days
6	16-32 hr	$\geq$ 150 days	4-9 days	$\geq$ 1000 days

It is clear that R6 is not an important SO_2 loss path if current estimates of SO_2 residence times and K6 are all correct. The residence times for H_2S, particularly in the summer, appear to be in harmony with earlier estimates (see Sections II.D and IV.A), which suggests that the reaction with OH, R12, is an important H_2S loss path. Currently the only SO_2 loss path that appear to be fast enough are either in the liquid phase or heterogeneous. No lifetimes were calculated for the upper troposphere, although it is clear that they would be considerably longer. As has been mentioned previously R6a, if K6a is as large as has been suggested, would be the dominant SO_2 removal path and would give a τ_{SO_2} of a few days depending on the season.

D. Nitrogen Compounds (HNO_3, NO, NO_2, NO_3, N_2O_5, and HNO_2)

The nitrogen photochemistry is coupled to that of the radical and carbon compounds since, as has already been discussed, the major nitrogen species, HNO_3, is strongly coupled to OH via mutual loss and production reactions.

1. Nitric Acid (HNO_3)

Altitude profiles for both winter and summer with and without rainout are given in Figure 9. The source of HNO_3 at all altitudes is the reaction with OH, R33, and the important loss paths are reaction with OH, photodissociation, and rainout. In the lower troposphere, HNO_3 has a residence time $\tau_{HNO_3} \geq 14$ and 70 days in the summer and winter, respectively, with rainout excluded; the inclusion of rainout gives $\tau_{HNO_3} = 4$ days in the summer and 5 days in the winter. Since HNO_3 is soluble and its chemical lifetime is so long, it might be expected to be rainted out. In this case, essentially the total

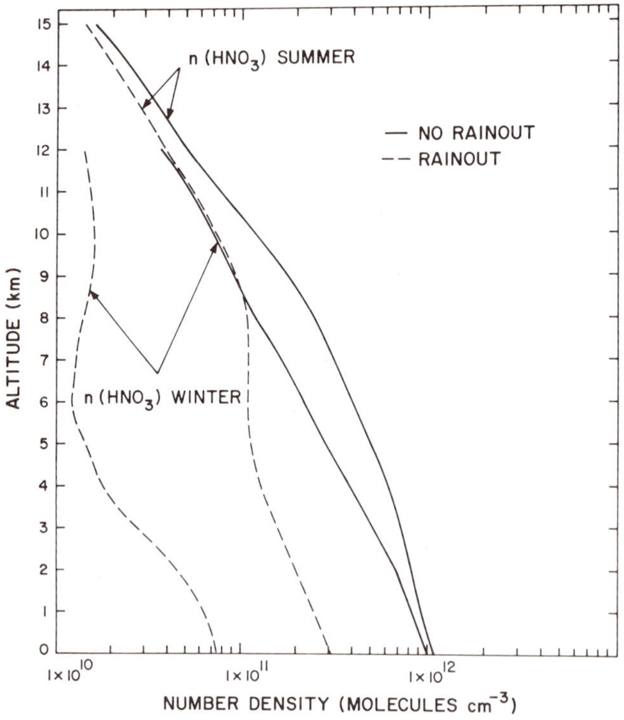

FIGURE 9. $n(HNO_3)$ altitude profiles.

production of HNO_3 in the first 5 km, 2.5×10^{11} and 5×10^{10} molecules cm^{-2} sec^{-1} in the summer and winter, respectively, is removed as nitrate. This calculated yearly average, 1.5×10^{11} molecules cm^{-2} sec^{-1}, is more than four times that measured by Eriksson (58).

As Figure 9 shows, rainout has the major influence on HNO_3 profiles. Without it, $n(HNO_3)$ will approach 40 ppb in the lower troposphere throughout the year, while its inclusion lowers $n(HNO_3)$ to 12 ppb in the summer and 3 ppb in the winter. HNO_3, however, is the major nitrogen species for all calculations. Its range of values at the tropopause

3 to 6 ppb, is in harmony with recent measurements of Murcray and Murcray (178).

In the summer, upper troposphere photodissociation is the dominant loss path, while in the winter, it is less, owing to the reduced amount of daytime, and becomes comparable to rainout, which itself is quite small. Residence times, chemical and rainout, run from 40 to 60 days in that region.

The usual behavior of the dashed lines in Figure 9 in the middle troposphere is caused by the sharp drop in the rainout rate above 5 km imposed by the calculation. While rainout should be much less in the middle and upper troposphere, the use of a 2-km scale height may not be the most realistic model.

2. Remaining Nitrogen Compounds (NO, NO_2, NO_3, N_2O_5, and HNO_2)

Summer and winter noontime profiles for the oxides of nitrogen as well as HNO_2 are given in Figure 10.

Because of the 2.4-kcal activation energy for R19, n(NO) remains relatively constant in the troposphere, while $n(NO_2)$ decreases steadily with increasing altitude (see Section IV.B.3 for further discussion). There is a slight seasonal variation due to small differences in seasonal temperature that affect R19, the reaction that converts NO to NO_2. These calculations were all performed with the sum n(NO) + $n(NO_2)$ fixed at 3 ppb throughout.

$n(NO_3)$ decreases with increasing altitude because its formation reaction has a 7-kcal activation energy, while $n(NO_2)$ is quite variable since both its production and its loss reactions have large activation energies. The oxide number densities of both remain quite small relative to n(NO) and $n(NO_2)$ during the daytime. At night (see Section IV.B.7) NO_3 becomes an important oxide in the lower troposphere, but it is rapidly

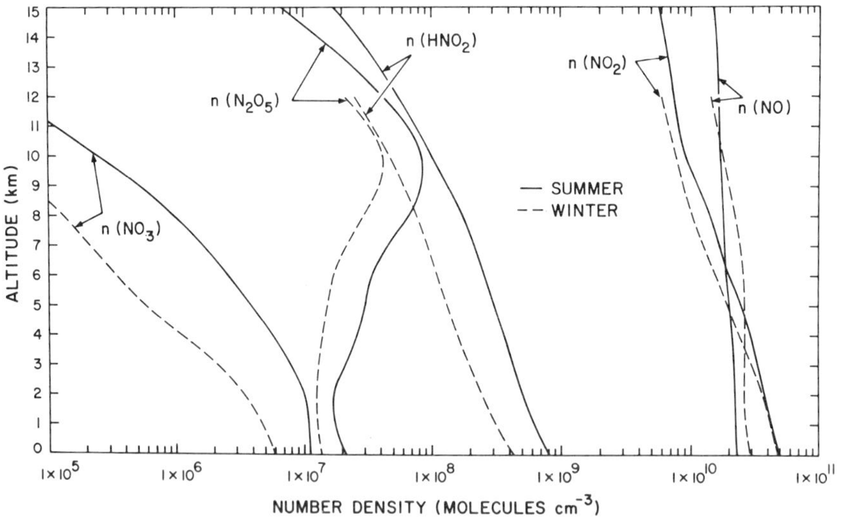

FIGURE 10. Noontime $n(NO_3)$, $n(N_2O_5)$, and $n(HNO_2)$ altitude profiles.

converted to NO and NO_2 in the early morning. There are seasonal variations in the NO_3 and N_2O_5 profiles, but they are not significant.

$n(HNO_2)$ decreases with increasing altitude because its source reaction involves OH while its loss rate is independent of altitude. Its reduced profile in the winter reflects a similar reduction in the n(OH) profile. Since, in general, HNO_2 forms a closed loop (see Figure 2) with its production requiring an OH and its destruction producing one, it has no effect on the tropospheric chemistry except near the surface, where HNO_2 production via R29 serves as a small radical source.

E. Methane and Related Species (CH_4, $H_2C{=}O$, CO, and H_2)

Altitude profiles of the average daily loss rates for CH_4 and CO, along with the average daily loss and production rates for H_2, are given in Table IX for both summer and winter. The summer and winter profiles for $n(H_2C{=}O)$ with and without rainout loss are shown in Figure 11.

1. Methane (CH_4)

An average of $\overline{n(OH)}$ with and without rainout loss of HNO_3 was used to calculate the summer and winter profiles of $\overline{L(CH_4)}$, with $n(CH_4) = 1.5$ ppm assumed. While the two winter profiles of n(OH) differ by as much as a factor of 2 in the middle and upper tropospheres, most of the CH_4 loss takes place in the lower troposphere because of the 3.8-kcal activation energy for R51.

The total summer and winter column loss rates are 3.5×10^{11} and 5×10^{10} molecules cm^{-2} sec^{-1}. This yields a yearly average column loss rate of 2×10^{11} molecules cm^{-2} sec^{-1}, in harmony with previous calculations [Levy (155) and Wofsy et al. (256)] and experimental estimates [see Ehhalt (52)]. The fact that the major CH_4 loss takes place at the same time of year as does the major anaerobic production might explain why the factor of 5 difference in seasonal loss rates does not cause a large seasonal fluctuation in $n(CH_4)$. The calculated average yearly column loss rate of 2×10^{11} would yield a CH_4 residence time of $\tau_{CH_4} = 3.7$ yr, which is in harmony with earlier estimates [Ehhalt (52) and Robinson and Robbins (214)].

2. Formaldehyde ($H_2C{=}O$)

The noon profiles for $n(H_2C{=}O)$ are shown in Figure 11. It is the first major product of the CH_4 oxidation chain (see Figure 5), and its most important loss reaction is photodissociation, although an OH reaction is also important. Its chemical residence time, except just after

TABLE IX
Average Daily Loss and Production Rates for CH_4 and Related Species

Altitude (km)	$\overline{L(CH_4)}$		$L(CO)/r(CO)$		$\overline{P(H_2)}$		$L(H_2)/r(H_2)$	
	Summer	Winter	Summer	Winter	Summer	Winter	Summer	Winter
0	1.16^{6}[a]	1.85^{5}	1.13^{13}	2.38^{12}	4.4^{5}	9.6^{4}	7.0^{11}	1.08^{11}
2	0.60^{6}	1.06^{5}	0.75^{13}	1.61^{12}	2.5^{5}	5.9^{4}	3.5^{11}	0.64^{11}
4	0.32^{6}	0.40^{5}	0.51^{13}	1.05^{12}	1.4^{5}	2.5^{4}	1.8^{11}	0.23^{11}
6	0.15^{6}	0.16^{5}	0.31^{13}	0.46^{12}	0.7^{5}	1.1^{4}	0.8^{11}	0.09^{11}
8	0.06^{6}	0.07^{5}	0.18^{13}	0.28^{12}	0.3^{5}	0.5^{4}	0.3^{11}	0.04^{11}
10	0.02^{6}	0.03^{5}	0.09^{13}	0.16^{12}	0.1^{5}	0.2^{4}	0.1^{11}	0.01^{11}
12	--	--	0.04^{13}	0.09^{12}	--	0.1^{4}	--	--
15	--	--	0.01^{13}	--	--	--	--	--

[a]The superscript designates the power of 10 by which each number is to be multiplied.

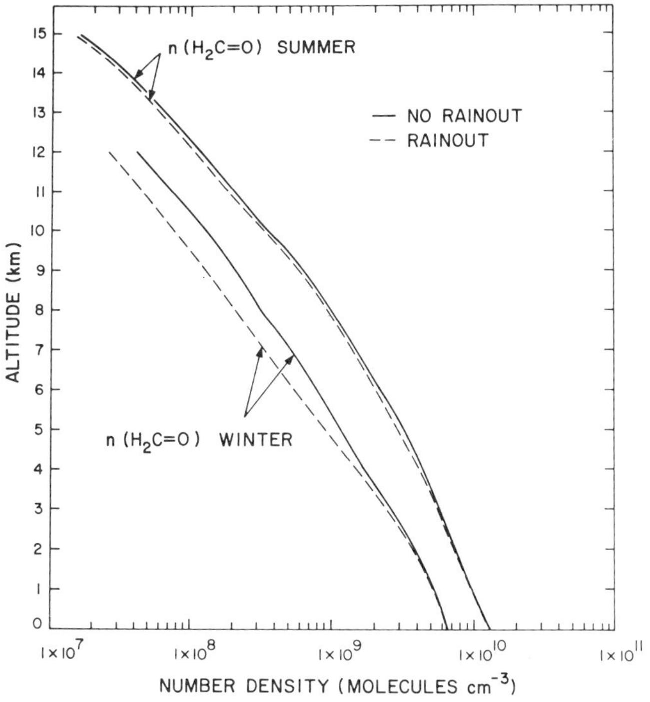

FIGURE 11. Noontime $n(H_2C{=}O)$ altitude profiles.

sunrise and just before sunset, is less than 1 hr, and it is treated as a short-lived molecule in the calculations.

The value of $n(H_2C{=}O)$ at the surface ranges from 0.5 ppb in the summer to 0.25 ppb in the winter, but it then rapidly drops with increasing altitude because its source, OH oxidation of CH_4, has a 3.8-kcal activation energy. The effects of a rainout loss path show up only through their influence on n(OH) and do not change $n(H_2C{=}O)$ in any important way.

3. Carbon Monoxide (CO)

Since all three loss reactions for $H_2C{=}O$ (R43a, R43b, and R48) result in the formation of CO, we can write

$\overline{P(CO)} = \overline{L(CH_4)}$. The profiles for $\overline{L(CO)}$, given in Table IX, were calculated for winter and summer; again $\overline{n(OH)}$ was averaged over the effects of rainout. If we assume that n(CO) is well mixed vertically, as suggested by its estimated residence time of 0.1 to 0.3 yr, we can solve for the CO mixing ratio

$$n(CO) = \int_0^{\text{tropopause}} \overline{P(CO)} \Big/ \int_0^{\text{tropopause}} \overline{L(CO)}/n(CO) \tag{49}$$

and find n(CO) = 0.07 and 0.055 ppm in the summer and winter, respectively. Therefore, while CH_4 is a large source of CO, it does not appear to be sufficient, and another large source (see Section IV.E.5) must be sought. If the activation energy for R51 were small, n(CO) produced by CH_4 would approach its surface value of 0.1 ppm, but current experimental data find the activation energy to be large.

If we assume n(CO) = 0.12 ppm, the calculated total column loss rates support chemical residence times of 0.1 yr in the summer and 0.6 yr in the winter, with an average yearly residence time of $\overline{\tau(CO)} = 0.2$ yr, in agreement with recent experimental estimates [Weinstock (250) and Weinstock and Niki (251)].

4. Molecular Hydrogen (H_2)

Another product of $H_2C{=}O$ photodissociation is H_2. Its production reaction is R43b and its loss reaction, which has a 4-kcal activation energy, is R49. The daily average production and loss rates given in Table IX were calculated with average winter and summer values of $\overline{n(OH)}$. If we assume $n(H_2) = 0.5$ ppm, the yearly average column loss rate given by

$$\frac{1}{2} \int_0^{\text{tropopause}} L(H_2)_{\text{summer}} + L(H_2)_{\text{winter}} \tag{50}$$

will provide a 5-yr residence time. As for CH_4, this residence time is long enough to permit both vertical and seasonal mixing. Using equation (50) to calculate yearly average production and loss rates, we find $n(H_2) = 0.75$ ppm, in reasonable agreement with reported measurements of 0.4 to 0.6 ppm.

F. Ozone

Average winter and summer profiles for photochemical O_3 production via

$$HO_2 + NO \rightarrow NO_2 + OH \tag{R20}$$

$$NO_2 + h\nu \xrightarrow{O_2} NO + O_3 \tag{R21}$$

and O_3 loss via

$$NO + O_3 \rightarrow NO_2 + O_2 \tag{R19}$$

$$NO_2 + O_3 \rightarrow NO_3 + O_2 \tag{R22}$$

$$NO_3 + h\nu \rightarrow NO + O_2 \tag{R23a}$$

are given in Table X.

With the major portion of production and loss taking place in the lower troposphere, where $n(HO_2)$ and $n(NO_3)$ are large, the effects of averaging rainout and nonrainout values are not important. Since the reaction of HNO_3 with OH is a source of NO_3 that results in the loss of only one O_3, the O_3 loss rate will be somewhat overestimated in this calculation.

TABLE X
Average Daily Loss and Production Rates for O_3

Altitude (km)	$L(O_3)$ Summer	$L(O_3)$ Winter	$\overline{P(O_3)}$ Summer	$\overline{P(O_3)}$ Winter
0	1.06^6 [a]	2.14^5	2.54^6	5.7^5
2	0.48^6	1.26^5	1.52^6	3.5^5
4	0.24^6	0.38^5	0.95^6	1.6^5
6	0.11^6	0.11^5	0.53^6	0.8^5
8	0.04^6	0.04^5	0.28^6	0.4^5
10	0.01^6	0.02^5	0.13^6	0.2^5
12	--	0.01^5	0.06^6	0.1^5
15	--		0.02^6	

[a] The superscript designates the power of 10 by which each number is to be multiplied.

The total column production rate for O_3 is $P(O_3) = 9.5 \times 10^{11}$ and 1.9×10^{11} molecules cm^{-2} sec^{-1} in the summer and winter, respectively, while the total column loss rates are $L(O_3) = 2.8 \times 10^{11}$ and 6×10^{10} molecules cm^{-2} sec^{-1} in the summer and winter.

In the lower troposphere, NO_2 is converted to NO_3 at night by R22, and that portion of the NO_3 that is photodissociated in the early morning represents a loss of O_3. R24 will become effective, however, as soon as $n(NO)$ reaches 5×10^9, so it is difficult to estimate the actual O_3 loss. It will have an upper limit of 3×10^{11}, but this is definitely an overestimate.

Those $O(^1D)$ atoms that react with H_2O also represent an O_3 loss, estimated at $L(O_3) \leq 4 \times 10^{11}$ in the summer and $\leq 5 \times 10^{10}$ in the winter.

It is possible that the large summer production will be balanced by nighttime and daytime NO_3 production followed by photodissociation and $O(^1D)$ production followed by reaction with H_2O, but it may also be that a net production or loss comparable to the much smaller O_3 flux estimates will result. Detailed diurnal time-dependent calculations coupled to a transport model will probably be necessary to resolve the question of atmospheric sinks and sources for O_3. As has been suggested by Crutzen (43a), R47 and R50b, with the new measurements and estimates of their rate coefficients, will provide a significant O_3 loss path and would have to be included in a detailed calculation.

VII. CONCLUSIONS

A. Summary

The driving force behind tropopsheric photochemistry is OH and, to a lesser extent, HO_2 and CH_3O_2. Their major source is the reaction of $O(^1D)$, itself produced by the photodissociation ($\lambda \leq 3150$ Å) of O_3, with H_2O. In the summer, n(OH) approaches 1.2×10^7 in the surface atmosphere and drops to 7×10^5 at the tropopause, while in the winter, surface levels reach 3×10^6 and drop to 5×10^5 at the tropopause. The total radical number density has a range of 0.3 to 1×10^9 at the surface and 1 to 2×10^7 at the tropopause.

The tropospheric chemistry of reactive carbon compounds has the anaerobic production of CH_4 as its major source. The OH reactions then convert CH_4 by a series of reactions to CO_2 and produce atmospheric $H_2C{=}O$, CO, and H_2 in the process. Atmospheric chemistry is the major source and sink for these three species, as well as the major sink for CH_4, and serves as the atmospheric link in the biosphere's carbon cycle.

The major source of tropospheric nitrogen chemistry is the natural production of NH_3. OH reactions then convert NH_3 to NO and NO_2, which, in turn, are converted to HNO_3 and removed by rainout either directly or as nitrate salts. The OH reactions provide the atmospheric link between the major source, NH_3 production, and sink rainout of NH_4^+ and NO_3^- salts and HNO_3. HNO_3, which is present in the surface atmosphere at 3 to 40 ppb, depending on the importance of rainout loss, and at 3 to 6 ppb at the tropopause, is a major atmospheric nitrogen species, along with NO and NO_2.

The tropospheric sulfur chemistry is different. Unlike the nitrogen and carbon chemistry, where combustion is an insignificant source, the combustion source of SO_2 appears to be very important. While OH reactions can be shown to convert sulfides to SO_2, it is not clear that normal atmospheric chemistry is important in the next step—the conversion of SO_2 to H_2SO_4, which is then removed from the atmosphere by rainout. It has also been suggested that a large amount of SO_2 is removed directly by rainout. Unfortunately we have the fewest data, both kinetic and atmospheric, on sulfur compounds. Most of the kinetic data we do have are at high temperatures, and most of the atmospheric data are for polluted environments.

A number of species that have not yet been directly observed—HNO_3, $H_2C{=}O$, H_2O_2, and CH_3O_2H—are believed to be present in the troposphere with number densities in the ppb range.

B. Approximations and Uncertainties

The important approximations used in the chapter are the following:

1. The maintaining of certain altitude profiles at fixed values during the calculations. While this approximation is valid for those species well represented by a single mixing ratio—CH_4, H_2, and CO—it is less so for the highly variable species O_3 and H_2O. Therefore, these calculations are representative of average latitudinal, altitudinal, and seasonal values, but not necessarily quantitatively correct at any given point in the troposphere.

2. The fixed value of 3 ppb given to $n(NO) + n(NO_2)$. The value chosen is in harmony with available data, but it is not yet known whether the sum of $n(NO)$ and $n(NO_2)$ is well mixed with altitude.

3. The use of hourly steady-state values to simulate time dependence. Because of the short chemical residence times, this approximation should be accurate for most of the day. Near sunrise and sunset, the decrease in number density will be overestimated, but this will not affect the average daily values.

4. The assumption that yearly average production and loss rates at 30°N are representative of the globe. However, given the lack of atmospheric data on the southern hemisphere, quantitative predictions may not be reliable.

The major uncertainties in the calculations are as follows:

1. The quantum yield for $O(^1D)$ production by $\lambda >$ 3000 Å. This number is crucial to the whole photochemistry. The values used in these calculations are believed to be quite conservative.

2. The actual reaction path for CH_3O_2. Unfortunately no gas-phase rates have been accurately measured for CH_3O_2 reactions, nor is the reaction path for CH_3O_2 + NO, NO_2, understood. While all the proposed paths produce $H_2C{=}O$, there is the possibility of CH_3OH production, as well as that of CH_3ONO_2. If the reaction with NO is conclusively shown not to go, more chemistry will need to be included in the CH_4 chain.

3. Lack of measurements of a number of activation energies—in particular, R22 and R20—in the tropospheric temperature range. Many of the rate coefficients—R6, R28, R20—and all reactions involving CH_3O_2 or CH_3O_2H are either not yet accurately known or not known at all. The absorption cross sections for NO_3 and HNO_2 are not well known.

4. Insufficient knowledge of the tropospheric values of $n(NO) + n(NO_2)$ and its altitude profiles.

C. Future Research

Many future areas for productive research were outlined in the preceding discussion of uncertainties and approximations involved in tropospheric chemistry. The needs of this field divide into two areas—kinetic data and mechanisms and atmospheric measurements.

1. Kinetic Data and Mechanisms

There is a great need for determination of activation energies in the tropospheric temperature range; absorption cross sections for NO_3 and HNO_2; reaction rates and mechanisms for CH_3O_2 and CH_3O_2H; more accurate rate coefficients for many reactions, such as R6, R28, and R20; and the $O(^1D)$ quantum yield for $\lambda > 3000$ Å.

From the modeling point of view, diurnal time-depende

models are needed, as well as an accurate coupling of transport and chemistry, in order to treat $n(H_2O)$ and $n(O_3)$ properly.

2. Atmospheric Measurements

To begin with, very little is known about the southern hemisphere, and more data of all kinds are necessary. This is particularly true if we wish to calculate true global production and loss rates. The values reported here are, at best, northern-hemisphere estimates, although they should also be qualitatively correct for the southern hemisphere.

Much more information about the altitude profile for $n(NO) + n(NO_2)$ is required, as well as measurements of $n(NH_3)$ and $n(NO) + n(NO_2)$ taken at the same time. Measurements of $n(Me_2S)$, $n(H_2S)$, and $n(SO_2)$ in the ambient atmosphere are also needed.

In order to determine the importance of rainout, measurements of $n(HNO_3)$ in the surface atmosphere are necessary.

The question of background levels of reactive hydrocarbons, which may be important to CO production and O_3 loss and production, is almost unanswered at this time.

Since the driving force for all the photochemistry is OH a measurement of its number density would be both useful and hopefully reassuring.

The current level of atmospheric photochemical modeling has advanced to the point that a coordinated monitoring of the atmosphere for $n(H_2O)$, $n(CH_4)$, $n(CO)$, $n(O_3)$, $n(NO)$ or $n(NO_2)$, $n(HNO_3)$, and reactive hydrocarbons would be of great importance. Many questions concerning sinks and sources of atmosphere species remain to be answered.

ACKNOWLEDGMENTS

I wish to acknowledge the many helpful discussions with Drs. Niki, Johnston, Dalgarno, Crutzen, Stuhl, Wofsy, McConnell, Davis, and several others in the preparation of this chapter. I thank Drs. Davis, Johnston, Stuhl, and Niki for providing experimental data before publication. This work was supported in part by a research grant from the Smithsonian Research Foundation.

REFERENCES

1. M. Ackerman in Mesopheric Models and Related Experiments, D. Reidel, Ed., Dordrecht (1971), p. 149.

2. A. Adel, Astrophys. J., 90, 627 (1939).

3. L. Aldaz, J. Geophys. Res., 74, 6943 (1969).

4. A. P. Altshuller and J. J. Bufalini, J. Environ. Sci. Tech., 5, 39 (1971).

5. J. G. Anderson and F. Kaufman, Chem. Phys. Lett., 16, 375 (1972).

5a. J. G. Anderson and F. Kaufman, Chem. Phys. Lett., 19, 483 (1973).

6. P. W. Arnold, J. Soil Sci., 5, 116 (1954).

7. P. G. Ashmore and M. G. Burnett, Trans. Faraday Soc., 58, 253 (1962).

8. A. E. Bainbridge and L. E. Heidt, Tellus, 18, 221 (1966).

9. E. J. Bair, private communication.

10. N. Basco, D. James, and F. James, Chem. Phys. Lett., 8, 265 (1971)

11 D. R. Bates and P. B. Hays, Planet. Space Sci., 15, 189 (1967).

12. D. R. Bates and A. Witherspoon, Mon. Not. Roy. Astron. Soc., 112, 101 (1952).

13. D. L. Baulch, D. D. Drysdale, and A. C. Lloyd, High Temperature Reaction Rate Data, University of Leeds, England, No. 1 (1968).

14. D. L. Baulch, D. D. Drysdale, and A. C. Lloyd, High Temperature Reaction Rate Data, University of Leeds, England, No. 2 (1968).

15. D. L. Baulch, D. D. Drysdale, and A. C. Lloyd, High Temperature Reaction Rate Data, University of Leeds, England, No. 3 (1969).

16. D. L. Baulch, D. D. Drysdale, D. G. Horne, and A. C. Lloyd, High Temperature Reaction Rate Data, University of Leeds, England, No. 4 (1969).

17. D. L. Baulch, D. D. Drysdale, and D. G. Horne, High Temperature Reaction Rate Data, University of Leeds, England, No. 5 (1970).

18. D. L. Baulch, D. D. Drysdale, and D. G. Horne, High Temperature Reaction Rate Data, University of Leeds, England, No. 6 (1970).

19. S. Beilke and H. Georgii, Tellus, 20, 435 (1968).

20. D. Biedenkapp, L. G. Hartshorn, and E. J. Bair, Chem. Phys. Lett., 5, 379 (1970).

21. K. F. Bishop, H. J. Delafield, A. E. J. Eggleton, C. O. Peabody, and B. T. Taylor, in Symposium on the Detection and Use of Tritium in the Physical and Biological Sciences, Vienna (1961).

22. R. D. Bojkov, J. Geophys. Res., 72, 5585 (1967).

23. B. Bolin and W. Bischof, Tellus, 22, 431 (1970).

24. B. Bolin and C. D. Keeling, J. Geophys. Res., 68, 3899 (1963)

25. J. N. Bradley and D. C. Dobson, J. Chem. Phys., 46, 2865 (1967).

26. R. H. Burgess and J. C. Robb, Chem. Soc. Spec. Publ., 9, 167 (1957).

27. R. H. Burris, Ann. Rev. Plant Physiol. 17, 155 (1966).

28. R. D. Cadle and E. R. Allen, Science, 167, 243 (1970).

29. R. D. Cadle, W. H. Fischer, E. R. Frank, and J. P. Lodge, Jr., J. Atmos. Sci., 25, 100 (1968).

30. R. D. Cadle and M. Ledford, Air and Water Poll. Int. J., 10, 25 (1966).

31. R. D. Cadle, J. Pate, and J. P. Lodge, Jr., paper presented to Amer. Chem. Soc., Chicago (1967).

32. R. D. Cadle and R. C. Robbins, Discuss. Faraday Soc., 30, 155 (1960).

33. J. Caldwell and A. R. Back, Trans. Faraday Soc., 61, 1939 (1965).

34. J. G. Calvert, J. A. Kerr, K. L. Demerjian, and R. D. McQuigg, Science, 175, 751 (1972).

35. E. Castellano and H. J. Schumacher, Z. Phys. Chem. Neue Folge, 65, 62 (1969)

36. H. Cauer, Compendium of Meteorology, T. F. Malone, Ed., Baltimore, 1126 (1951).

37. L. A. Cavanagh, C. F. Schadt, and E. Robinson, Environ. Sci. Tech., 3, 251 (1969).

38. A. C. Chamberlain, Int. J. Air Poll., 3, 63 (1960).

39. F. D. Colegrove, W. B. Hanson, and F. S. Johnson, J. Geophys. Res., 70, 4931 (1965).

40. F. D. Colegrove, W. B. Hanson, and F. S. Johnson, J. Geophys. Res., 71, 2227 (1966).

41. COSPAR International Reference Atmosphere, North-Holland, Amsterdam (1965).

42. P. J. Crutzen, J. Geophys. Res., 76, 7311 (1971).

43. P. J. Crutzen, Report AP-10, Inst. of Meteorol., U. of Stockholm, Sweden (1972).

43a. P. J. Crutzen, Tellus (1973), to be published.

44. R. Dalmon, Mem. Serv. Chim. Etat., 30, 141 (1943).

45. E. F. Danielsen, J. Atmos. Sci., 25, 502 (1968).

46. E. F. Danielsen, R. Bleck, J. Shedlovsky, A. Wartburg, P. Haagenson, and W. Pollock, J. Geophys. Res., 75, 2343 (1970)

46a. D. D. Davis, J. Prusacyzk, M. Dwyer, and P. Kim, AIAA Meeting, Denver, Colorado (1973).

46b. D. D. Davis, W. A. Payne, and L. J. Stief, Science, 179, 280 (1973).

46c. D. D. Davis, R. Schiff, and S. Fischer, AIAA Meeting, Denver, Colorado (1973).

47. K. L. Demerjian, J. A. Kerr, and J. G. Calvert, Adv. Environ. Sci. (1973).

48. W. B. Demore and O. F. Raper, J. Chem. Phys., 44, 1780 (1966).

48a. W. B. Demore, Science, 180, 735 (1973).

49. C. R. Detwiler, D. L. Garrett, J. D. Purcell, and R. Tousey, Ann. Geophys., 17, 9 (1961).

50. N. R. Dhar and A. Ram, J. Ind. Chem. Soc., 10, 287 (1939).

51. H. U. Dutsch, J. Geophys. Res., 78, 1707 (1970).

52. D. H. Ehhalt, presented at "Carbon and the Biosphere," Brookhaven Symposium in Biology (1972).

53. D. H. Ehhalt and L. E. Heidt, J. Geophys. Res., 78, 5265 (1973).

54. L. Elterman, Report AFCRL-64-740, AFCRL, Bedford, Massachusetts, (1964).

55. L. Elterman, Report AFCRL-68-0153, AFCRL, Bedford, Massachusetts, (1968).

56. L. Elterman, Report AFCRL-70-0200, AFCRL, Bedford, Massachusetts, (1970).

57. N. Endow, G. J. Doyle, and J. L. Jones, J. Air Poll. Control Assoc., 13, 141 (1963).

58. E. E. Eriksson, Tellus, 4, 215 (1952).

59. E. E. Eriksson, J. Geophys. Res., 68, 4001 (1963).

60. P. Fabian, W. F. Libby, and C. E. Palmer, J. Geophys. Res., 73, 3611 (1968).

61. G. J. Fergusson, Proc. Roy. Soc., 243A, 561 (1957).

62. W. H. Fischer, J. P. Lodge, Jr., A. F. Wartburg, and J. B. Pate, Environ. Sci. Tech., 2, 464 (1968).

63. S. N. Foner and R. L. Hudson, J. Chem. Phys., 36, 2681 (1962).

64. S. H. Fonselius, J. Geophys. Res., 68, 4009 (1963).

65. H. W. Ford, G. J. Doyle, and N. Endow, J. Chem. Phys., 26 1337 (1957).

66. C. W. Francis, G. Chester, and L. A. Haskin, Environ. Sci. Tech., 4, 586 (1970).

67. T. Frankiewicz and R. S. Berry, Environ. Sci. Tech., 6, 365 (1972).

68. I. Gabally, J. Geophys. Res., 74, 6869 (1969).

69. J. W. Gall, presented at "Physico-Chemical Transformation of Sulfur Compounds in the Atmosphere and the Formation of Acid Smogs," Mainz (1967).

70. J. W. Gall, D. H. Slater, and J. G. Calvert, presented at "Chemical Reactions in Urban Atmosphere," Warren, Michigan (1969).

71. F. E. Gartell, F. W. Thomas, and S. B. Carpenter, Amer. Ind. Hyg. Assoc. J., 24, 113 (1963).

72. D. Garvin and L. H. Gevantman, NBS Report no 867, National Bureau of Standards (1972).

73. C. W. Gear, Proceedings IFIPS Conference, A81 (1968).

74. C. W. Gear, IEEE, CT-18, No. 1, January 1971.

75. R. J. Gelinas, J. Comp. Phys., 9, 222 (1972).

76. H. W. Georgii, J. Geophys. Res., 75, 2365 (1970).

77. H. W. Georgii and D. Jost, Pure Appl. Geophys., 59, 216 (1964).

78. E. R. Gerhard and H. F. Johnstone, Ind. Eng. Chem., 47, 972 (1955).

78a. J. A. Ghormley, R. L. Ellsworth, C. J. Hochanadel, J. Phys. Chem., 77, 1341 (1973).

79. W. Glasson and C. Tuesday, GMR-584, 1-32, General Motors, Corporation (1966).

80. E. Glueckauf, Compendium of Meteorology, T. F. Malone, Ed., Baltimore (1951), p. 3.

81. R. M. Goody, Planet. Space Sci., 17, 1319 (1969).

82. R. M. Goody and C. D. Walshaw, Quart. J. Roy. Meteorol. Soc., 79, 496 (1953).

83. D. Gray, E. Lissi, and J. Heicklen, J. Phys. Chem., 76, 1919 (1972).

84. N. R. Greiner, J. Phys. Chem., 72, 406 (1968).

85. N. R. Greiner, J. Chem. Phys., 51, 5049 (1969).

86. N. R. Greiner, J. Chem. Phys., 53, 1070 (1970).

87. F. T. Gucker and S. Basu, Sci. Rep. No. 1, Contract AF 19(122)-400, U. of Indiana, Bloomington (1953).

88. P. H. Gudiksen, A. W. Fairhall, and R. J. Reed, J. Geophys. Res., 73, 4461 (1968).

89. H. L. Hamilton, J. J. B. Worth, and L. A. Ripperton, U. S. Army Res. Off. Contract No. DA-HC19-67-C-0029, Research Triangle Park, North Carolina (1968).

90. R. F. Hampson, R. L. Brown, D. Garvin, J. T. Herron, R. E. Huie, J. D. McKinley, and W. Tsang, NBS Report 10 692, National Bureau of Standards (1972).

91. R. F. Hampson, W. Braun, D. Garvin, J. T. Herron, R. E. Huie, M. J. Kurylo, A. H. Laufer, H. Okabe, and W. Tsang, NBS Report 10 828, National Bureau of Standards (1972).

91a. A. B. Harker and H. S. Johnston, J. Phys. Chem., 77, 1153 (1973).

92. P. B. Hays and J. J. Olivero, Plant. Space Sci., 18, 1729 (1970).

93. J. Heicklen, Adv. Chem. Ser., 76, 23 (1968).

94. J. Heicklen and H. S. Johnston, J. Amer. Chem. Soc., 84, 4394 (1962).

95. D. G. Hendry, private communication (1972).

96. W. S. Hering and T. R. Borden, Jr., AFCRL-64-30 (II), AFCRL, Bedford, Massachusetts (1964).

97. W. S. Hering and T. R. Borden, Jr., AFCRL-64-30 (III), AFCRL, Bedford, Massachusetts (1965).

98. W. S. Hering and T. R. Borden, Jr., AFCRL-64-30 (IV), AFCRL, Bedford, Massachusetts (1967).

99. J. T. Herron and R. E. Huie, J. Chem. Phys., 51, 4164 (1969).

100. C. J. Hochanadel, J. A. Ghormley, and P. J. Ogren, J. Chem. Phys., 56, 4426 (1972).

101. J. A. Howard, Adv. Free Radical Chem., 4, 49 (1972).

102. J. A. Howard and K. U. Ingold, J. Amer. Chem. Soc., 90, 1056 (1968).

103. R. E. Huie, J. T. Herron, and D. D. Davis, J. Phys. Chem., 76, 2653 (1972).

104. B. G. Hunt, J. Geophys. Res., 71, 1385 (1966).

105. B. G. Hunt, Mon. Weather Rev., 97, 289 (1969).

106. D. Hussain and R. G. W. Norrish, Proc. Roy. Soc., 273A, 165 (1963).

107. R. B. Ingersoll and R. E. Inman, presented at "Sources, Sinks, and Concentration of CO and CH_4 in the Earth's Environment," joint meeting of Amer. Geophys. Union and Amer. Meteorol. Soc., St. Petersburg, Florida (1972).

108. R. E. Inman, R. B. Ingersoll, and E. A. Levy, Science, 172, 1229 (1971).

109. E. C. Y. Inn and Y. Tanaka, J. Opt. Soc. Amer., 43, 870 (1953).

110. International Symposium on Trace Gases and Radioactivity, J. Geophys. Res., 68, (1963).

111. International Sympoisum on Trace Gases and Radioactivity, Tellus, 18, (1966).

112. International Symposium on Trace Gases and Radioactivity, J. Geophys. Res., 75 (1970).

113. W. Jacobi and K. Andre, J. Geophys. Res., 68, 3799 (1963).

114. S. Jaffe and H. W. Ford, J. Chem. Phys., 71, 1832 (1967).

115. F. S. Johnson, J. Meteorol., 11, 431 (1954).

116. H. Johnston, National Standard Reference Data Series, National Bureau of Standards, 20 (1968).

117. H. Johnston, Project Clean Air, Task Force No. 7, U. of California, Berkeley (1970).

118. H. Johnston, Lawrence Radiation Lab. Rep. UCRL-20568 (1971).

119. H. Johnston, private communication (1972).

120. H. Johnston and H. Crosby, J. Chem. Phys., 22, 689 (1954).

121. H. Johnston and D. Yost, J. Chem. Phys., 17, 386 (1949).

122. H. F. Johnstone and D. R. Coughanowr, Ind. Eng. Chem., 50, 1169 (1958).

123. E. J. Jones and O. R. Wulf, J. Chem. Phys., 5, 873 (1937).

124. I. T. N. Jones and R. P. Wayne, Proc. Roy. Soc. London, A319, 273 (1970).

125. C. E. Junge, Tellus, 8, 127 (1956).

126. C. E. Junge, J. Geophys. Res., 65, 227 (1960).

127. C. E. Junge, Tellus, 14, 363 (1962).

128. C. E. Junge, Air Chemistry and Radioactivity, Academic Press, New York (1963).

129. C. E. Junge, J. Geophys. Res., 68, 3840 (1963).

130. C. E. Junge and G. Czeplak, Tellus, 20, 422 (1968).

131. C. E. Junge and J. Hahn, J. Geophys. Res., 76, 8143 (1971).

132. C. E. Junge and T. G. Ryan, Quart. J. Roy. Meteorol. Soc., 84, 46 (1958).

133. C. E. Junge, W. Seiler, and P. Warneck, J. Geophys. Res., 76, 2866 (1971).

134. M. Katz and G. A. Ledingham, NCR No. 815, National Research Council of Canada, Ottawa (1939).

135. F. Kaufman, Can. J. Chem., 47, 1917 (1969).

136. F. Kaufman, private communication (1972).

137. C. D. Keeling, Tellus, 12, 200 (1960).

138. W. W. Kellogg, R. D. Cadle, E. R. Allen, H. L. Lazrus, and E. A. Martell, Science, 175, 587 (1972).

139. J. J. Kelly and J. D. McTaggart-Cowan, J. Geophys. Res., 73, 3228 (1968).

140. G. Kockarts, Mesospheric Models and Related Experiments, D. Reidel, Ed., Dordrecht, 160 (1971).

141. J. L. Kroening and E. P. Ney, J. Geophys. Res., 67, 1867 (1962).

142. T. Koyama, J. Geophys. Res., 68, 3971 (1963).

143. H. Kuhme, [see Ref. (214)].

144. R. H. Kummler and T. Baurer, J. Geophys. Res., 78, 5306 (1973).

145. R. H. Kummler and M. H. Bortner, J. Geophys. Res., 75, 3115 (1970).

146. R. H. Kummler, M. H. Bortner and T. Baurer, Environ. Sci. Tech., 3, 248 (1969).

147. D. Lal and Rama, J. Geophys. Res., 71, 2865 (1966).

148. T. Langley and W. D. McGrath, Planet. Space Sci., 19, 413 (1971).

149. P. A. Leighton, Photochemistry of Air Pollution, Academic Press, New York (1961).

150. C. B. Leovy, J. Geophys. Res., 74, 417 (1969).

151. H. Lettau, Compendium of Meteorology, T. F. Malone, Ed., Baltimore (1951), p. 320.

152. H. Levy II, Science, 173, 141 (1971).

153. H. Levy II, Planet. Space Sci., 20, 919 (1972).

154. H. Levy II, Planet. Space Sci., 21, 575 (1973).

155. H. Levy II, J. Geophys. Res., 78, 5325 (1972).

156. E. Lissi and J. Heicklen, J. Photochem., 1, 39 (1972).

157. A. C. Lloyd, NBS Report 10 447, National Bureau of Standards (1970).

158. J. P. Lodge, Jr., private communication (1972).

159. J. P. Lodge, Jr., A. J. MacDonald, and E. Vihman, Tellus, 12, 184 (1960).

160. J. P. Lodge, Jr. and J. B. Pate, Science, 153, 408 (1966).

161. J. E. Lovelock, R. J. Maggs, and R. A. Rasmussen, Nature, 237, 452 (1972).

162. J. D. Mahlman and S. Manabe, DOT-TSC-OST-72-13, Dept. of Transportation (U.S.) (1972).

163. S. Manabe and B. G. Hunt, Mon. Weather Rev., 96, 477 (1968).

164. S. Manabe, J. Smagorinsky, and R. F. Strickler, Mon. Weather Rev., 93, 769 (1965).

165. E. A. Martell, J. Atmos. Sci., 25, 113 (1968).

166. H. J. Mastenbrook, J. Appl. Meteorol., 5, 737 (1966).

167. H. J. Mastenbrook, J. Atmos. Sci., 25, 299 (1968).

168. H. J. Mastenbrook and D. R. Purdy, NRL Report 6891, Naval Research Laboratory, Washington (1969).

169. H. J. Mastenbrook and D. R. Purdy, NRL Report 6974, Naval Research Laboratory, Washington (1969).

170. J. C. McConnell, J. Geophys. Res., 78, 7812 (1973).

171. J. C. McConnell, M. B. McElroy, and S. C. Wofsy, Nature, 233, 187 (1971).

172. M. B. McElroy, private communication (1972).

173. M. B. McElroy and J. C. McConnell, J. Atmos. Sci., 28, 1095 (1971).

174. M. V. Migeotte, Phys. Rev., 73, 519 (1948).

175. M. V. Migeotte and L. Neven, Physica, 16, 423 (1950).

176. T. Mill, F. Mayo, H. Richardson, K. Irwin, and D. L. Allard, J. Amer. Chem. Soc., 94, 6802 (1972).

176a. C. Morley and I. W. Smith, J. Chem. Soc., 68, 1016 (1972).

177. E. D. Morris and H. Niki, J. Chem. Phys., 55, 1991 (1971).

177a. M. F. R. Mulcahy and R. H. Smith, J. Chem. Phys., 54, 5215 (1971).

178. F. H. Murcray, A. Goldman, D. G. Murcray, A. Csoeke-Poeckh, W. J. Williams, and R. N. Stocker, J. Geophys. Res., 78, 7033 (1973).

179. M. Nicolet, J. Geophys. Res., 70, 679 (1965).

180. M. Nicolet, Ann. Geophys., 26, 531 (1970).

181. M. Nicolet, Inst. d'Aeronomie Spatiale de Belgique, Aeronomica Acta A-91, Brussels (1971).

182. M. Nicolet, DOT-TSC-OST-72-13, Dept. of Transportation (U.S.) (1972).

183. M. Nicolet and W. Peetermans, presented at "Sources, Sinks, and Concentration of CO and CH_4 in the Earth's Environment," joint meeting of Amer. Geophys. Union and Amer. Meteorol. Soc., St. Petersburg, Florida (1972).

184. H. Niki, C. H. Wu, E. D. Morris, J. Phys. Chem., 77, 2507 (1973)

184a. H. Niki, L. Brietenbrach, and E. Morris, Am. Chem. Soc. Meeting, New York (1972).

185. J. F. Noxon, J. Chem. Phys., 52, 1852 (1970).

186. R. Nydal, J. Geophys. Res., 73, 3617 (1968).

187. T. C. O'Connor, Contract No. DA-91-591-EUC-2126, University College, Galway, Ireland (1962).

188. H. G. Ostlund and J. Alexander, J. Geophys. Res., 68, 3995 (1963).

189. J. C. Pales and C. D. Keeling, J. Geophys. Res., 70, 6053 (1965

190. G. Paraskevopoulos and R. J. Cvetanovic, Chem. Phys. Lett., 9, 603 (1971).

191. T. T. Paukert and H. Johnston, J. Chem. Phys., 56, 2824 (1972).

191a. W. A. Payne, L. J. Stief, and D. D. Davis, J. Am. Chem. Soc., to be published.

192. R. Penndorf, J. Opt. Soc. Amer., 47, 176 (1957).

193. J. N. Pitts, Jr., Adv. Environ. Sci., 1, 289 (1969).

194. J. N. Pitts, Jr., J. H. Sharp, and S. I. Chan, J. Chem. Phys., 40, 3655 (1964).

195. S. E. Poet, H. E. Moore, and E. A. Martell, J. Geophys. Res., 77, 6515 (1972).

196. J. Pressman and P. Warneck, J. Atmos. Sci., 27, 155 (1970).

197. M. A. Price and E. R. Stephens, private communication (1971).

198. E. Rashke, Tellus, 19, 215 (1967).

199. R. A. Rasmussen, J. Air Poll. Control Assoc., 22, 537 (1972).

200. R. A. Rasmussen and F. W. Went, Proc. Nat. Acad. Sci., 51, 215 (1965).

201. V. H. Regener, Z. Phys., 109, 642 (1938).

202. V. H. Regener, J. Geophys. Res., 62, 221 (1957).

203. V. H. Regener, J. Geophys. Res., 65, 3975 (1960).

204. E. R. Reiter, Atmospheric Transport Processes, Part 1 - Energy Transfers and Transformations, AEC Critical Review Series (1969).

205. E. R. Reiter, Atmospheric Transport Processes, Part 2 - Chemical Tracers, AEC Critical Review Series (1971).

206. E. R. Reiter, Atmospheric Transport Processes, Part 3 - Hydrodynamic Tracers, AEC Critical Review Series (1972).

207. E. R. Reiter, M. E. Glasser and J. D. Mahlman, Atmos. Sci. Paper No. 107, Colorado State U., Fort Collins (1967).

208. N. A. Renzetti and G. J. Doyle, Int. J. Air Poll., 2, 327 (1960).

209. L. A. Ripperton, J. Geophys. Res., 70, 5009 (1965).

210. L. A. Ripperton, H. Jeffries and J. J. B. Worth, Environ. Sci. Tech., 5, 246 (1971).

211. L. A. Ripperton, J. J. B. Worth, and H. E. Jeffries, presented at Technician Symposium, New York (1967).

212. L. A. Ripperton, J. J. B. Worth, and L. Kornreich, presented at 61st Annual Meeting of Air Poll. Control Assoc. (1968).

213. L. A. Ripperton and F. M. Vukovich, J. Geophys. Res., 76, 7329 (1971).

214. E. Robinson and R. C. Robbins, Stanford Res. Proj. PR-6755, Amer. Petrol. Inst., Washington, D. C. (1968).

215. E. Robinson and R. C. Robbins, APCA J., 5, 303 (1970).

216. E. Robinson and R. C. Robbins, SRI Proj. PR-6755, Amer. Petrol. Inst., Washington, D. C. (1971).

217. H. I. Schiff, Ann. Geophys. 28, 67 (1972).

218. K. Schofield, Planet. Space Sci., 15, 643 (1967).

219. T. G. Scholz, D. H. Ehhalt, L. E. Heidt, and E. A. Martell, J. Geophys. Res., 75, 3049 (1970).

220. G. Schott and N. Davidson, J. Amer. Chem. Soc., 80, 1841 (1958).

221. E. A. Schuck and G. J. Doyle, Air Poll. Foundation Rep. No. 29 (1959).

222. W. C. Schumb, C. N. Gatterfield, and R. L. Wentworth, Hydrogen Peroxide, Reinhold, New York (1955).

223. K. Schutz, C. Junge, R. Beck, and B. Albrecht, J. Geophys. Res., 75, 2230 (1970).

224. P. M. Scott and R. J. Cvetanovic, J. Chem. Phys., 54, 1440 (1971).

225. W. Seiler, private communication (1972).

226. W. Seiler and C. Junge, J. Geophys. Res., 75, 2217 (1970).

226a. R. Simonaitis and J. Heicklen, J. Phys. Chem., 77, 1096 (1973).

226b. R. Simonaitis and J. Heicklen, private communication (1973).

227. H. W. Sidebottom, C. C. Badcock, G. E. Jackson, J. G. Calvert, G. W. Reinhardt, and E. K. Damon, Environ. Sci. Tech., 6, 72 (1972).

228. R. Simonaitis and J. Heicklen, Int. J. Chem. Kinetics, 4, 529 (1972).

229. N. Sissenwine, D. D. Grantham, and H. A. Salmela, J. Atmos. Sci., 25, 1129 (1968).

230. A. F. Smith, D. G. Jenkins, and D. E. Cunningworth, J. Appl. Chem., 11, 317 (1961).

231. R. M. Smith, Tellus, 20, 76 (1968).

232. D. R. Snelling and E. J. Bair, J. Chem. Phys., 47, 228 (1967).

233. C. W. Spicer, A. Villa, H. A. Wiebe, and J. Heicklen, CAES Report No. 223-71, Dept. of Chem., Penn. State (1972).

234. L. Stief, W. Pain, and D. Davis, Science (1972).

235. E. R. Stephens and W. E. Scott, Proc. Amer. Petrol. Inst., 42, 665 (1962).

236. F. Stuhl, J. Chem. Phys., 59, 635 (1973).

237. F. Stuhl and H. Niki, J. Chem. Phys., 55, 3943 (1971).

238. F. Stuhl and H. Niki, J. Chem. Phys., 57, 3677 (1972).

239. J. W. Swinnerton, V. J. Linnenbom, and C. H. Cheek, Environ. Sci. Tech., 3, 836 (1969).

240. J. W. Swinnerton, V. J. Linnenbom, and R. A. Lamontagne, Science, 167, 984 (1970).

241. S. S. Thomas and J. G. Calvert, J. Amer. Chem. Soc., 84, 4207 (1962).

242. R. Tousey, Space Sci. Rev., 2, 3 (1963).

243. U. S. Standard Atmosphere Supplements, 1962, Govt. Printing Office, Washington, D.C. (1962).

244. U.S. Standard Atmosphere Supplements, 1966, Govt. Printing Office, Washington, D.C. (1966).

245. H. E. Van den Bergh and A. B. Callear, Trans. Faraday Soc., 67, 2017 (1971).

246. E. Vigroux, Contribution a l'Etude Experimentale de l'Ozone, Marson et Cie, Paris (1953).

247. K. Watanabe, E. C. Y. Inn, and M. Zelikoff, J. Chem. Phys., 21, 1026 (1953).

248. L. G. Wayne and D. M. Yost, J. Chem. Phys., 19, 41 (1951).

249. E. Weber, J. Geophys. Res., 75, 2909 (1970).

250. B. Weinstock, Science, 166, 224 (1969).

251. B. Weinstock and H. Niki, Science, 176, 290 (1972).

252. F. W. Went, Nature, 187, 641 (1960).

253. K. Westberg and N. Cohen, Aerospace Corp. Rep. ATR-70-(8107)-1, El Segundo, California (1969).

253a. A. A. Westenberg and N. deHaas, J. Chem. Phys., 57, 5375 (1972).

253b. A. A. Westenberg and N. deHaas, J. Chem. Phys., 58, 4061 (1973).

253c. A. A. Westenberg and N. deHaas, J. Chem. Phys., 58, 4066 (1973).

253d. A. A. Westenberg and N. deHaas, J. Phys. Chem., 76, 1586 (1972).

253e. A. A. Westenberg, Science, 177, 255 (1972).

254. W. E. Wilson, J. Phys. Chem. Ref. Data, 1, No. 2 (1972).

255. W. E. Wilson and A. A. Westenberg, presented at 11th International Symposium on Combustion, U. of California, Berkeley (1966).

256. S. C. Wofsy, J. C. McConnell, and M. B. McElroy, J. Geophys. Res., 77, 4477 (1972).

256a. W. Wong and D. D. Davis, Intl. J. Chem. Kinetics (in press) (1973)

257. R. A. Young, G. Black, and T. G. Slanger, J. Chem. Phys., 49, 4758 (1968).

258. L. Zafonte and H. Johnston, private communication (1971).

ADDENDUM

Recently Mahlman (Proceedings of the Second Conference on the Climatic Impact Assessment Program, November 14-17, 1972, Department of Transportation Report No. DOT-TSC-OST-73-4 (April 1973), pp. 321-337) reported on the simulation of a stratified tracer with a crudely simulated photochemical equilibrium at the upper level and a precipitation-dependent sink at the lowest levels. The tracer experiment was run with three-dimensional output from the GFDL general circulation model. The results as far as latitude, longitude, and altitude distribution as well as seasonal dependence exhibited many features common with those observed for tropospheric ozone. The net transport into the troposphere was about 1×10^{11} molecules cm^{-2} sec^{-1}.

On the other hand Chameides and Walker (J. Geophys. Res., to be published [1974]) have recently presented a photochemical steady state calculation for tropospheric oxone which shows production and loss rates of about 1×10^{12} molecules cm^{-2} sec^{-1} and implies a 10-day chemical lifetime for ozone.

They also claim to find qualitative agreement with both the altitude and seasonal dependence of ozone at 30°N and observed day-to-day variations of the number density. They used the photochemical model developed by Levy (154) to calculate the production and loss of ozone as first suggested by Crutzen (43a).

Thus it should be clear that the question of tropospheric ozone is far from settled. The results of Mahlman would suggest that there is a net sink for ozone in the troposphere and that the net source is transport from the stratosphere while the results of Chameides and Walker suggest that atmospheric ozone is in a photochemical stationary state.

A possible explanation is that, while the photochemical production and loss of ozone is much larger than the transport contribution, the sum of the production and loss provides a small net sink to balance the transport source. At this time the accurate data necessary for such calculations are not available.

Author Index

Numbers in parentheses are reference numbers and show that an author's work is referred to although his name is not mentioned in the text. Numbers in *italics* indicate the page on which the full references appear.

Subject Index